S0-ASW-716

ELECTRONIC PROPERTIES OF INORGANIC QUASI-ONE-DIMENSIONAL COMPOUNDS

Part II – Experimental

PHYSICS AND CHEMISTRY OF MATERIALS WITH LOW-DIMENSIONAL STRUCTURES

Series B: Quasi-One-Dimensional Materials

ELECTRONIC PROPERTIES OF INORGANIC QUASI-ONE-DIMENSIONAL COMPOUNDS

Part II – Experimental

Edited by

PIERRE MONCEAU

Centre de Recherches sur les Très Basses Températures, CNRS, 38000 Grenoble, France

D. REIDEL PUBLISHING COMPANY

DORDRECHT / BOSTON / LANCASTER

Library of Congress Cataloging in Publication Data
Main entry under title:

Electronic properties of inorganic quasi-one-dimensional compounds.

(Physics and chemistry of materials with low-dimensional structures. Series B, Quasi-one-dimensional materials)
Includes indexes.
Bibliography: p.
Contents: pt. 1. Theoretical – pt. 2. Experimental.
1. One-dimensional conductors. 2. Chemistry, Inorganic. 3. Chemistry, Physical and theoretical. I. Monceau, Pierre, 1942– II. Series.
QC176.8.E4E37 1985 530.4'1 84-27565
ISBN 90-277-1801-6 (set)
ISBN 90-277-1789-3 (pt. 1)
ISBN 90-277-1800-8 (pt. 2)

Published by D. Reidel Publishing Company.
P.O. Box 17, 3300 AA Dordrecht, Holland.

Sold and distributed in the U.S.A. and Canada
by Kluwer Academic Publishers,
190 Old Derby Street, Hingham, MA 02043, U.S.A.

In all other countries, sold and distributed
by Kluwer Academic Publishers Group,
P.O. Box 322, 3300 AH Dordrecht, Holland.

Printed in The Netherlands

TABLE OF CONTENTS

TABLE OF CONTENTS TO PART I

PREFACE TO PART II

Intensive experimental and theoretical work has been undertaken these last few years in the understanding of the chemical and physical properties of quasi one-dimensional compounds. This designation is ascribed to materials which exhibit large anisotropy ratios in some physical properties when measured along one direction with regard to orthogonal directions. The recently-acquired possibility for chemists to synthetize real, inorganic as well as organic, one-dimensional systems with fascinating physical properties has been one of the essential motivations which have attracted many new research workers. The first two volumes in the present series "Physics and Chemistry of Materials with Quasi-One-Dimensional Structures" are intended to review the experimental and the theoretical progress made in this field in the last years. These volumes deal with electronic properties of inorganic compounds.

In Part I, concepts and theoretical models developed to understand the properties of one-dimensional systems have been analysed. In the present volume, the physical properties of three families of one-dimensional materials are described, namely the platinum chain compounds, the polymetic sulfur nitride and the transition metal tri- and tetrachalcogenides. Some of the latter compounds have been considered as prototypes for a novel collective mechanism of electrical conduction induced by the motion of a charge density wave, as proposed by Fröhlich in 1954. Two chapters are devoted to optical reflectivity and lattice dynamic study measurements which reveal the low-dimensional structure of these chain-like materials. This book is intended to be of interest to solid state physicists -both students as well as advanced researchers-, specialists in synthetic chemistry as well as scientists concerned with new materials.

I wish to express my sincere gratitude to the authors for their collaboration in this book and I warmly thank Prof. F. Lévy, the managing editor, for his suggestions and his constant encouragement.

PIERRE MONCEAU Grenoble, January 1984

P. Monceau (ed.), Electronic Properties of Inorganic Quasi-One-Dimensional Materials, II, xi.

PROPERTIES OF CONDUCTING PLATINUM CHAIN COMPOUNDS

KIM CARNEIRO

Physics Laboratory I,
University of Copenhagen,
Universitetsparken 5,
DK 2100 Copenhagen,
Denmark

"It is therefore likely that a one-dimensional model could never have metallic properties. However . . . . "
R. E. Peierls, 1954.

1. Introduction

Given the steadily increasing volume of models dealing with one-dimensional metals, the statement of Peierls [1] cited above should to some extent be considered a stroke of serendipity. However, it is generally accepted as the first discovery of the fact that a hypothetical one-dimensional metal is unstable towards the formation of a semiconductor, often referred to as a Peierls insulator. It is noteworthy that in those early days of solid state physics thermal energies were considered to be small compared to excitation energies (by analogy to atomic and nuclear physics) so that only the ground state of the one dimensional conductor, the insulator, was supposed to exist in thermal equilibrium. But part of the reason why the Peierls instability has attracted so much interest is as a matter of fact related to the fact that several chain compounds exhibit one dimensional metallic behavior at room temperature and undergo a phase transition into the insulating state at some lower temperature. This allows both the high temperature state as well as the ground state to be studied and, what is important, to build the connection between the two states.

Another interesting feature of the Peierls argument is that it can be generalized to allow for other types of ground states such as superconductivity and magnetism in one-dimensional conductors.

The compounds which Peierls predicted not to exist had in fact been reported in the chemical literature more than 100 years before [2]. In studies of compounds where chains of platinum occurred with the different oxidation states Pt-II and Pt-IV, intermediate or mixed cases were found where crystals showed bronze to gold lusters, indicative of metallic character. However, it was not until Krogman and Hausen undertook a detailed study of some of these salts in solving their crystallographic structures that the connection between the mixed valence or partially oxidized Krogman salts and the Peierls instability was made [3, 4]. But once it was made, it spurred intense efforts to explore and understand the behavior of the conducting platinum atom chain salts. In this chapter we review what has come out of these efforts. By chemical modifications the important physical parameters have been varied, and this has allowed systematic studies of the Peierls instability.

What are the important physical parameters? As shown in Figure 1 the instability

P. Monceau (ed.), Electronic Properties of Inorganic Quasi-One-Dimensional Materials, II, 1–68.

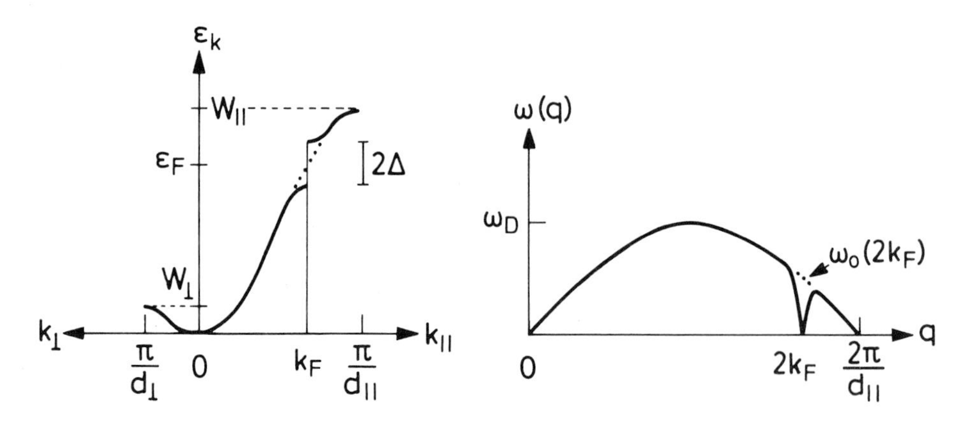

Fig. 1. To the left is shown the electronic band structure ϵ_k for the quasi-one-dimensional conductor. Bandwidths are $W_{\|}$ and $W_{\perp}$ in parallel and perpendicular directions respectively. The gap 2Δ at the Fermi level (k_F, ϵ_F) is shown. To the right are shown the phonons $\omega_0(q)$ of the conductor exhibiting a softening at $q = 2k_F$ according to the Peierls distortion. $\omega_0(2k_F)$ is the unperturbed phonon frequency at $q = 2k_F$. Note that the characteristic electronic energies are much bigger than the phonon energies $(\epsilon_F \gg \hbar\omega_D)$.

involves a gap in the electronic spectrum and a 'soft phonon' lattice distortion. It comes about because a lattice distortion will lower the electronic energy so much in a one-dimensional system that it will always overcome the elastic energy cost of the ionic distortion, provided that the wavelength of the lattice distortion is $2\pi/(2k_F)$. Therefore the important parameters are those characterising the electronic band structure ϵ_k of the high temperature metal, the elastic properties of the ionic lattice as given by the phonon frequencies $\omega_0(q)$ and the Fermi vector k_F. This is quantified if one considers the Fröhlich Hamiltonian [5]:

$$\mathcal{H} = \sum_{k,\sigma} \epsilon_k c_{k,\sigma}^+ c_{k,\sigma} + \sum_q \hbar\omega_0(q) b_q^+ b_q + \sum_{q=\pm 2k_F} \sum_{k,\sigma} \frac{g(q)}{\sqrt{N}} c_{k+q,\sigma}^+ c_{k,\sigma} (b_q + b_{-q}^+), \qquad (1)$$

where $c_{k,\sigma}^+$ and b_q^+ are creation operators for one-dimensional Bloch electrons and phonons, g is the electronic energy gain per relative ionic displacement, and N is the number of atoms in the chain. As first shown by Rice and Strässler the dimensionless electron–phonon coupling constant [6]:

$$\lambda = g^2 \frac{N(\epsilon_F)}{\hbar\omega_0(2k_F)}, \qquad (2)$$

accounts for the insulating ground state in terms of the high temperature metallic parameters. In (2), $N(\epsilon_F)$ is the density of electrons at the Fermi surface. The importance of λ is for instance seen in the expression for the activation energy of the low temperature semiconducting state:

$$\Delta = 8C\epsilon_F e^{-1/\lambda}, \tag{3}$$

where the parameter C depends on the electronic band structure (given by ϵ_k) and the band filling (given by k_F) [6–8].

Another important characteristic of the one-dimensional conductor is its degree of one-dimensionality. Landau [9] has shown that if it was strictly one-dimensional, thermal fluctuations would exclude that any phase transition occurs at finite temperatures, leaving only these fluctuations to be studied. But real compounds are not strictly one-dimensional and therefore a finite transition temperature is observed. It is important to characterize the three-dimensional character of a real 'quasi-one-dimensional' conductor in order to understand its behavior a finite temperatures. What happens is that the observed transition temperature T_P gets suppressed below its mean field or BCS value [6, 7]:

$$T_P^{MF} = \frac{\Delta}{1.76k_B}, \tag{4}$$

and the suppression depends on the character and strength of the three-dimensional coupling between the chains in the conductor. Several authors have dealt with this problem [10–16] which we return to in Section 5. Here we quote the work of Horowitz, Gutfreund and Weger, who assumed a transverse bandwidth $W_\perp$ (cf. Figure 1) and found that within certain limits of the bandwidth anisotropy $W_\perp/W_\parallel$ the observed transition temperature T_P may be related to T_P^{MF} as follows [14]:

$$\frac{T_P}{T_P^{MF}} = \exp[-2.5k_B T_P^{MF}/(\eta\epsilon_F)], \tag{5}$$

where η is a parameter related to the bandwidth anisotropy. As we shall see Equation (5) provides a useful means of deriving the anisotropy of platinum compounds, since T_P, T_P^{MF} and ϵ_F may be derived independently.

2. Platinum Compounds and Their Crystal Structures

Substantial chemical efforts have been given to the synthesis and crystallographic characterization of the conducting platinum salts, e.g., as described in two recent reviews by Williams *et al.* [17, 18] for the tetracyano-platinates (sometimes referred to as *CP's*) and by Underhill *et al.* [19] for the bisoxalatoplatinates (sometimes referred to as *OP's*). Below we will give a summary of what seems important for the understanding of the physical behavior of the salts, bearing in mind that it is only via this systematic chemical work that the detailed physical insight has become available.

2.1. LIGANDS WHICH GIVE PARTIALLY OXIDIZED SALTS

The existence of one-dimensional platinum chain conductors is based on the ability of Pt to exist in nonintegral valence state between Pt^{+2} and Pt^{+4} as well as the ability to form columnar structures. Both Pt^{+2} and Pt^{+4} crystallize in chain structures when surrounded by planar ligand systems and in some of these the Pt ion may be partially oxidized (or reduced) to give conducting salts. As an example let us consider the prototype conductor $K_2[Pt(CN)_4]Br_{0.30}\cdot H_2O$, abbreviated as KCP or KCP(Br), first prepared in 1842 by Knop [20]. It is a partially oxidized tetracyano-platinate which may be thought of as being derived from the unoxidized Pt-II salt $K_2[Pt(CN)_4]\cdot 3H_2O$ by inclusion in the lattice of the extra anion Br^{-1}. Both salts form the columnar structure indicated in Figure 2; but upon oxidation two things happen: (i) the conductivity is raised from

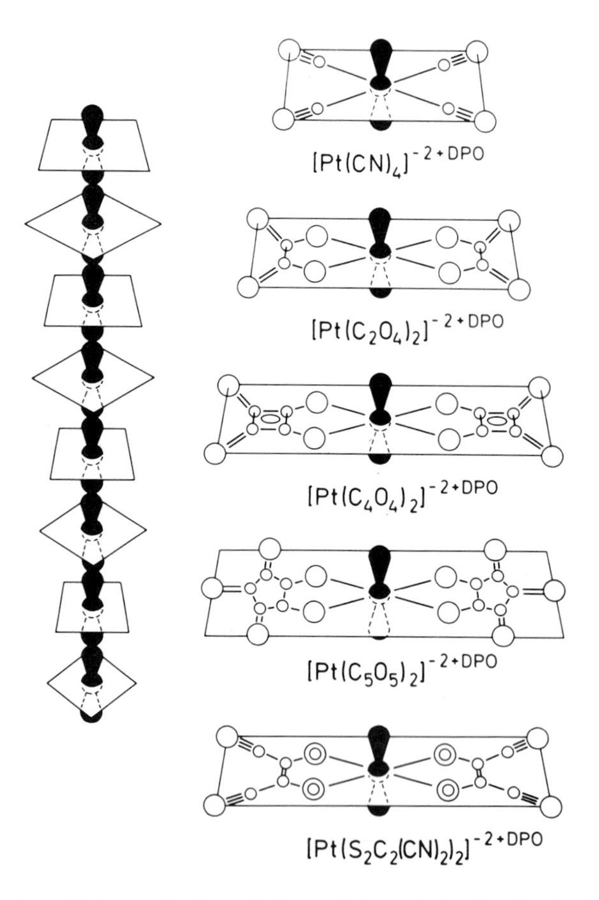

Fig. 2. Perspective view of the conducting platinum chain and the ligands that have been used in partially oxidized salts.

5×10^{-6} $(\Omega\ \text{cm})^{-1}$ to 200 $(\Omega\ \text{cm})^{-1}$ and (ii) the intrachain Pt–Pt distance $d_{\parallel}$ is changed from 3.478 Å to 2.89 Å. This allows substantial overlap between the d_{z^2} orbitals of neighboring platinum atoms along a chain, thereby increasing the conductivity.

Fortunately KCP is not unique in that several modifications have been made leaving the conducting Pt-chain intact. Out of the substantial amount of tetra-cyanoplatinates that have been made Table I lists some which have been well characterized and to which we will refer in the following [17, 18, 21–34]. As is seen both anions and cations may be

TABLE I

Partially oxidized tetracyano-platinates

Compound	Abbreviation	Reference	Space group
$Rb_2[Pt(CN)_4](FHF)_{0.40}$	RbCP(FHF)	21	*I4/mcm*
$Rb_2[Pt(CN)_4][H(OSO_3)_2]_{0.49} \cdot H_2O$	RbCP(DSH)	22	$P\bar{1}$
$Cs_2[Pt(CN)_4](FHF)_{0.40}$	CsCP(FHF)	23	*I4/mcm*
$Cs_2[Pt(CN)_4]Cl_{0.30}$	CsCP(Cl)	24	*I4/mcm*
$K_2[Pt(CN)_4]Cl_{0.30} \cdot 3H_2O$	KCP(Cl)	25	*P4mm*
$Cs_2[Pt(CN)_4](N_3)_{0.25} \cdot 3H_2O$	$CsCP(N_3)$	26	$P\bar{4}b2$
$K_2[Pt(CN)_4]Br_{0.30} \cdot 3H_2O$	KCP(Br)	17,27–29	*P4mm*
$Pb_{0.27}K_{1.73}[Pt(CN)_4]Cl_{0.30} \cdot 3H_2O$	Pb/KCP(Cl)	30	*P4mm*
$Rb_2[Pt(CN)_4]Cl_{0.30} \cdot 3H_2O$	RbCP(Cl)	31	*P4mm*
$[(NH_2)_3]_2[Pt(CN)_4]Br_{0.25} \cdot H_2O$	GCP(Br)	32	*I4mm*
$(NH_4)_2[Pt(CN)_4]Cl_{0.42} \cdot 3H_2O$	ACP(Cl)	33	*P4mm*
$K_{1.75}[Pt(CN)_4] \cdot 1.5H_2O$	K(def)TCP	34	$P\bar{1}$

substituted. This has the effect of expanding or contracting the lattice dependent on the size of the ions 'in use' resulting in substantial changes of the physical parameters [35]. Table I also shows the space groups in which the compounds crystallize to demonstrate that a given structure can only incorporate changes within certain limits. But irrespective of this change in crystal symmetry the salts listed in Table I behave to a high degree analogously with respect to the Peierls instability [36].

Not only may the environment of the $[Pt(CN)_4]$-moiety be varied, but also the ligand itself in which the platinum is embedded. Ligands that have successfully been used to host partially oxidized Pt are shown in Figure 2. Most work has been done with the bis-oxalatoplatinate salts based on partial oxidation of the $[Pt(C_2O_4)_2]$-ion. In some respect this ligand is more versatile than $[Pt(CN)_4]$. For instance, both divalent as well as monovalent metal cations have been employed, whereas no divalent cation CP-salt has been made (with the exception of the *partially* divalent Pb/KCP of Table I) and also bulky organic cations may be incorporated. On the other hand OP-salts are restricted to the *cation deficient* type, i.e., where the cation content is in between what corresponds to Pt^{+2} and Pt^{+4}. Here CP's are more flexible since both cation deficient and *anion deficient* salts have been made, e.g., K(def)TCP and KCP(Br) in Table I, respectively. Although the substantial chemical modifications of the $[Pt(C_2O_4)_2]$-salts do leave the conducting Pt-chain intact, at least some counterions play a direct role in determining the physical behavior of these salts. It is therefore becoming increasingly clear that the OP's do not at large exhibit simple Peierls transitions, but that their behavior is dominated by the competition between more than one instability. The bis-oxalatoplatinates that will be referred to later are listed in Table II [4, 19, 37–40].

Whereas the $[Pt(CN)_4]$-ion has no natural extension, one of the interesting features of the $[Pt(C_2O_4)_2]$-ion is that can be modified. Two examples are shown in Figure 2,

TABLE II

Partially oxidized bis-oxalatoplatinates

Compound	Abbreviation	Reference	Space group
$Mn_{0.81}[Pt(C_2O_4)_2]\cdot 6H_2O$	MnOP	37	*Cccm*
$Zn_{0.81}[Pt(C_2O_4)_2]\cdot 6H_2O$	ZnOP	38	*Cccm*
$Co_{0.83}[Pt(C_2O_4)_2]\cdot 6H_2O$	CoOP	39	*Cccm*
$Ni_{0.84}[Pt(C_2O_4)_2]\cdot 6H_2O$	NiOP	40	*Cccm*
$Mg_{0.82}[Pt(C_2O_4)_2]\cdot 5.3H_2$	MgOP	4	*Cccm*

namely the bis-squaratoplatinate-ion $[Pt(C_4O_4)_2]$ and the bis-croconatoplatinate-ion $[Pt(C_5O_5)_2]$. Partially oxidized salts of these ions have been made, but they have not been extensively characterized [41, 42]. They are listed in Table III and will be discussed

TABLE III

Conducting salts with large ligands

Compound	Abbreviation	Reference	Space group
$K_{1.6}[Pt(C_4O_4)_2]\cdot 2H_2O$	KSQAR	41	–
$K_{1.2}[Pt(C_5O_5)_2]\cdot 2H_2O$	KCROC	42	–
$Li_x[Pt(S_2C_4N_2)_2]\cdot 2H_2O$	LiPt(mnt)	43, 44	$P\bar{1}$

later because they show an interesting new development regarding the versatility of the conducting Pt-chain, and possibly demonstrate new physical phenomena not encountered in the above mentioned platinum salts. The same is true for the recently prepared $Li_x[Pt(S_2C_4N_2)_2]\cdot 2H_2O$, which marks a case of an 'in-between' between the Pt-chain conductors and the organic charge transfer salts [43, 44]. Also this compound is included in Table III.

2.2. CRYSTAL STRUCTURES OF $[Pt(CN)_4]$-SALTS

Figure 3 shows the tetragonal crystal structure of KCP(Br) [27]. It contains two $[Pt(CN)_4]$-units per unit cell but the two unequivalent platinum distances are equal, indicating that the Pt-bond length $d_{\parallel}$ is determined by the d_{z^2} overlap and not influenced by the fact that the structure is noncentrosymmetric. The K-cations are located in the upper part of the unit cell in between the planes defined by the $[Pt(CN)_4]$ groups, whereas the crystal water is situated in the bottom part connecting neighboring Pt-chains via hydrogen bonds. Here, there are two different site symmetries for the H_2O. The Br^{-1} is in the center with a 60% occupancy, distributed over two sites, one exactly in this center and one slightly displaced [17, 45]. It appears that water occupies the center of

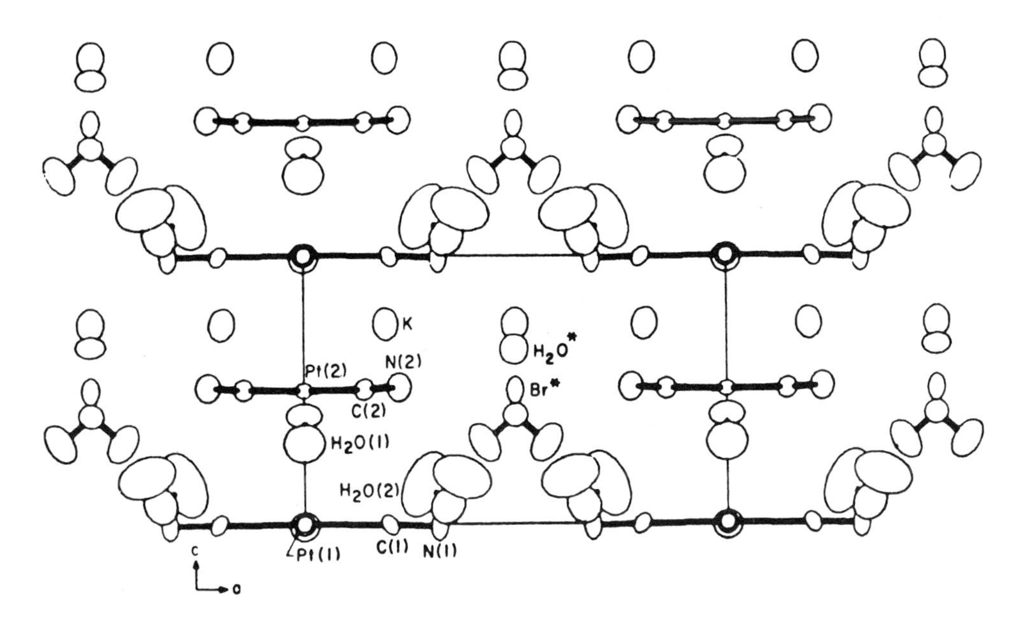

Fig. 3. View of the unit cell of the KCP-structure. There is one Pt-chain with two units of [Pt(CN)$_4$] in the tetragonal cell (*P4mm*). The metal cation occupies the upper part and the crystal water the lower part. The anion that partially oxidizes the Pt is in the center of the unit cell. This structure prevails for intermediate intrachain platinum distances $d_{\|}$.

the cell when Br is absent, giving a total of 3.2 water molecules per formula unit as shown in Table I. The concentration of Br^{-1} determines the nominal oxidation state for the Pt-ion to be 2.30 and one of the issues that we will address below is how this number appears to be determined by a combination of the sizes of the ions and the electronic structure of the [Pt(CN)$_4$]-salts.

Although the structure of KCP(Br) is probably well given by the above description, there are still some unclear points regarding its details, in particular the water content. Thermogravitometric analyses indicate that in protonated KCP(Br) only 3.0 of the 3.2 H_2O-molecules are released [17] whereas deuterated KCP(Br) releases all its D_2O [29], indicating a difference in binding energy of the central water molecule in the two cases. This may seem as a minute detail to the uninitiated eye. However, there is solid indication both from comparison between inelastic neutron scattering as well as from Raman scattering results that protonated and deuterated crystals have significant different bonding energies, as measured by their phonon spectra $\omega(q)$. In consequence, since $\omega_0(2k_F)$ enters the electron–phonon coupling according to (2), λ is not necessarily the same in protonated and in deuterated KCP, a point to be discussed in Section 4. The uncertainty in the accurate structural characterization of even the most studied platinum conductor is just one of many lessons which tell that these compounds are complicated creatures, and the more details we try to unravel the less clearcut do the answers to our questions often come out.

Crystalline homologues of KCP(Br) have been made by varying both cations and anions, resulting into a stretching or compression of the one dimensional Pt-chains which appear to behave very much as accordions. Empirically one finds that this particular crystal structure prevails when the room temperature platinum separation satisfy the relation:

$$2.874 \text{ Å} \leqslant d_{\|} \leqslant 2.920 \text{ Å}. \tag{6}$$

What happens as $d_{\|}$ is further diminished is shown in Figure 4 for the case of RbCP(FHF)

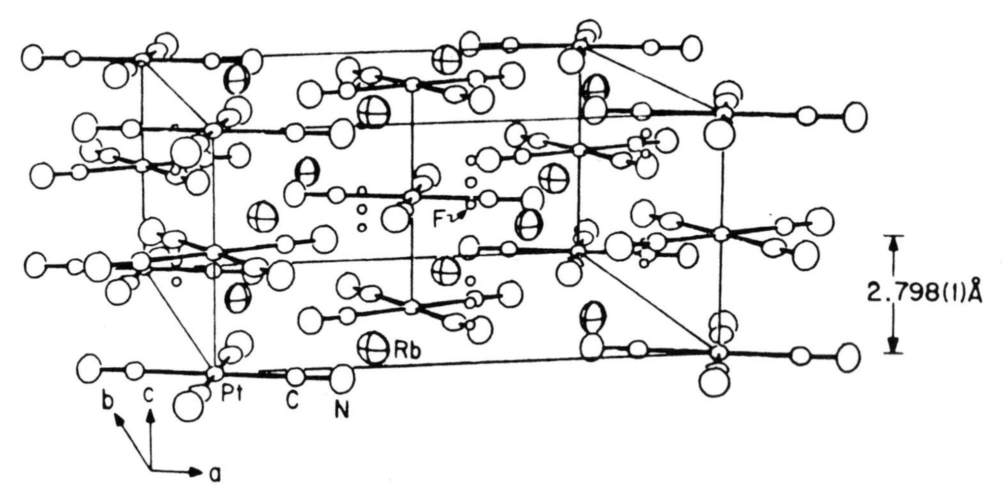

Fig. 4. Perspective view of the unit cell of RbCP(FHF)-structure, which is tetragonal (*I4/mcm*). It represents the shortest Pt–Pt distance achieved. There are two Pt-chains in the cell, but unlike the KCP-structure this structure is anhydrous and the cations are located in the planes of the $[Pt(CN)_4]$-ions.

[21]. The RbCP(FHF) structure is body-centered tetragonal (space goup *I4/mcm*) and it has the shortest reported platinum separation of 2.800 Å, not much longer than the bond length in elementary Pt which is 2.775 Å. Compounds have been prepared with this structure in the range:

$$2.800 \text{ Å} \leqslant d_{\|} \leqslant 2.859 \text{ Å}. \tag{7}$$

Compared with the KCP structure the shortening has been achieved in the following way. The cations are now situated *in the planes* of the $[Pt(CN)_4]$-groups, and in the case of $(FHF)^{-1}$ these anions are small compared to the halide ions. Also these crystals are anhydrous; a fact that of course contributes to the diminishing of the unit cell as a whole. With respect to the physical behavior the interesting feature is that Pt here has nominal oxidation states as high as 2.4, which results in significant changes in the solid state properties compared to KCP(Br).

Although the two structures described above seem to be able to accommodate major changes in the sizes and shapes of both cations and anions, some ions do require separate structures. This is true for disulfatohydrogen ion $[H(OSO_3)_2]^{-3}$ [22], the azide ion $(N_3)^{-1}$ [26], as well as the guanidinium cation $[(NH_2)_3]^+$ [32]. It is also true in the case of the cation deficient salts. In Figure 5 we show the structure of the cation deficient K(def)TCP [34]. This salt is interesting for several reasons. It has the longest $d_{\|}$ reported for the $[Pt(CN)_4]$-compounds, the Pt chain is slightly zig-zagged, and finally related to

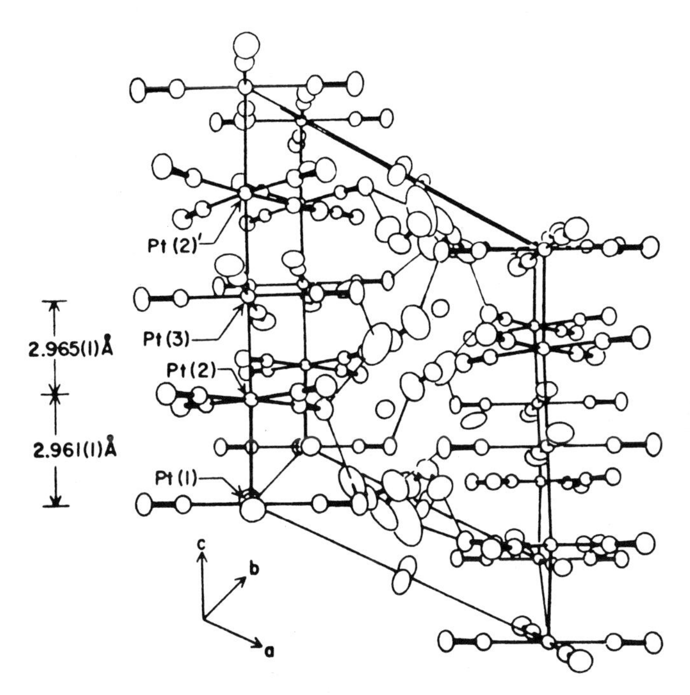

Fig. 5. Perspective view of the unit cell of K(def)TCP, which represents the longest $d_{\|}$ of $[Pt(CN)_4]$-compounds. This structure is triclinic ($P\bar{1}$).

the fact that is has only 1.5 water molecule per formula unit the hydrogen-bonding network is quite different from that of the KCP-structure of Figure 3. Further, neighboring $[Pt(CN)_4]$'s are nearly eclipsed in K(def)TCP as opposed to the other salts, where they are rotated 45° with respect to each other. These structural characteristics are likely to be responsible for some of the anomalous behavior of K(def)TCP, but it is at present not clear how to comprehend in detail the observed phenomena in this compound.

As we shall see below, it is for many purposes sufficient to consider the conducting $[Pt(CN)_4]$-salts as consisting of the 'platinum accordion' sketched in Figure 2, where all physical parameters may be related to the interchain distance $d_{\|}$ which may be varied between:

$$2.80\ \text{Å} \leqslant d_{\|} \leqslant 2.96\ \text{Å}. \tag{8}$$

2.3. CRYSTAL STRUCTURES OF $[Pt(C_2O_4)_2]$-SALTS

From a crystallographic point of view the conducting bis-oxalatoplatinates show a similar richness as the tetra-cyanoplatinates, and the occurrence of different kinds of superstructures make them fascinating objects to study [19, 46]. In this review we will, however, limit ourselves to the isostructural series of *divalent cation deficient* conducting salts. The structure is shown in Figure 6 for the case of CoOP [37]. It is orthorhombic

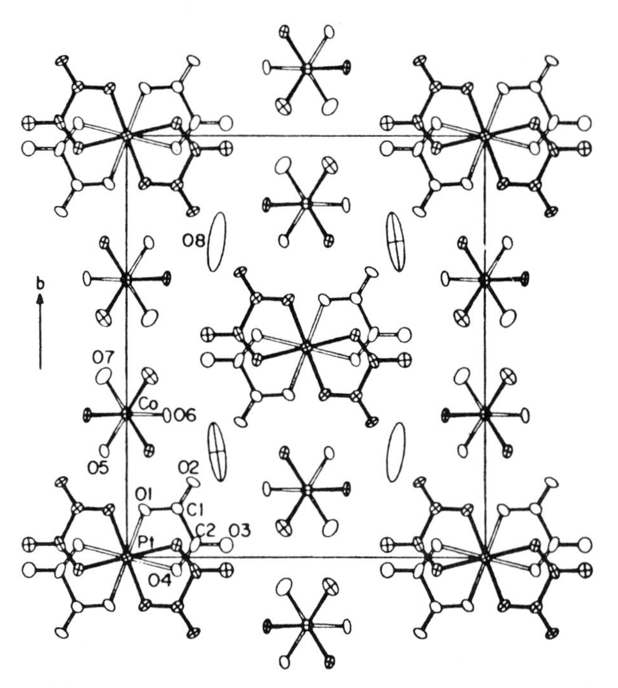

Fig. 6. The unit cell of CoOP, representing the structure of the divalent cation deficient $[Pt(C_2O_4)_2]$-compounds. The metal cation is positioned between the planes of the ligands and surrounded by six water molecules. Some extra H_2O is located in these planes. This structure is orthorhombic (*Cccm*).

(*Cccm*) with two chains per unit cell and with the metal cations at approximately 40% occupancy coordinated with six water molecules situated in between the planes of the ligands. Some extra H_2O is found at special sites so that the total water content is 6 per formula unit. This may also be true for MgOP although the extra water site was not found originally. But lack of this extra water *may* also be responsible for the somewhat unique behavior of MgOP. Regarding the conducting Pt-chain, it seems far less stretchable within a given structure here than in the case of $[Pt(CN)_4]$-salts. This is clearly seen when considering the range of observed platinum distances of the compounds of Table II:

$$2.835\ \text{Å} \leqslant d_{\parallel} \leqslant 2.85\ \text{Å}, \tag{9}$$

which is evidently a much smaller range than given by (8). This indicates that interactions between the $[Pt(C_2O_4)_2]$-ligands play a greater role in establishing the platinum bond than is the case for the $[Pt(CN)_4]$-ligands. This is consistent with the fact that the bis-oxalatoplatinates do not display the Peierls transition as clearly as the tetracyanoplatinates. However, it is astonishing that the compounds of Table II exhibit a greater diversity in physical behavior than their $[Pt(CN)_4]$-counterparts since they are structurally so similar. This will be discussed in Section 6 with emphasis on the specific role of the cations.

2.4. Crystal Structures of Other Conducting Salts

As mentioned above the information available on other conducting platinum salts is more scarce. Only the full structure of LiPt(mnt) is known [44]. It is triclinic (space group $P\bar{1}$) with the $[Pt(S_2C_4N_2)_2]$ ligands very nearly eclipsed. The Pt separation is $d_{\|} = 3.639$ Å. For the two other compounds in Table III only the nominal oxidation states of platinum are known to be 2.4 and 2.8 for KSQAR and KCROC, respectively [41, 42]. In the latter compound $d_{\|}$ has been determined to be 3.11 Å. This distance is just greater than the shortest $d_{\|}$ of 3.09 Å reported for the 'unoxidized' Pt-II salt $Sr[Pt(CN)_4] \cdot H_2O$ [46] and 3.10 Å for $Mg[Pt(CN)_4] \cdot 7H_2O$ under a pressure of 12 kbar [48].

2.5. Summary

The above review of conducting platinum compounds and their crystal structures tells that a conducting platinum chain may be manufactured in several ways. Several ligand systems have been successfully used and there is no reason why those shown in Figure 2 should exhaust the possibilities.

The smaller ligands $[Pt(CN)_4]$ and $[Pt(C_2O_4)_2]$ give rise to very short intrachain Pt–Pt bonds $d_{\|}$ so here we expect a big overlap between d_{z^2}-orbitals along the chain. An interesting feature of the former class is that $d_{\|}$ may be varied from 2.798 Å to 2.963 Å, quite a stretch of the Pt-chain with important changes of the physical properties in consequence. Another intriguing point is that the conducting chain remains unexpectedly intact in spite of big changes in crystallographic environments; but on the other hand a small change in KCP(Br) such as deuteration has a significant effect which has to be dealt specifically with. The latter class of small ligands, the oxalatoplatinates, also form a rich variety of crystals but here the intrinsic properties of the conducting chain are changed more profoundly.

For conducting crystals with larger ligands the direct role of the Pt is less clear. As $d_{\|}$ exceeds 3 Å the d_{z^2}-overlap diminishes as evidenced by the occurrence of nonconducting $[Pt(CN)_4]$-salts, but still conducting salts can be made. In LiPt(mnt) $d_{\|} = 3.639$ Å which is similar to the S–S distances in organic conductors so here one has probably reached a distance where the direct role of the Pt-ion is small.

3. The Metallic Phase

As mentioned in the introduction, the metallic phase of a one-dimensional conductor is stable at some high temperature, whereas a semiconducting state prevails at lower temperatures. The scale temperature, separating the two phases is T_P^{MF}, so in order to study the metal we need in principle to be well above this temperature. Here 'well above' has a rather delicate meaning, namely that we want not only to be above T_P^{MF}, but also outside the region of fluctuations. And the more one dimensional the system is, the more extended is this temperature region. This is sketched in Figure 7, where we show the phase diagram of a quasi one dimensional conductor as a function of temperature and the parameter η, characterising the interchain coupling. In Section 5 we discuss how

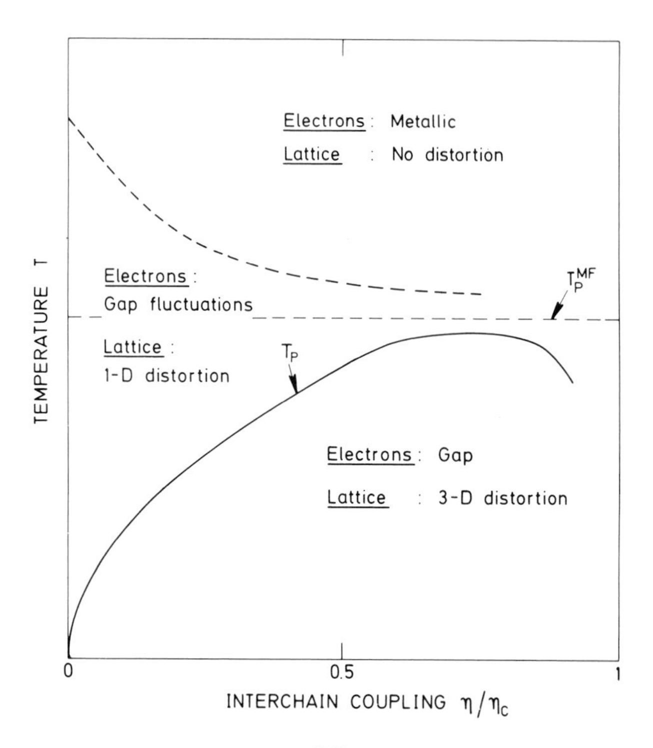

Fig. 7. Phase diagram for the Peierls instability. T_P^{MF} is the mean field scale-temperature and T_P is the real transition temperature.

η is expressed when specific models are considered. The actual transition temperature T_P is calculated from (5) at intermediate and large values of interchain coupling [14], and supposed to follow an $\sqrt{\eta}$ dependence for small η's [49]. The separation between the undistorted metal and the fluctuating metal is not well defined, but for $\eta = 0$ the onset of fluctuations has been set to their Ginzburg value of 1.6 [50].

Given these considerations it quickly becomes obvious that the high temperature metallic phase of conducting platinum compounds is experimentally inaccessible. Measured energy gaps are typically between 50 and 100 meV leading to value of T_P^{MF} well above room temperature from (4). Since the crystals decompose at temperatures not much above room temperature one has to resort to more sophisticated ways of deducing the high temperature characteristics of the metallic phase than simply raising the temperature. In practice the temperature can only be raised to somewhere between the actual transition temperature T_P and T_P^{MF} where the compound is literally in the metallic phase, but influenced by the Peierls instability. We address this problem in the following way: Given an experimental result which reflects a metallic property, care is taken to assess how the measurement is influenced by the fact that semiconducting fluctuations are present at all practical temperatures. In this respect the conducting platinum compounds are different from the organic charge transfer salts, where the genuine metallic phase may often be studied over an extended temperature range.

Nevertheless, as we shall see below, it is possible to give a fairly complete characterization of the metallic phase of the platinum compounds, in particular deriving the parameters of importance for Peierls instability.

3.1. THE ROOM TEMPERATURE CONDUCTIVITY

The first indication that the salts listed in Tables I–III are metallic stems from their room temperature conductivity in the chain direction $\sigma_{\parallel}$, as measured by dc- or very low frequency ac-techniques. Table IV lists the reported values [18, 22, 30, 32, 37, 39, 41, 43, 53–62] as well as the distance $d_{\parallel}$. As one would expect these two parameters vary oppositely each other. Also shown is the conductivity of Pt-metal and the unoxidized Pt-II salts $Mg[Pt(CN)_4] \cdot 7H_2O$ and $K_2[Pt(CN)_4] \cdot 3H_2O$ [60]. From the conductivity values of Table IV the metallic character of the partially oxidized platinum compounds is striking, in particular when it is kept in mind that the density of electrons in the salts is only about 10% of Pt-metal; but it must of course be kept in mind that the

TABLE IV

One-dimensional metallic characteristics. References are given in brackets.

Compound	$d_{\parallel}$ Å	$\sigma_{\parallel}$ $(\Omega\ \text{cm})^{-1}$	ω_p $10^{15}\ \text{s}^{-1}$	k_F $\pi/d_{\parallel}$	$\omega_0(2k_F)$ $10^{12}\ \text{s}^{-1}$
Pt-metal	2.775 [52]	9.4×10^4 [52]	–	–	–
RbCP(FHF)	2.798 [21]	2000 [53]	–	0.80	–
RbCP(DSH)	2.826 [22]	400 [22]	–	0.85 [22]	–
CsCP(FHF)	2.833 [23]	2000 [53]	–	0.8 [85]	–
CsCP(Cl)	2.859 [24]	200 [54]	–	0.85	–
KCP(Cl)	2.874 [25]	200 [18]	–	0.84	–
CsCP(N_3)	2.877 [26]	150 [55]	–	0.875	–
KCP(Br)	2.888 [27]	200 [56]	4.42 [56, 63]	0.86 [86]	12.2 [91]
Pb/KCP(Cl)	2.899 [30]	100 [30]	–	0.885 [22]	–
RbCP(Cl)	2.900 [31]	10 [57]	–	0.845	–
GCP(Cl)	2.910 [32]	11 [32]	–	0.875 [87]	–
ACP(Cl)	2.920 [33]	0.4 [58]	–	0.875 [57]	11.1 [57]
K(def)TCP	2.963 [34]	100 [59, 60]	4.49 [65]	0.875 [87]	11.1 [92]
SrTCP	3.09 [47]	–	–	1	–
MgTCP	3.155 [48]	10^{-5} [61]	–	1	–
KTCP	3.478 [51]	10^{-8} [61]	–	1	–
MnOP	2.835 [37]	30 [37]	–	0.875 [22]	–
NiOP	2.836 [38]	10 [37]	–	0.85 [81]	–
ZnOP	2.838 [39]	60 [39]	2.50 [38]	0.84 [85]	–
CoOP	2.841 [40]	15 [62]	–	0.85 [85]	–
MgOP	2.85 [4]	25 [62]	–	0.85 [85]	–
KSQAR	–	0.0005[a] [41]	–	0.8	–
KCROC	3.11 [42]	–	–	0.6	–
LiPt(mnt)	3.639 [44]	100 [43]	1.1 [66]	0.59 [88]	–

[a] Powder value similar to powder conductivities for OP's.

conductivity at one temperature is only a very coarse measure of the electronic properties of a solid. Another point with respect to the conducting salts is that room temperature falls in different places of the diagram in Figure 7 for different compounds, and therefore a comparison is not necessary meaningful. However, the temperature dependence of $\sigma_{\parallel}(T)$ shows a smooth maximum around room temperature for most of the compounds, and as such represents a common feature. Regarding the degree of one dimensionality the conductivity anisotropy lies in the range $60-10^4$ for the $[Pt(CN)_4]$-salts and is about 100 for the conducting $[Pt(C_2O_4)_2]$-compounds.

3.2. The Electronic Band Structure

More compelling evidence for the metallic nature of the conducting Pt-chain compounds comes from investigations of their electronic band structure. By analogy with three dimensional *d*-metals one expected the conduction band to be of the tight binding type and to have a rather narrow bandwidth. Given the band filling of KCP this material should have hole-like transport with the number of carriers n corresponding to $n_h = 0.3$ holes per Pt-ion. It was therefore surprising when infrared reflectivity measurements on KCP, reproduced in Figure 8, showed a very high plasma edge corresponding to a

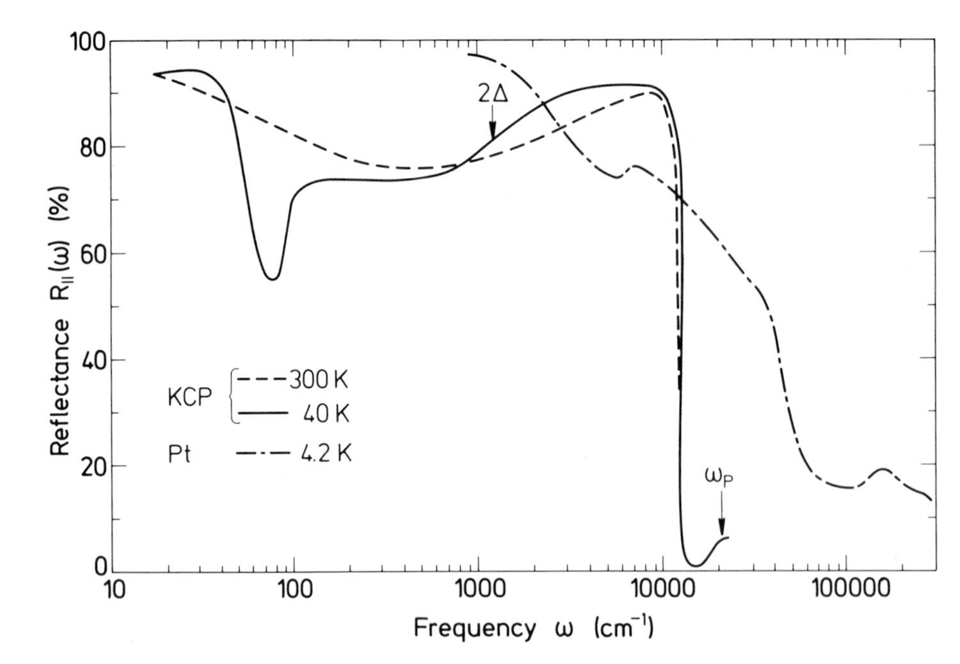

Fig. 8. Reflectance along the conducting axis of KCP(Br). The perpendicular reflectivity is about 5% and structureless. The plasma frequency ω_p derived from a Drude fit is shown together with the position of 2Δ where excitations across the Peierls set in. The reflectivity of Pt-metal is also shown. Frequencies are given in units of their corresponding wavenumber $\omega/2\pi c$ (in cm^{-1}). 1 cm^{-1} corresponds to an angular frequency 0.188×10^{12} s^{-1} or a frequency of 30 GHz.

plasma frequency of $\omega_p = 4.42 \times 10^{15}$ s^{-1} (i.e., a wavenumber $\omega_p/2\pi c = 23500$ cm^{-1}) [56, 63]. From the relation:

$$\omega_p^2 = \frac{ne^2}{\epsilon_0 m^*}, \tag{10}$$

where $\epsilon_0 = 8.85 \times 10^{-12}$ As/(Vm) is the vacuum dielectric constant. The measured ω_p corresponds to an effective hole-mass of $m^* = 0.18\ m$, or a very large bandwidth in contradiction with the assumed narrow bandwidth. This suggests that KCP is not adequately described within the tight-binding approximation. Alternatively, if one assumes a free electron approximation the density of carriers is $n = n_e = 1.7$ electrons per Pt-ion, and then (10) yields $m^* = 1.0\ m$. Comparing the result of the two analyses one arrives at the conclusion that the electronic of KCP is well described in the free electron approximation. Infrared studies of the compounds K(def)TCP [63, 64] and ZnOP [39] indicate that also these compounds are very nearly free electron like, an important fact for our analysis of the Peierls instability in these compounds. All measured plasma frequencies are given in Table IV. The composite evidence is that all the conducting salts based on [Pt(CN)$_4$] and [Pt(C$_2$O$_4$)$_2$] have the simple band structure, shown in Figure 9. The compounds with larger ligands shown in Table III are, however, likely to be well described within the tight binding approximation. Only LiPt(mnt) has been investigated by IR, and the bandwidth

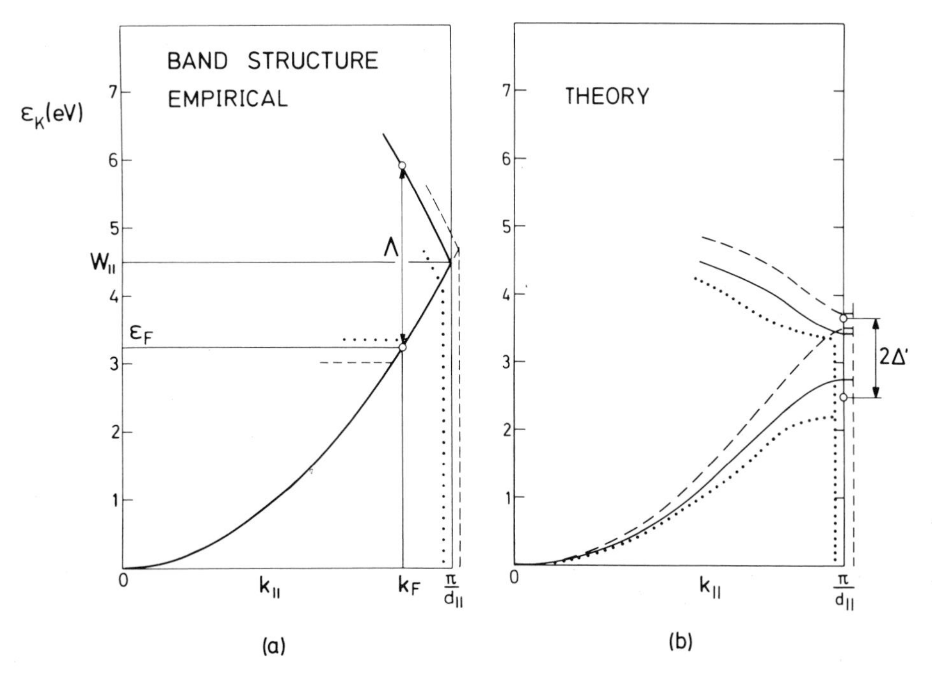

Fig. 9. Electronic band structure ϵ_k of conducting [Pt(CN)$_4$]-compounds in their undistorted metallic phase. (a) shows the experimentally derived free electron structure with full lines corresponding to KCP(Br). Dashed and dotted lines indicate the outer extremes for the series of Table I. (b) shows the almost free electron like band from recent calculations [73]. Two Δ′ represent an extrapolated hypothetical gap as discussed in the text. Energies are given in eV (1 eV = 1.60 × 10^{-19} J).

appears to be of the order of 0.4 eV [66]. In order to compare in a rough way the characteristics of a tight binding metal and a free electron metal we define the bandwidth for the latter case as:

$$W_{\parallel} = \tfrac{1}{4}(\hbar d_{\parallel}^{*})^2/(2m). \tag{11}$$

In (11) we have defined the reciprocal lattice vector $d_{\parallel}^{*} = (2\pi)/d_{\parallel}$. This corresponds to a unit cell containing one Pt-ion, a convenient definition when discussing the solid state properties. (11) gives $W_{\parallel} = 4.5$ eV for KCP, and similar values for the other salts of Tables I and II.

Independent evidence of the free electron band structure from luminescence [67, 68] and infrared absorption [63, 39] stemming from the interband transition of energy Λ in Figure 9. These transitions have been identified in KCP [63] and ZnOP [39] from shoulders on the plasma edge in IR experiments, and as luminescence in KCP [67] and K(def)TCP [68]. The observed transition energies agree within a few percent with the free electron value for Λ:

$$\Lambda = 4W_{\parallel}(1 - k_F d_{\parallel}/\pi). \tag{12}$$

A third experimental result indicating that KCP has a free electron band structure comes from thermopower measurements. The theoretical value for the one-dimensional free electron gas is:

$$S_{\parallel} = -\frac{\pi^2 k_B}{6e}\left(\frac{k_B T}{\epsilon_F}\right), \tag{13}$$

which, with $\epsilon_F = 3.25$ eV appropriate for KCP, gives $S_{\parallel} = -1.13$ μV/K at room temperature. The measured value is $S_{\parallel} = -1.5$ μV/K [69]. In the oxalatoplatinates $S_{\parallel}$ is also small at room temperature consistent with a large bandwidth [70]. However, as the temperature dependence of the thermopower also varies from compound to compound around room temperature it is hard to use this value for quantitative measurements of the band structure ϵ_k.

The theoretical understanding of the band structure of the conducting platinum chain compounds has also brought us from the original tight binding picture to – almost – the free electron picture. In an early calculation for KCP the bandwidth of the $5d_{z^2}$ electron-band was found to be $W_{\parallel} = 1.13$ eV, i.e., similar to results for the three dimensional *d*-metals at that time [71]. However, it was soon realized that the overlap in the one-dimensional array of Figure 2 is markedly different from Pt-metal, leading to a modified band structure. Hence later calculations have shown that although starting from basically a tight binding picture one gets a large bandwidth, and the calculated ϵ_k approximates fairly well the free electron dispersion curve found in the experiments. However, even the latest calculations [72, 73] still gives hole characteristics for KCP, as indicated in Figure 9, and this of course limits the usefulness of the theoretical results in quantitative comparison with experiments. Calculations of the band structure in conducting $[Pt(C_2O_4)_2]$-salts as well as IR-measurements indicate that $W_{\parallel}$ is narrower than in the $[Pt(CN)_4]$-compounds [74, 75].

It is interesting to compare our current understanding of the band structure of one-dimensional Pt-chain compounds to that of Pt-metal. Photoemission spectroscopy [76] infrared reflectance studies [77] as well as band structure calculations [78] show that in Pt-metal 4 bands have d-like character and a combined bandwidth of about 7 eV. However, none of these bands bear resemblance to the free electron d_{z^2}-band characterizing ϵ_k for the Pt-chain metals. As a demonstration of their different electronic structure Figure 8 also shows the reflectivity of Pt-metal [77]. Hence, apart from the fact that it has become acknowledged that both types of compounds have larger bandwidth than originally anticipated, the development described here has shown that one-dimensional Pt-conductors should definitely not be thought of as representing a particular direction in ordinary Pt-metal.

When investigating the metallic properties of the Pt-salts, how did we deal with the above mentioned problem of separating metallic and semiconducting features in our analysis of the experiments? In the IR-spectrum of Figure 8 we relied on the plasma edge as representative of the high temperature metal, but we neglected the indentation in $R(\omega)$ around wavenumbers of 200 cm^{-1}, which we will later relate to the Peierls gap Δ. Hence the reflectivity at room temperature shows features of both the high as well as the low temperature phase, but they naturally separate, thereby allowing an unambiguous determination of parameters. Regarding the interband transition Λ, this transition is only allowed since the compound is in fact distorted even above the transition temperature, so here a determination of a metallic characteristic depends upon the fact that we are in the fluctuating regime of Figure 7. With respect to the transport parameters $\sigma_{\parallel}$ and $S_{\parallel}$, these are harder to analyze unambiguously, since the competing tendencies of the metal and the semiconductor cannot be separated at any given temperature. Therefore these measurements should be considered 'supplementary' as far as characterizing the metallic state.

Although the evidence for the free electron nature of the tetra-cyanoplatinates and the bis-oxalatoplatinates seems convincing, it should be mentioned that two alternative pictures have been put forward to explain the former class. Firstly Yersin *et al.* have proposed that ϵ_k for the conducting $[Pt(CN)_4]$-salts can be derived by extrapolation from ϵ_k for the nonconducting salts [79]. This in particular implies a zone boundary gap Δ', which has been accurately measured in a series of nonconducting Pt-II compounds, by a mechanism similar to the interband transition Λ in (12), since here one may formally set $k_F = d_{\parallel}^*/2$. Extrapolation suggests that in KCP the gap is $\Delta' \cong 0.6$ eV which is close to the result of the recent band structure calculations reproduced as in Figure 9. But as just mentioned this gap must be severely overestimated since it implies hole nature for the transport properties of KCP, apart from the fact that there is no experimental evidence for a second gap in the conducting platinum salts. So the relation between the electronic structure of the conducting compared with the nonconducting $[Pt(CN)_4]$-salt must be more complicated than suggested by Yersin *et al.* A more realistic approach in this regard would be to take the enormous screening into account caused by the large dielectric constant of KCP. This would yield $\Delta' \approx 0.6\ \text{eV}/\epsilon_{\parallel}^2$ which become negligible with the measured dielectric constant [90] $\epsilon_{\parallel} \gtrsim 1000$. A different possible connection between metallic and nonmetallic Pt-chain compounds would be that Δ' in the Pt-II salts could (in part or fully) stem from a large Peierls distortion in these would be half filled band conductors, similar to what appears to be the case in *trans*-polyacetylene $(CH)_x$. This is, however, not the case since Δ' would then be accompanied by a dimerization of the intra-chain Pt-separation, of which there is no evidence.

Secondly, Nagasawa has recently proposed that the $[Pt(CN)_4]$-chain of KCP should not be thought of as a band conductor with delocalized electrons, but rather as a chain containing localized electrons on alternating ions of Pt^{+2} and Pt^{+4} appropriately mixed to give the correct average oxidation state [81]. Within this framework, he can explain certain experimental results in particular those which do not reflect the extended nature of the electrons. Here we shall continue the 'band picture' because it, in my opinion, most clearly and consistently display the interesting physics.

3.3. Relation between Pt-Separation and Band Filling

An interesting systematic behavior of the compounds in Table I which may be brought in direct relation with their electronic band structure is the connection between the platinum separation $d_{\|}$ and the nominal oxidation state of the Pt-ion. Defining the degree of partial oxidation DPO as the deviation of the oxidation state from 2, Williams [82] first noticed that $d_{\|}$ followed the empirical relation:

$$d_{\|}(\mathrm{DPO}) = d_{\|}(1) + \delta \cdot \log(\mathrm{DPO}), \tag{14}$$

with values for $d_{\|}$ and δ closely resembling those derived by Pauling for conducting platinum compounds ('metallic resonance') based on integral valence states [83]. Since the DPO is related to the Fermi vector in the following way:

$$k_F = \frac{\pi}{d_{\|}}\left(1 - \tfrac{1}{2}\mathrm{DPO}\right). \tag{15}$$

the systematic behavior reflects a 'controlled' variation of the band filling as given by k_F. Figure 10 illustrates the relation between k_F and $d_{\|}$ for the salts in Tables I–III,

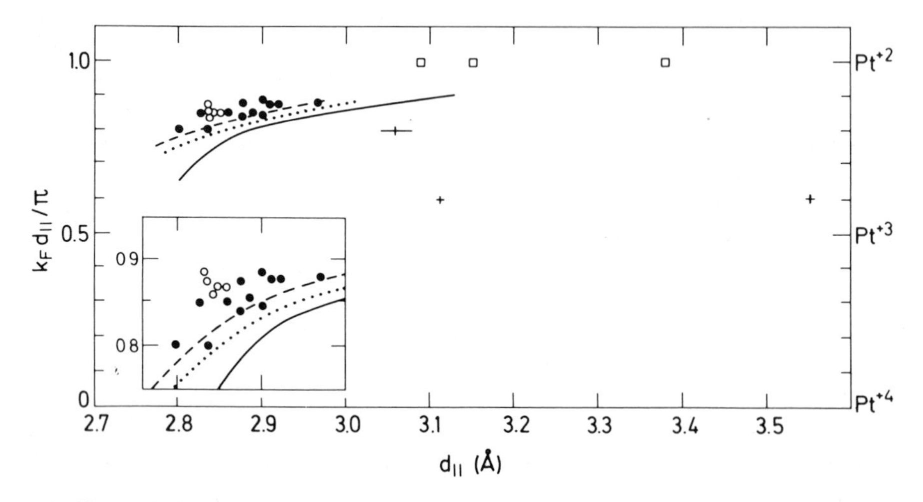

Fig. 10. Electronic band filling *vs* platinum distance $d_{\|}$. Full line: Theoretical result by Whangbo and Hoffman [73]. Dashed line: Pauling's empirical result for 'metallic resonance'. Dotted line: 'non-metallic' resonance. Full circles: $[Pt(CN)_4]$-conductors, open circles: $[Pt(C_2O_4)_2]$-conductors, and crosses: other conductors. Squares: nonconducting $[Pt(CN)_4]$-salts.

together with Pauling's empirical rules both for metallic behavior as well as nonmetallic behavior [84]. Values for $d_{\parallel}(1)$ and δ are listed in Table V.

TABLE V

Parameters of (15) derived as described in the text. The results of [73] have been fitted in the region $0.2 < \text{DPO} < 0.5$.

	$d_{\parallel}(1)$ (Å)	δ
Non-metallic Pt-complexes [84]	2.59	0.71
Metallic Pt-complexes [83]	2.59	0.60
Partially oxidized Pt-conductors	2.56	0.62
Theory [73]	2.48	0.45

Apart from demonstrating that the conducting platinum chain compounds are safely characterized as metallic by the criteria of Pauling, the qualitative behavior of the compounds may be understood from the band structure calculations of Whangbo and Hoffmann [73]. They considered the energy of a chain with a given DPO as a function of $d_{\parallel}$ to find the minimum energy, which then determines the optimum $d_{\parallel}$. As shown in Figure 10 their calculated curve resembles that of Pauling as well as the experimental behavior. Given the known deficiency of the calculated band structure described above it is not surprising that the agreement between the theoretical curve and the experimental is not quantitative, but this *qualitative* agreement is a solid indication that the intra-chain Pt–Pt bond in the conducting $[Pt(CN)_4]$-compounds is a genuine metallic bond. One important consequence of the relation between k_F and $d_{\parallel}$ is that it explains the chemical composition of the compounds, e.g., the concentration of Br^{-1} in KCP(Br). According to the above conclusions it is governed by the delicate balance between accommodating counterions and water in the unit cell (thereby determining $d_{\parallel}$) and accommodating an appropriate number of electrons in the Pt-bond of length $d_{\parallel}$ to maintain its metallic character.

The bandfilling may be determined in three ways from experiments. *Firstly*, platinum oxidation titration yields the nominal oxidation state of the Pt-ions directly; but this technique is not always applicable. *Secondly*, quantitative chemical analysis determines the concentration of oxidizing anions (or the cation-deficiency). Assuming that only Pt is oxidized, the DPO (or the bandfilling) follows from this concentration. However, if H_3O^+ is present in the lattice DPO from chemical analysis gives different results from titration, a problem which first arose in the analysis of ACP(Cl) [58]. This compound has DPO = 0.25 and should therefore strictly be written as $(NH_4)_2(H_3O)_{0.17}[Pt(CN)_4]Cl_{0.42} \cdot 2.83H_2O$. Similarly RbCP(DSH) incorporates nominally 0.17 H_3O^+ per formula unit. The *third* and most reliable method to determine the bandfilling is diffuse X-ray scattering at room temperature. Owing to the Peierls instability the lattice distorts along the Pt-chains with a period of $2k_F$, giving rise to characteristic diffuse scattering lines. DPO's from this technique have always been found to be in accord with titration measurements, when the latter were possible. The values for k_F listed in Table IV are taken from X-ray experiments [22, 58, 85–88] except in the few cases, where only the 'chemical' k_F is available.

Figure 10 also shows the k_F as a function of $d_\parallel$ for the compounds in Tables II and III. For the $[Pt(C_2O_4)_2]$-salts the situation is less clear than for the $[Pt(CN)_4]$'s, since they fall roughly within the regular behavior, but without displaying the same extended systematic behavior. For the conductors of Table III, their position in Figure 10 makes it quite clear that they must have electronic structures which are very different from the other compounds.

3.4. THE LATTICE DYNAMICS

In principle it is enough to know the bare phonon frequency at high temperatures $\omega_0(2k_F)$, in order to treat the Peierls instability, since only this particular phonon couples strongly to the electrons in the Fröhlich Hamiltonian (cf. (1) and (2)). In practice several complications may occur. A rather mundane reason for the lack of extended knowledge about the lattice dynamics in one-dimensional conductors is that they are produced as very small crystals. So only in a few cases have crystals been produced big enough for inelastic neutron scattering, the only method that gives detailed insight into the lattice dynamics. Regarding conducting platinum chain compounds only KCP(Br) is well studied [89–91], but ACP(Cl) [58] and K(def)TCP [92] have been also investigated. Another complication is the already mentioned that high temperatures are not attainable, so that we expect the lattice dynamics to be modified at all practical temperatures by the electron-phonon coupling. Thirdly, a fundamental point is that although the compounds are one-dimensional conductors with the conduction electrons associated only with the Pt-ions, they are genuinely three-dimensional from a lattice dynamical point of view with many degrees of freedom. This leads to a multitude of three-dimensional phonons $\omega_i(\mathbf{q})$, all of which may couple to the electrons if the component of $\mathbf{q}$ along the conducting axis is $2k_F$.

Inelastic neutron scattering experiments do, however, reveal that the lattice dynamics of conducting platinum compounds are relatively simple with respect to the Peierls instability. The prevailing picture is that only a single longitudinal phonon branch is active in the coupling to the conduction electrons, as it is sketched in Figure 1. Figure 11 shows the measured phonon spectrum of deuterated KCP(Br). One finds a 'giant' Kohn anomaly at all temperatures in the phonon branch associated with the longitudinal acoustic phonon $\omega(q_\parallel)$ propagating along the conducting **c**-direction of the crystal, but the anomaly is so narrow that it allows determination of $\omega_0(2k_F)$ by interpolation. Moreover, its value proves to be very closely independent of the perpendicular component of $\mathbf{q}$. However, the derived $\omega_0(2k_F)$ does depend on the temperature reflecting the general anharmonic nature of the crystal not associated with the electron–phonon coupling at $2k_F$. In order not to incorporate these effects in the analysis of the Peierls instability it seems therefore most correct to attribute the low temperature interpolated value of 8.0 meV to $\hbar\omega_0(2k_F)$ for KCP(Br). Similar analysis of the dispersion relations in the two other studied compounds suggests that the unperturbed phonon frequency of the tetra-cyanoplatinates appropriate for use in (2) is described by:

$$\omega_0(2k_F) = \omega_D\,|\sin(k_F d_\parallel)|. \qquad (16)$$

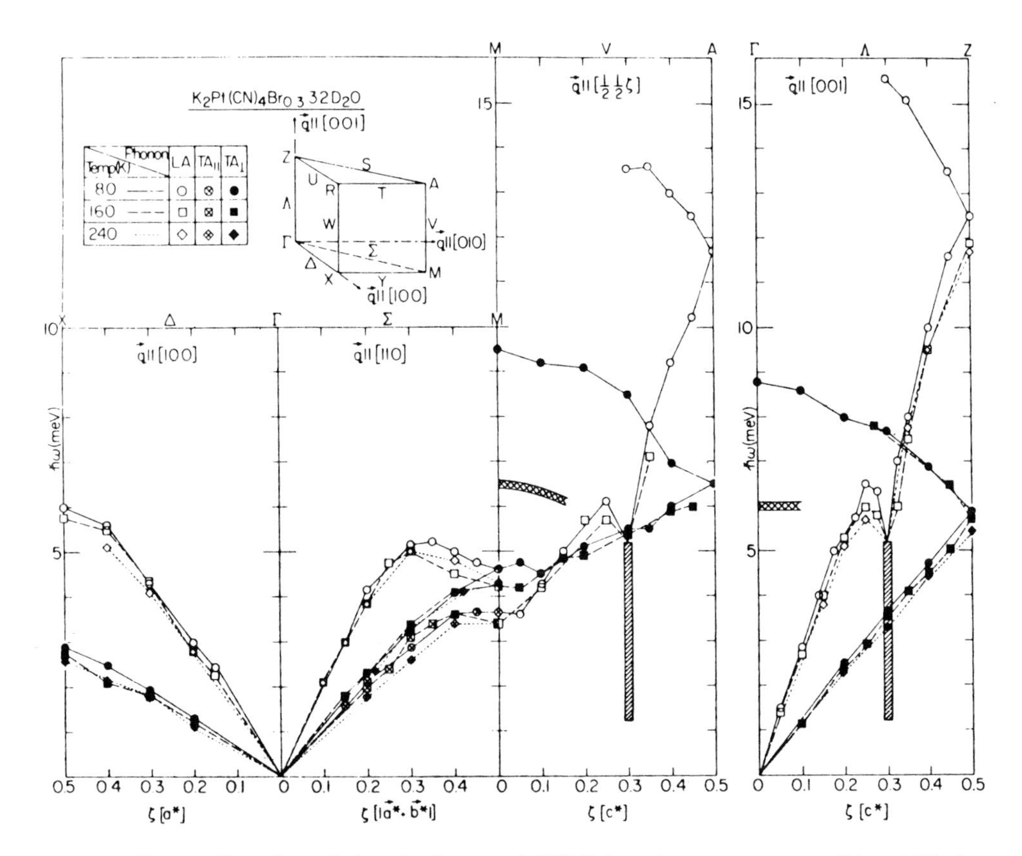

Fig. 11. Phonon dispersion relations in deuterated KCP(Br) at three temperatures. LA and TA designate longitudinal and transverse acoustic modes respectively. The shaded areas correspond to $q_{\|} = 2k_F$ where the strong electron–phonon coupling is active. The cross-hatched area indicates a possible optical phonon. Energies are given in meV and 1 meV corresponds to an angular frequency of 1.52×10^{12} s^{-1} or a frequency of 159 GHz.

This 'universal' phonon dispersion relation $\omega_0(q_{\|})$ is shown in Figure 12, and specific values for $\omega_0(2k_F)$ are given in Table IV. In (16) ω_D is the Debye frequency corresponding to $\hbar\omega_D$ = 18 meV which seems to be a good common value for the compounds in Table I.

Compared to their organic counterparts the electron–phonon coupling in the $[Pt(CN)_4]$-salts is remarkably simple, since no low lying – external or intramolecular – modes seem to contribute significantly to λ (with the possible exception of K(def)TCP [65]). This is of course related to the fact that the intramolecular modes in the Pt-salts are of high frequency, as for instance the CN-stretching at a wavenumber $\omega/2\pi c$ = 1800 cm^{-1} ($\hbar\omega$ = 200 meV).

Nothing is known at present about the unperturbed phonon spectrum in the bis-oxalatoplatinates or the compounds in Table III, and this limits the degree of sophistication to which the Peierls instability can be investigated.

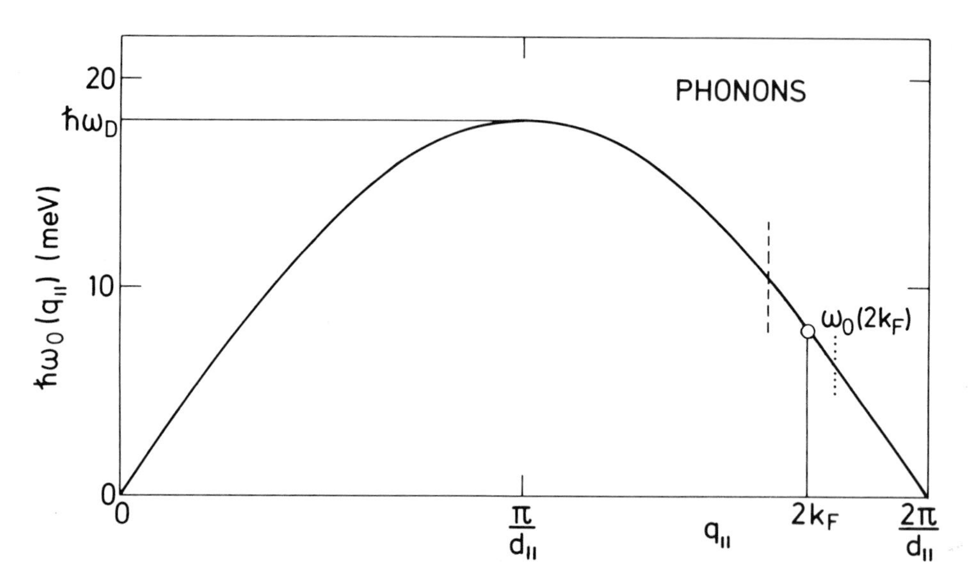

Fig. 12. Phonons $\omega(q_{\|})$ of conducting $[Pt(CN)_4]$-compounds in their undistorted metallic phase. Band fillings are indicated by $2k_F$ for KCP (solid line and circle) and for the outer extremes for the series of Table I (dashed and dotted lines).

3.5. The Electron–Phonon Coupling

Having characterized the high temperature metallic phase of the conducting $[Pt(CN)_4]$-salts it is possible to insert specific expressions for the density of states, the energy gain g, and the unperturbed phonon frequency into the dimensionless electron–phonon coupling constant λ as given by (2). The density of electronic states of the one dimensional free electron is easily derived:

$$N(\epsilon_F) = \frac{\sqrt{2m}}{\hbar d_{\|}^*}\frac{1}{\sqrt{\epsilon_F}} = \frac{1}{2\sqrt{W_{\|}\epsilon_F}}, \tag{17}$$

whereas g may be calculated by modifying earlier results [6, 93, 94] for the tight binding model to the free electron band structure [95]:

$$g = \pm i\frac{\partial W_{\|}}{\partial \alpha}\sqrt{\frac{\hbar}{2M\omega_0(2k_F)}}\,(k_F d_{\|}), \tag{18}$$

where α is the amplitude of the distortion with wave vector $2k_F$. M is the effective ionic mass which has been shown to be the mass of $[Pt(CN)_4]$, i.e., 199 atomic mass numbers. We estimate the magnitude of $\partial W_{\|}/\partial\alpha$ by calculating $\partial W_{\|}/\partial d_{\|}$ from (11) which correspond to a static deformation ($q = 0$) instead of one at $q = 2k_F$. In this way we get:

$$|g| \approx \frac{\partial W_{\parallel}}{\partial d_{\parallel}} \sqrt{\frac{\hbar}{2M\omega_0(2k_F)}} \cdot k_F d_{\parallel}$$

$$= 2W_{\parallel} \sqrt{\frac{m}{M} \frac{\epsilon_F}{\hbar\omega_0(2k_F)}}$$

$$= \frac{1}{2} \frac{(\hbar d_{\parallel}^*)^2}{2m} \sqrt{\frac{m}{M} \frac{\epsilon_F}{\hbar\omega_0(2k_F)}} . \qquad (19)$$

Inserting (16), (17), and (19) into (22) we get the following expression for the dimensionless electron–phonon coupling constant:

$$\lambda = \frac{1}{4} \left[\frac{\hbar^2 d_{\parallel}^{*2}}{2m} \right]^{3/2} \frac{m}{M} \sqrt{\epsilon_F} \frac{1}{\hbar^2 \omega_D^2 \sin^2(k_F d_{\parallel})}$$

$$= 2 \frac{W_{\parallel}^{3/2} \epsilon_F^{1/2}}{(\hbar\omega_D)^2} \frac{m}{M} \frac{1}{\sin^2(k_F d_{\parallel})} . \qquad (20)$$

With respect to the Peierls instability (20) is the key formula for this section. From the microscopic characteristics of the high temperature metallic phase we can calculate the physical parameter λ which determines the macroscopic behavior of the low temperature semiconducting phase by using (3). As may be seen from Figures 9 and 12, the dominating effect of varying k_F, as may be done in the $[Pt(CN)_4]$-conductors, is to vary $\omega_0(2k_F)$ and since this enters squared in (20), λ changes significantly with k_F. It should be noted that (20) has a parallel in superconductivity as first derived by McMillan [96]:

$$\lambda = \frac{\langle I^2 \rangle}{M\langle \omega^2 \rangle} . \qquad (21)$$

In both cases does one find an ω^{-2} dependence, making λ very dependent on the phonon spectrum. But in contrast to the three-dimensional case relevant for superconductivity where a nontrivial average of the phonon density of states has to be considered, complicating the determination of λ from the McMillan formula, in one dimension one only has to consider the phonon $\omega_0(2k_F)$. Therefore (19) and (20) are more directly applicable than (21). Hence, the observed variation of λ among the $[Pt(CN)_4]$ compounds is expected to be well accounted for by (20), as shown in the next section. But the absolute value of g (and therefore λ) as calculated from (18) is uncertain owing to the way in which $\partial W_{\parallel}/\partial \alpha$ was evaluated.

3.6. SUMMARY

In the above treatment we have demonstrated how one with confidence can characterize the metallic state of the conducting Pt-chain compounds. Despite the fact that they never show genuine metallic behaviors it is nevertheless possible to establish that they have large enough electronic band widths along the chains to be characterized was one-dimensional metals. This is of course important since if they were not metals they would not be susceptible to the interesting instabilities of the one dimensional conductor, and then our subsequent analysis of their low temperature behavior would not be justified.

The fact that Pt-chain conductors do not clearly display metallic behaviors above a well defined metal-insulator transition temperature, like many organic conductors do, has two reasons. Firstly, their scale temperature T_P^{MF} is high, which is related to the large Fermi energies (or bandwidths) according to (3) and (4). In fact this temperature is typically above room temperature and cannot be achieved experimentally because crystals disintegrate upon heating. Secondly, Pt-chain conductors are very one-dimensional and therefore fluctuations occur even at temperatures well above T_P^{MF} as indicated in Figure 7 for low η's. An exception is LiPt(mnt) which because of its low bandwidth and rather two-dimensional structure has a relatively low T_P^{MF} and shows relatively weak fluctuation effects.

Of particular interest are the conducting $[Pt(CN)_4]$-salts. Here the variation in $d_{\parallel}$ is associated with a variation in band filling. Hence, the compounds listed in Table I together with detailed results obtained from KCP(Br) allow a systematic study of the effect of changing k_F. In particular the formula (20) for the electron phonon coupling λ suggests that this parameter will change appreciably with k_F.

4. The Simple Charge Density Wave Phase

As a result of the Peierls instability the low temperature state of a one-dimensional conductor is a semiconductor. Here 'low temperature' means below the actual transition temperature T_P which may be determined from experiments. From (5) it is seen that T_P is given *in part* by the Peierls instability and *in part* by the three-dimensionality of the compound. However, below the transition temperature the system is in its ground state which is solely determined by the one-dimensional properties, so that these may be studied unambiguously as long as we stay at temperatures below T_P. The fact that it does not have a fixed relation to T_P^{MF} as seen from Figure 7 will later be used to derive the interchain coupling parameter η, but it is important to keep in mind that it is T_P which shows up in experiments.

The Peierls semiconductor has some interesting characteristic features. As discussed in the introduction (cf. Figure 1), the electronic gap is associated with a distortion of the ionic lattice, leading to a *charge density wave*. The $T = 0$ state of a one dimensional conductor is therefore often referred to as a charge density wave state, abbreviated CDW. Below we will study both the gap Δ, the lattice distortion and their relation by varying k_F, analyzing different compounds in Table I. But we will also look into more details of the electron–phonon coupling by reviewing the lattice dynamics as they have been studied by inelastic neutron scattering, infrared measurements, and Raman scattering.

As it turns out, most of the compounds in Table I behave according to the simple picture of the charge density wave in a free electron like metal, and these are the ones to be treated here. *A Priori* one might expect all partially oxidized tetracyanoplatinates to behave regularly, since Figure 10 demonstrates that they all obey the same regular dependence of the band filling on the Pt-spacing $d_{\|}$. But the compound with the largest $d_{\|}$ K(def)TCP does not behave like the others and will be treated separately.

4.1. The Electronic Gap

The gap 2Δ which occurs in the electronic spectrum as a result of the Peierls instability may be experimentally studied by several techniques. Transport properties such as conductivity and magnetic susceptibility will show activated behavior, and spectroscopy such as IR-reflectivity and luminescence will reveal spectral characteristics that can be related to Δ.

The temperature dependence of the conductivity below the transition has been studied in all compounds treated here. Figure 13 shows the temperature dependent conductivity

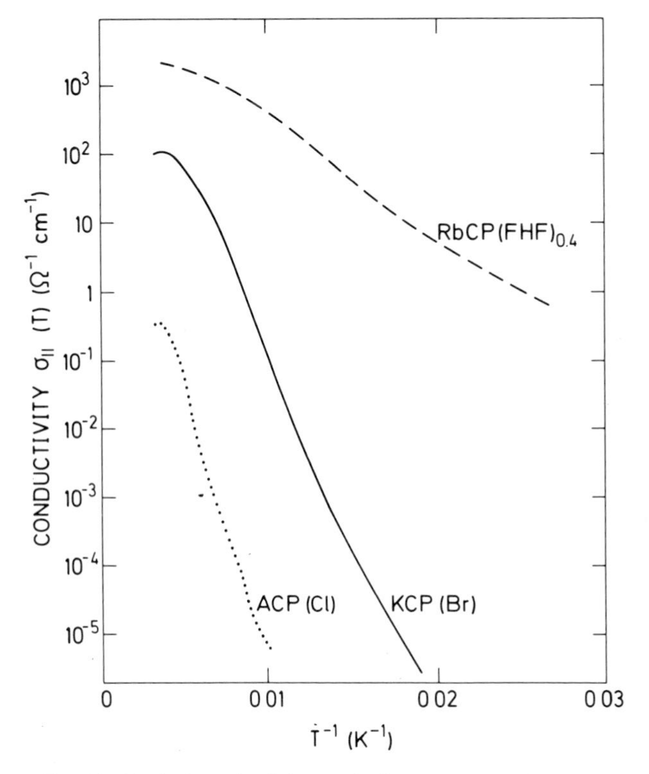

Fig. 13. Typical conductivity results for tetracyanoplatinates.

$\sigma_{\|}(T)$ for the three compounds ACP(Cl), KCP(Br), and RbCP(FHF), representative of the $[Pt(CN)_4]$-series [53, 56, 58]. It is seen that at low temperatures the conductivity follows an activated behavior from which Δ can be deduced. Using the fact that the band structure is free electron like one can derive a specific relation between the gap Δ and the electron–phonon coupling constant λ [7]:

$$\Delta = 8 \frac{1 - k_F d_{\|}/\pi}{1 + k_F d_{\|}/\pi} \epsilon_F e^{-1/\lambda}, \tag{22}$$

which together with the fact that the Fermi energy is given by:

$$\epsilon_F = \hbar^2 k_F^2/(2m) \tag{23}$$

makes it possible to compute λ from the measured Δ. Both are given in Table VI for the regularly behaving tetracyanoplatinates. It may at first sight seem astonishing that varying $d_{\|}$ from 2.80 Å to 2.92 Å has such a big effect on the low temperature conductivity and hence on the parameters Δ and λ. However this is immediately explained when λ is computed from (20). What varies rapidly in (20) is the bare phonon frequency as given by the relation $\omega_0(2k_F) = \omega_D |\sin(k_F d_{\|})|$ since the values of k_F is rather close to $\pi/d_{\|}$. Setting $k_F = \pi/d_{\|}$ makes $\lambda \to \infty$. The agreement between the two different ways of deriving λ is demonstrated in Figure 14, where we compare the λ calculated from (22), i.e., from the measured low temperature activation energy to the calculated λ from (20). In the latter case we use the relation (15) to convert $d_{\|}$ to k_F and we have adjusted the absolute value of λ. The agreement in variation is a good indication of the consistency of our picture.

Other transport properties than conductivity should of course also show the signature of the semiconducting state at low temperatures. This is demonstrated in the case of the paramagnetic susceptibility [81, 97, 98] for KCP which shows activated behavior with an activation energy very close to what is found from conductivity. This agreement is inherent in the simple picture of the charge density wave in a metal with perfect screening; but if Coulomb correlations are of importance Δ measured by susceptibility becomes smaller than measured by conductivity, a feature which frequently occurs in organic conductors. Hence the correspondence between susceptibility and conductivity is a manifestation of the fact that Coulomb correlations seem to be negligible in conducting Pt-compounds.

TABLE VI

Zero temperature gap parameter Δ and electron–phonon coupling λ deduced from conductivity measurements. References are given in Table IV.

Compound	Δ (meV)	λ
RbCP(FHF)	18	0.20
RbCP(DSH)	43	0.25
CsCP(FHF)	27	0.22
CsCP(Cl)	30	0.23
KCP(Cl)	70	0.29
$CsCP(N_3)$	41	0.26
KCP(Br)	72	0.30
RbCP(Cl)	75	0.30
GCP(Cl)	45	0.27
ACP(Cl)	125	0.38

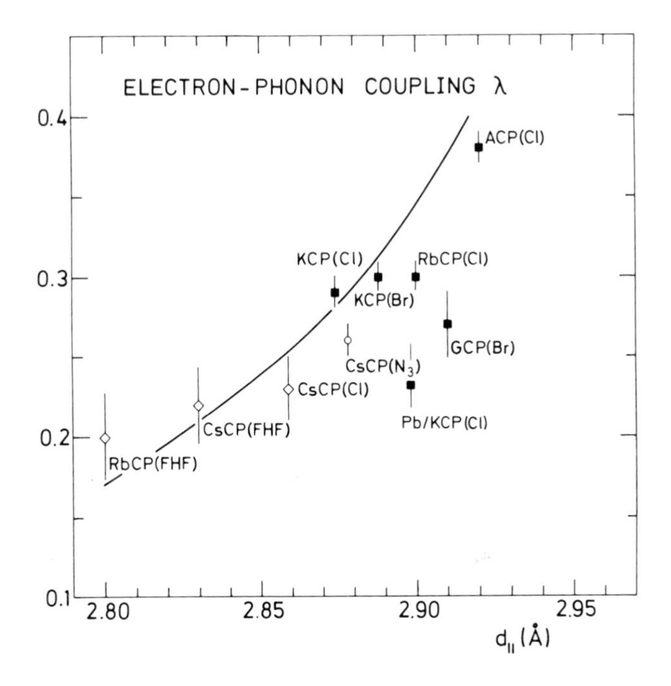

Fig. 14. Comparison of the electron–phonon coupling constant λ as determined from the activation energy of the low temperature conductivity (points) and the microscopic properties of the high temperature metallic phase (line). One parameter has been adjusted.

Turning now to measurements where electrons are directly excited across the Peierls gap we analyze the reflectivity and luminescence to get information about Δ. The theory for the frequency dependent conductivity $\sigma(\omega)$ was worked out by Lee, Rice, and Anderson [99], who showed that it is zero when $\hbar/\omega < 2\Delta$ and has a square root singularity when $\hbar\omega > 2\Delta$:

$$\sigma(\omega) = \frac{ne^2}{m}\,\frac{\pi\hbar}{4\Delta}\left(\frac{2\Delta}{\hbar\omega}\right)^2 \frac{1}{\sqrt{(\hbar\omega/2\Delta)^2 - 1}}. \tag{24}$$

Only in the case of KCP has the infrared reflectivity been measured extensively enough so that the frequency dependent conductivity $\sigma(\omega)$ could be deduced [100]. This quantity is shown in Figure 15 together with the result of (24). One recognizes the increase in conductivity as $\hbar\omega$ approaches 2Δ, which from the conductivity data corresponds to 144 meV or $\omega/2\pi c = 1145\ \mathrm{cm}^{-1}$; but one also immediately realizes that $\sigma(\omega)$ does not behave according to a simple picture of a gap semiconductor since it is finite within the gap and peaks at a frequency well above $\omega = 2\Delta/\hbar$. There are two aspects of this apparent lack of consistency. One is the finite density of states in the gap and the rounding of the observed conductivity peak. Qualitatively, the finite mass of the charge density wave (discussed below) rounds the square root singularity, but this effect is very small, given the actual parameters in KCP. Quantitatively the rounded peak in $\sigma(\omega)$ could

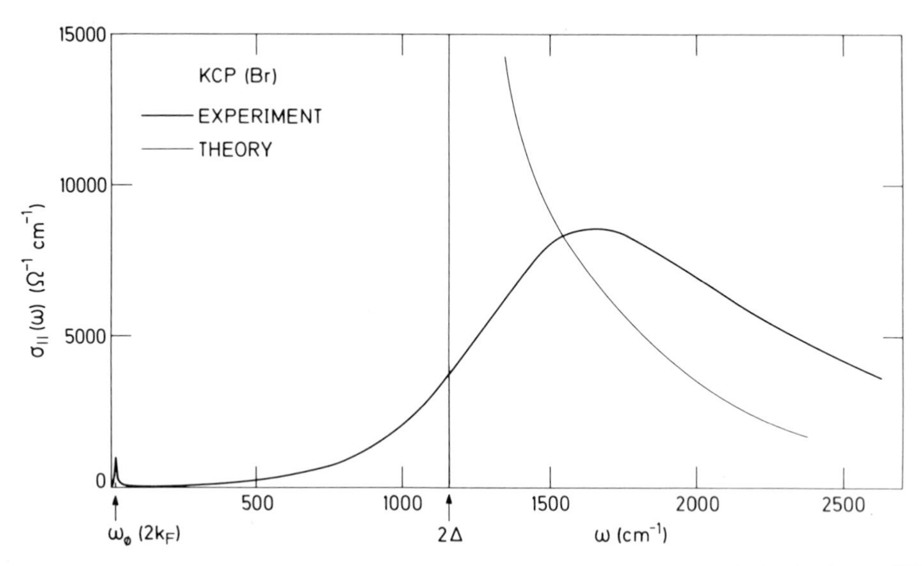

Fig. 15. Measured frequency dependent conductivity (full curve) of KCP and simple theory [99] (thin curve). Frequencies are measured in wavenumbers (cf. Figure 8).

stem from either disorder and from one-dimensional localization. Since we are at present dealing with low temperatures, where KCP is effectively three-dimensional and where, from other evidence, we conclude that disorder effects are important this aspect seems to be associated with disorder. The other discrepancy between the simple gap model and the measured $\sigma(\omega)$ is that the observed peak is shifted upwards in frequency towards $\omega/2\pi c = 1600 \text{ cm}^{-1}$. Part of this is probably due to the above-mentioned rounding of the peak, so if one takes the characteristic frequency to be where $\sigma(\omega)$ has its steepest descent the value of $2\Delta/2\pi\hbar c$ from reflectivity is 1200 cm^{-1}, i.e. only 10% higher than the value inferred from dc conductivity. To get an accurate value for the difference is difficult since it depends on the model one chooses for the disorder effect; but the qualitative feature that Δ derived from dc conductivity is smaller than Δ derived from optical studies is indeed what is to be expected. Owing to the strong coupling to the lattice we expect the electrons to deform it when travelling in an electric field of frequency less than or comparable to phonon frequencies. This is known as the *polaron-effect*, and due to the binding energy of the lattice distortion, the polaron, the effective gap becomes less than the real one. On the other hand when exciting electrons optically across the gap the lattice does not have time to relax and therefore one should then measure the genuine 2Δ. Both the quantitative estimates given here as well as recent theoretical calculations for the half filled band case [101] indicate that the polaron binding energy is rather small, so we shall proceed using the conductivity results listed in Table VI as if the polaron effect was negligible.

As one would expect from Figure 9 the luminescence peak splits into two peaks separated by 2Δ, when ϵ_k is distorted. This effect has been seen in KCP [67] and K(def)TCP [68], and the values for 2Δ are a little higher than those determined by conductivity, consistent with the above conclusions.

In summary the dc conductivity below the transition temperature T_P follows an activated behavior as predicted from the microscopic properties of the high temperature metallic phase. The substantial variation in measured activation energies from 18 to 120 meV among the compounds of Table VI is accounted for by the variation in the phonon frequency $\omega_0(2k_F)$, caused by the variation in $2k_F$. Other values of the charge density wave gap Δ in KCP, both from transport measurements and from optical excitations, agree with results from conductivity experiments. From this it may be concluded that neither Coulomb correlations nor polaron effects are significant in this class of compounds.

4.2. THE LATTICE DISTORTION

The specific mechanism of the Peierls instability that brings about the above-mentioned metal-insulator transition involves a concomitant lattice distortion. The one-dimensional character of this distortion leads to characteristic Bragg-*planes* perpendicular to the highly conducting axis of the crystal as opposed to the usual three dimensional Bragg spots. These planes are seen in X-ray as diffuse scattering. The notation 'diffuse scattering' relates both to the characteristic distribution of the intensity, as well as to the fact that a special diffuse scattering set up is often used to see this very weak scattering. Below the phase transition T_P one should observe a condensation into regular superlattice spots of the diffuse scattering, but the width is often finite because of the disorder in the crystals. The lattice distortion caused by the Peierls instability has unique features that separate it from other superstructures, and it has been studied extensively in conducting Pt-salts following the first work on KCP by Comès *et al.* [86, 102]. To get insight into the nature of the ionic displacement let us consider a single chain. In the absence of any distortion the position of the nth ion is given by $x_n = nd_\parallel$, and the X-ray intensity is given by the static structure factor $S(\kappa)$

$$S_0(\kappa) = e^{-2W} \sum_H \delta(\kappa - Hd_\parallel^*), \tag{25}$$

where $d_\parallel^* = 2\pi/d_\parallel$, H is the one-dimensional Miller index of the reflection, and $e^{-2W} = e^{-u^2\kappa^2}$ is the Debye Waller factor that accounts for the thermal vibrations of root mean square amplitude u. $S(\kappa)$ determines the intensity of the X-rays (or neutrons) scattered in the direction 2θ given by $\kappa = 2k_0 \sin\theta$, where $k_0(=2\pi/\lambda)$ is the wave vector of the scattering wave and θ is the scattering angle. Hence (25) may be considered as a reformulation of the usual Bragg law $2d_\parallel \sin\theta = H\lambda$ which is equivalent to $\kappa = Hd_\parallel^*$.

In the presence of the sinusoidal distortion the position of the nth ion is given by:

$$x_n = nd_\parallel + \alpha(x, t) \cdot \sin[n \cdot 2k_F d_\parallel + \phi(x, t)]. \tag{26}$$

Here the amplitude $\alpha(x, t)$ and the phase $\phi(x, t)$ of the charge density wave are both assumed to vary slowly in space and time. From (26) one can calculate the static structure factor [103]:

$$S(\kappa) = \sum_{k=-\infty}^{\infty} S^k(\kappa) \tag{27}$$

with

$$S^k(\kappa) = e^{-2W} \sum_{H=-\infty}^{\infty} \delta(\kappa - Hd_{\parallel}^* - k \cdot 2k_F)\, |J_k(\kappa \cdot \alpha)|^2 .$$

Here $\alpha = \langle \alpha(x, t) \rangle$ is the average amplitude of the lattice distortion, $J_k(x)$ is the kth Bessel function of the first kind. Expanding these in their small argument $(\kappa \cdot \alpha)$ we get the following explicit expressions for the partial structure factors $S^k(\kappa)$:

$$k = 0: \quad S^0(\kappa) = e^{-2W} \sum_H \delta(\kappa - Hd_{\parallel}^*)[1 - \tfrac{1}{2}\alpha^2\kappa^2 + \ldots]. \tag{28}$$

This term gives rise to scattered intensity at the Bragg points of the original lattice. However, this intensity is diminished corresponding to an increase in u^2 of $\frac{1}{2}\alpha^2$, i.e., exactly the root mean square displacement of the ions according to (26). However, several other deviations from an ideal crystal lattice give extra contributions to the Debye-Waller factor, so the influence on the original Bragg intensities of the periodic displacement is not of practical use when investigating the characteristics of the charge density wave. The first interesting term in (27) is:

$$k = \pm 1: \quad S^{\pm 1}(\kappa) = e^{-2W} \sum_H \delta(\kappa - Hd_{\parallel}^* \pm 2k_F)[\tfrac{1}{4}(\kappa\alpha)^2 + \ldots]. \tag{29}$$

This term is often referred to as the $2k_F$-diffuse scattering. In our one-dimensional model lattice it gives rise to satellite reflections around the Bragg spots of intensity $\frac{1}{2}(\kappa\alpha)^2$ relative to these. In general this intensity is quite small since the amplitude of the distortion α is small relative to the average distance between Pt-ions, $d_{\parallel}$. In KCP(Br) a neutron scattering study was used to determine that α at low temperatures is 0.027 ± 0.003 Å, corresponding to $\alpha/d_{\parallel} = 0.0094$, or approximately 1% [103]. It should be noted that the $2k_F$ scattering has the characteristic wave vector dependence $\kappa^2 e^{-2W}$, i.e., it rises quadratically with H^2, the order of the closest Bragg reflection, until it gets damped by the Debye–Waller factor. The second interesting term in (27) is:

$$k = \pm 2: \quad S^{\pm 2}(\kappa) = e^{-2W} \sum_H \delta(\kappa - Hd_{\parallel}^* \pm 4k_F)[\tfrac{1}{64}(\kappa\alpha)^4 + \ldots]. \tag{30}$$

This term, which gives rise to second order satellites at $4k_F$ should not be confused with satellites at the same wave vector stemming from Coulomb correlations between electrons on a chain. We show it here to point out that the presence of a sinusoidal charge density wave does give rise to a very weak $4k_F$ X-ray scattering. Its intensity may be calculated from the observed $2k_F$ scattering and when this is compared to the observed $4k_F$ scattering (if observed) one is led to the conclusion that other mechanisms than the CDW cause this scattering. In $Pt(CN)_4$ conductors $4k_F$ scattering is not seen. Similarly $k > 2$ gives negligible contribution to $S(\kappa)$.

In three-dimensional compounds below T_P we would expect a three-dimensional ordering of the charge-density wave on each chain. In the tetragonal lattice of KCP(Br) one expects an antiferroelastic ordering perpendicular to the chains, both if chains order as a result of Coulombic repulsion between charge density waves on neighbouring chains as well as if electrons can hop between chains. Retaining only $k = 0, \pm 1$ from above the static structure factor takes the form:

$$S(\kappa) = e^{-2W} J_0^2(\kappa \cdot \alpha) \sum_{\tau} \delta(\kappa - \tau) \pm$$

$$\pm e^{-2W} J_1^2(\kappa \cdot \alpha) \sum_{\tau} \delta(\kappa - \tau \pm \mathbf{Q}_{\mathrm{CDW}}) \tag{31}$$

where τ is a reciprocal lattice vector, $\boldsymbol{\alpha}$ is a vector in the conducting direction of amplitude α, and $\mathbf{Q}_{\mathrm{CDW}} = (Q_\perp, 2k_F)$ with $\mathbf{Q}_\perp = \frac{1}{2}(\mathbf{a}^* + \mathbf{b}^*)$.

With the following modifications (31) describes what is observed in KCP(Br) [86, 96, 97, 98]: the unit cell contains two Pt-ions along the conducting axis so that $\mathbf{c} = 2\mathbf{d}_\parallel$ (i.e., $\mathbf{c}^* = \frac{1}{2}\mathbf{d}_\parallel^*$) and the presence of the other chemical species in the cell adds a geometrical structure factor to $S(\kappa)$, which may be used to identify the distorted entity as $[Pt(CN)_4]$ from elastic neutron scattering [90]. However, the most important deviation between the simple charge density wave static structure factor $S(\kappa)$ and that observed in KCP is the lack of long range order. The $2k_F$ satellites never develop into sharp superlattice spots but retain a finite correlation length, at least perpendicular to the chains $\xi_\perp$, and presumably also along the chains $\xi_\parallel$. Both these correlation lengths should go to infinity at T_P and long range order should exist below the observed transition temperature T_P, but $\xi_\perp$ never exceeds 80 Å corresponding to 8 lattice spacings, $d_\perp$, perpendicular to the chains, whereas $\xi_\parallel$ if finite exceeds 300 Å at low temperatures. Sham and Patton [16] have calculated the limiting effects on correlation lengths from disorder, as present in KCP(Br) by the random Br-ions; but below we will argue that both commensurability pinning and impurity pinning play a role in establishing a limiting case between a charge density glass (finite $\xi_\parallel$) and a charge density wave (infinite $\xi_\parallel$) in KCP at low temperatures.

The solution to the Fröhlich Hamiltonian (1) gives a relation between the gap Δ and the magnitude of the distortion α [104]

$$\alpha = \frac{2\Delta}{g} \sqrt{\frac{\hbar}{2M\omega_0(2k_F)}}\,. \tag{32}$$

The observed magnitude of α may be used to evaluate g from (32), taking Δ from conductivity, and $\omega_0(2k_F)$ and M from neutron scattering. In the case of KCP this gives $g = 0.16$ eV. This is in excellent agreement with the evaluation of $g = 0.14$ eV from (2) using λ from conductivity results. The estimates of g according to (19) is 0.3, i.e., somewhat too large, as expected from the discussion of this formula.

4.3. Soft Phonons. The Amplitude Mode.

As mentioned in the Introduction, the Peierls transition belongs to the class of 'soft mode' transitions where in the simplest picture the phonon frequency at wave vector $2k_F$ goes to zero at the phase transition. This softening has its parallel in three-dimensional metals as first pointed out by Kohn, and is therefore often referred to as a giant Kohn anomaly. The important difference between the Kohn effect in one and three dimensions is that in ordinary (three-dimensional) metals the Kohn anomaly is very small and is not associated with a lattice instability. As originally pointed out by Lee, Rice and Anderson [99], and later confirmed by several theoretical papers [49, 105–112] it is convenient to describe the lattice dynamics below the transition (strictly speaking at $T = 0$, but in reality $T < T_P$) as the dynamics of the charge density wave. As shown in Figure 16 there are two excitations: (i) the *amplitudon* ω_α associated with oscillations in the amplitude $\alpha(x, t)$ and (ii) the *phason* ω_ϕ associated with oscillations in the phase $\phi(x, t)$. The dispersion relations $\omega_\alpha(q)$ and $\omega_\phi(q)$ are given by [99, 113]:

$$\omega_\alpha^2(q) = \omega_\alpha^2(2k_F) + \frac{1}{3}\frac{m^*}{M^*}\,[v_F(q - 2k_F)]^2 \tag{33}$$

where $\omega_\alpha(2k_F) = \sqrt{\lambda}\omega_0(2k_F)$ and

$$\omega_\phi(q) = \sqrt{\frac{m^*}{M^*}}\, v_F(q - 2k_F). \tag{34}$$

In (33) and (34) M^* denotes the mass of the charge density wave which in this model accounts for the slow response of the lattice to the electronic perturbation, not to be confused neither with the effective mass of the electrons m^* (which we may set equal to m corresponding to the free electron picture developed above) nor with the mass of the ionic lattice (which we set to 199 atomic masses for $[Pt(CN)_4]$). It is related to the parameters of the Fröhlich Hamiltonian in the following way:

$$\frac{M^*}{m^*} = 1 + \frac{4}{\lambda}\left[\frac{\Delta}{\hbar\omega_0(2k_F)}\right]^2. \tag{35}$$

The symmetries of the two modes are such that the amplitude mode is Raman active and the phase mode is infrared active. Both modes are, as all density fluctuations, seen by neutron scattering.

Several neutron scattering experiments [89, 90, 91, 111, 113, 115], Raman scattering [116, 117], and infrared reflectivity studies [100, 108], as well as later theoretical work has illuminated the dynamical properties of the charge density wave in KCP. The amplitude mode energy at $q = 2k_F$ has been accurately determined in deuterated KCP(Br) by Raman scattering to be $\hbar\omega_\alpha(2k_F) = 4.7$ meV corresponding to a wavenumber of 38 cm^{-1}. Since $\omega_0(2k_F)$ is known from neutron scattering we can use (33) to estimate $\lambda = 0.35$ in good agreement with values derived above. The neutron scattering intensities of $\omega_\alpha(q)$ around $q = 2k_F$ are shown in Figure 17, but because of the relatively poor resolution of this technique one cannot unambiguously identify the scattering with the theoretical dispersion relations (33) and (34), and shown in Figure 16. This ambiguity has given rise to several interpretations [91, 102, 111], but here we follow one [91] that establishes a close connection with the theory outlined above. We identify the intensity at $q = 2k_F$ for frequencies $\hbar\omega \gtrsim 4$ meV with the amplitude mode and this then allows a determination of both $\hbar\omega_\alpha(2k_F) \simeq 6$ meV and $M^*/m^* \simeq 1200$ from (33). Hence the amplitude mode is well characterized experimentally and found both to be in qualitative accordance with (33) as well as in good agreement with subsequent theoretical work.

CONFIGURATION

ENERGY

METAL

Equilibrium: $x_n = nd_{\|}$

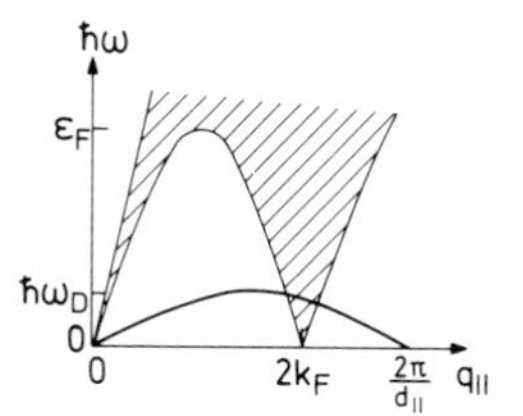

Excitations: Phonons

$x_n = nd_{\|} + u_0 \sin(q_{\|}x - \omega t)$

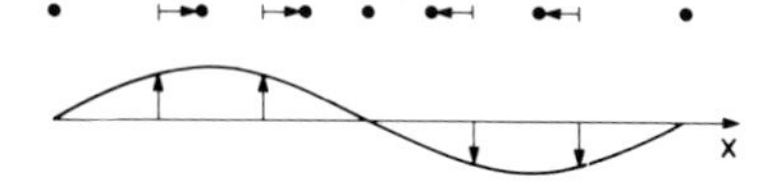

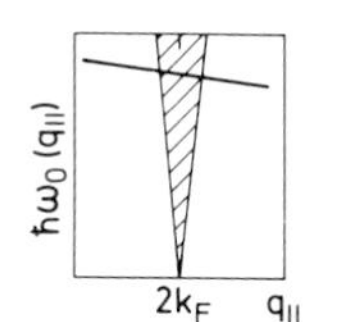

CHARGE DENSITY WAVE

Equilibrium: $x_n = nd_{\|} + \alpha \sin(2k_F nd_{\|} - \varphi)$

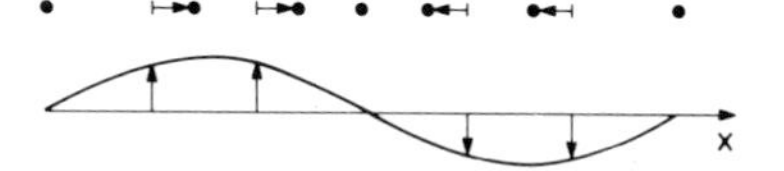

Excitations: Amplitudons

$\alpha(x, t) = \alpha + u_\alpha \sin(q_{\|}x - \omega t)$

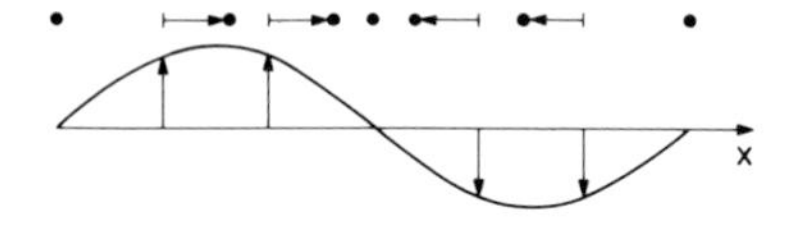

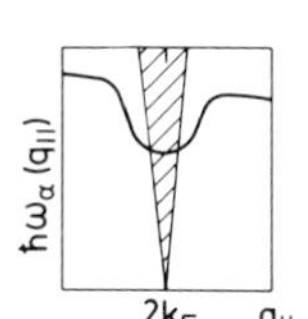

Excitations: Phasons

$\varphi(x, t) = \varphi + u_\varphi \sin(q_{\|}x - \omega t)$

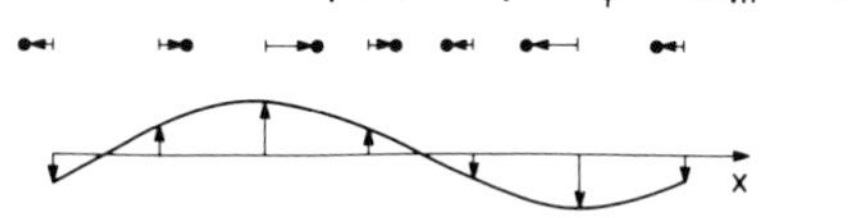

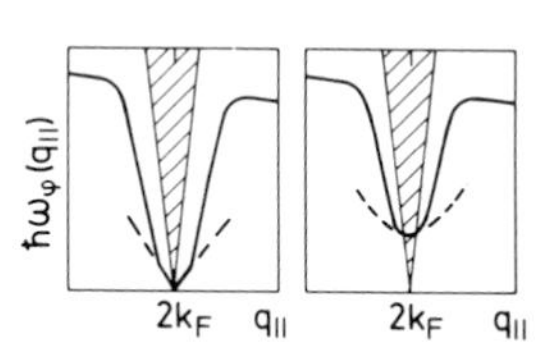

Fig. 16. Illustration of the lattice dynamics of the one-dimensional conductor. In the metallic phase the equilibrium configuration is an equidistant ionic lattice and the normal modes are periodic modulations with period $q_{\|}$ and energy $\hbar\omega_0(q_{\|})$ (phonons). For the charge density wave state the equilibrium configuration is distorted with period $2k_F$ and the normal modes are modulations of either amplitude (amplitudon with energy $\hbar\omega_\alpha(q_{\|})$) or phase (phason with energy $\hbar\omega_\phi(q_{\|})$). The phase node is shown both in the unpinned case ($\omega_\phi(2k_F) = 0$) and the pinned case ($\omega_\phi(2k_F) \neq 0$), and the extension of the theory of Lee, Rice and Anderson is indicated (dashed lines). The hatched region is the region of allowed electron-hole excitations in the metallic phase.

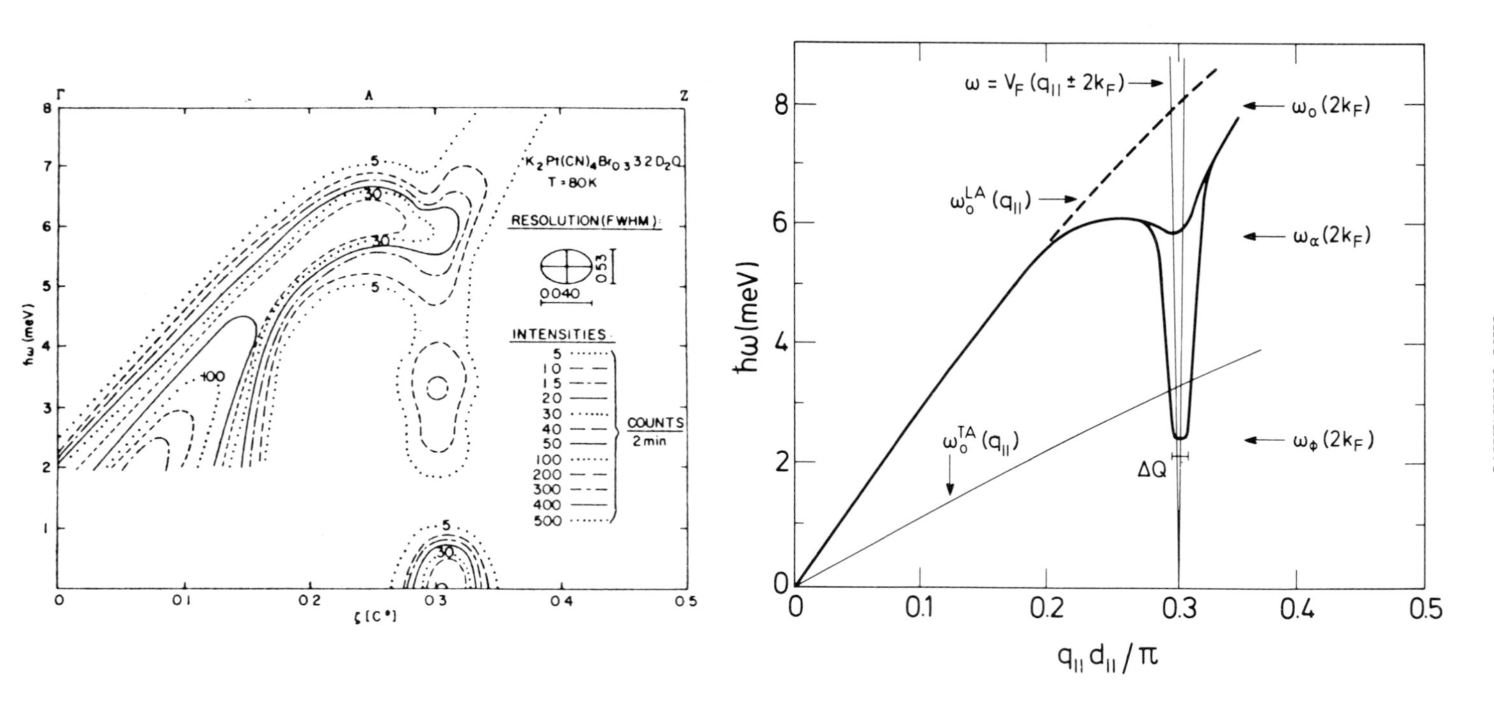

Fig. 17. Neutron scattering intensities around $\mathbf{q}$ = (0, 0, $2k_F$) in KCP(Br) (left), and the interpretation according to the concepts discussed in the text. The maximum at ($q_{\|}$, $\hbar\omega$) = ($2k_F$, 3.5 meV) can be attributed to the crossing of a transverse acoustic phonon as discussed in [91]. Energies are given in meV (cf. Figure 11).

An interesting extension of the picture developed above regards the lifetime of the amplitude mode. An anomalously large linewidth γ_α (half width at half maximum) of the Raman-active amplitude mode observed in KCP [116, 117] led Kurihara to investigate the nonlinear coupling between this mode and the phason [111]. As shown in Figure 18,

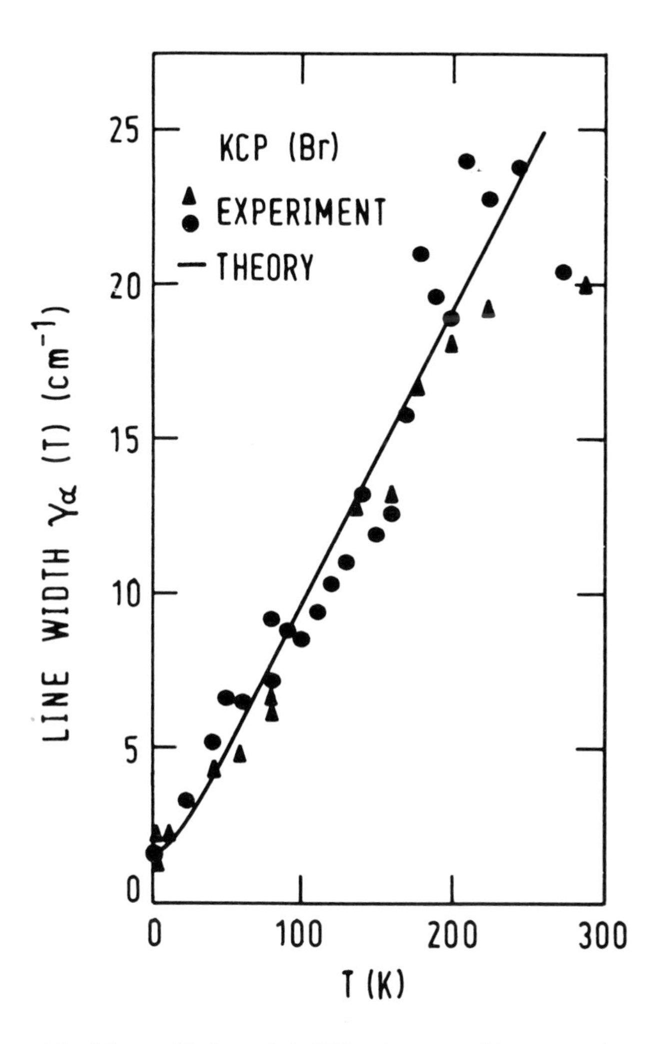

Fig. 18. Linewidth of the amplitude mode in KCP. γ is measured in wavenumbers (cf. Figure 8).

the experimental $\gamma_\alpha(T)$ goes from 0.38×10^{12} s^{-1} (2 cm^{-1}) at $T = 5$ K to 4.7×10^{12} s^{-1} (25 cm^{-1}) at $T = 270$ K, which is well reproduced when the decay of one amplitudon into two phasons is considered. The theoretical result is:

$$\gamma_\alpha(T) = \gamma^0 \coth(T^0/T), \tag{36}$$

where

$$\gamma^0 = \frac{\pi\omega_\alpha(2k_F)}{4\sqrt{M^*/m^*}} \frac{1}{\sqrt{\left[1 - \frac{2\omega_\phi(2k_F)}{\omega_\alpha(2k_F)}\right]^2}}$$

and

$$k_B T^0 = \tfrac{1}{4}\hbar\omega_\alpha(2k_F).$$

Since all parameters determining γ^0 and T^0 are known independently there are no adjustable parameters, and the agreement between theory and experiment is therefore very satisfying. However, literally the theory only applies below T_P, i.e., $T < 100$ K. The fact that (36) describes the observed $\gamma_\alpha(T)$ so well even at high temperatures is to some extent fortuitous and it is not astonishing that certain aspects of the theory (e.g., the temperature dependence of the frequency $\omega_\alpha(2k_F)$) does not account for the experimental results. However, at low temperatures the linewidth of the amplitude mode as a result of its coupling to the phase mode is a dramatic demonstration of the fact that even at $T = 0$ these two modes cannot be considered independently in a one-dimensional conductor.

In K(def)TCP [67] and in ZnOP and CoOP [119] one finds that the amplitudon linewidth γ_α is much smaller than in KCP, and that Kurihara's theory cannot account for its observed magnitude. One reason for the lack of general applicability of the Kurihara theory is that interchain coupling is not considered. Since interchain coupling will diminish the amplitude of the phase oscillations and hence the nonlinear coupling to the amplitudons, a likely explanation for the fact that only KCP exhibits the characteristic $\gamma_\alpha(T)$ given by (36) is that only this compound is sufficiently one-dimensional.

Finally, we consider the difference between deuterated and protonated KCP(Br). As mentioned above they appear to bind differently the water molecules that substitute for the deficient Br-occupation in the center of the unit cell. As a pronounced effect of deuteration, Raman scattering investigations of the amplitude mode showed that $\hbar\omega_\alpha(2k_F) = 4.7$ meV (38 cm^{-1}) in deuterated KCP(Br) at low temperatures [117] as compared to 5.5 meV for the protonated salt [116]. This was taken as evidence for a strong coupling between conduction electrons and water vibrations, which would then enhance the electron phonon coupling constant λ in the protonated salt compared with deuterated. However, this conclusion is correct only if the bare phonon $\omega_0(2k_F)$ is the same in the two cases, and it was overlooked that this is not true. In fact comparison between the neutron scattering results from protonated [115] and deuterated KCP(Br) [90–91] indicate that $\omega_0(2k_F)$ is as different as $\omega_\phi(2k_F)$. One is therefore led to the conclusion that the overall binding in KCP(Br) changes upon deuteration, but λ is not much changed. This is in accordance with the fact that the electrical properties do not depend significantly of whether D_2O or H_2O is used as crystal water.

4.4. THE PHASE MODE

The phase mode in KCP was first studied by infrared reflectivity [100] as shown in the peak in $\sigma(\omega)$ at $\hbar\omega_\phi(2k_F) = 2$ meV (15 cm^{-1}) in Figure 15, and later neutron scattering substantiated this finding. The fact that $\omega_\phi(2k_F) \neq 0$ is in disagreement with (34) but is explained as being a result of pinning of the charge density wave. In this case it is natural to assume that the phase mode dispersion relation becomes:

$$\omega_\phi^2(q) = \omega_\phi^2(2k_F) + \frac{m^*}{M^*} v_F^2(q - 2k_F)^2 . \tag{37}$$

By analyzing the lineshape of the reflectivity from 10 to 100 cm^{-1} in terms of a plasma oscillation of the charge density wave Brüesch *et al.* [100] derived from an analysis similar to that of the metallic plasma frequency the CDW-mass is $M^*/m^* = 980$. This is in very good agreement with the neutron determination from (33). Neutron scattering studies of the phase mode for $\hbar\omega > \hbar\omega_\phi(2k_F)$ indicate, however, a phason 'velocity' much greater than corresponding to $M^* \approx 1000\ m$. In fact the phase mode rises so steeply that neutron scattering cannot resolve its velocity. This allows only an estimate of the effective mass to be $M^* < 30\ m$ from the data of Carneiro *et al.* [91] Analyzing the data of Käfer [110] in this way their improved revolution gives $M^* < 15\ m$. This paradox regarding the charge density wave mass M^* is resolved when the theory of Lee, Rice and Anderson is compared with the numerical calculations [105, 109]. Both find that (33) and (34) apply only in a very narrow region of $q \approx 2k_F$, a region monitored by visible light, whereas outside this region the phase velocity is not given by (37). Hence, in the analysis of the neutron scattering from the phase mode at $q \neq 2k_F$ (or $\hbar\omega > \hbar\omega_\phi(2k_F)$) the high velocity observed is not directly related to a small M^*.

Several detailed calculations have been made for the full dynamical structure factor $S(q, \omega)$ which in principle could be related to the measured neutron intensities of Figure 17. However, the fact that the full details of $S(q, \omega)$ cannot be revealed experimentally nor theoretically makes it a risky task to pursue the analysis further than attempted here.

The precise origin of the finite pinning frequency $\omega_\phi(2k_F) = 2$ meV in KCP is not well established. Lee, Rice and Anderson [99] originally stated that pinning might occur as a result of 'commensurability energy', 'three-dimensional ordering' or 'sticking at impurities'. In KCP(Br) it has been widely assumed that the Br^--disorder provides an impurity potential necessary for pinning. However, as shown by Fukuyama [120, 121] it is difficult to imagine a finite $\omega_\phi(2k_F)$ in the absence of a commensurate field. Defining the two lengths $l_0 = \sqrt{(m^*/M^*)} v_F/\omega_\phi(2k_F)$ and $d = \hbar v_F/\Delta$ (the latter being the width of a commensurate phase soliton), the ratio l_0/d determines whether the ground state is the charge density wave state or a charge density wave glass. CDW occurs when $l_0/d > 1.43$ and here one gets long range order and a pinning frequency determined by the commensurability pinning alone. When $l_0/d < 1.43$ a 'glassy' state without long-range order is stable and $\omega_\phi(2k_F)$ is enhanced over what commensurability pinning would give. In KCP $l_0/d \approx \sqrt{(m^*/M^*)} \times \Delta/[\hbar\omega_\phi(2k_F)] = 1.1$, i.e., close to the limiting value. It should be noted that the assumption $n_{Br} \cdot l_0 \gg 1$, where $n_{Br} = 0.6/5.7$ Å (=0.1 Å^{-1}) is the Br^{-1} number density ($l_0 \approx 100$ Å) is well satisfied, justifying the random walk treatment of the impurity scattering. Hence within this model, which treats the first and the third reason for pinning proposed by Lee, Rice and Anderson, one can explain the lack of true long range order in KCP(Br) as a result of disorder while attributing the finite pinning frequency $\omega_\phi(2k_F)$ almost solely to commensurability effects.

Such a picture, if applied to KCP requires a periodic potential commensurate with the charge density wave period $2k_F$. Since $2k_F$ is incommensurate with c^*, the crystallographic lattice cannot provide such a potential. But a closer look at the Br-occupation reveals that indeed the potential from these anions will have components *both* commensurate with the lattice as well as with $2k_F$, a fact which is related to the fact that it is the 60% occupancy frequency of the Br at a particular position in the crystallographic unit cell with one Br for each two Pt-ions which determines the degree of partial oxidation of the Pt and hence k_F. It is therefore not unlikely that the particular Br configuration in KCP provides both a commensurability potential and a disorder (or impurity) field exploring the pinning characteristics in this compound, the finite correlation length $\xi_{\|}$ and the finite phason frequency $\omega_0(2k_F)$.

4.5. The special behavior of Pb/KCP(Cl), GCP(Br) and LiPt(mnt)

Above we have established a detailed description of a one-dimensional conductor, both in its high temperature metallic phase and in its low temperature semiconducting phase. We have demonstrated that 9 of the compounds of Table I, namely the anion deficient compounds with monovalent cations, are model compounds for the Peierls instability. In particular the measured electron–phonon coupling constant λ varies from compound to compound according to (20) as demonstrated in Figure 14, and numerous other experimental results, in particular from KCP, all give a consistant picture. The only assumptions are that in the high temperature phase the electronic structure ϵ_k is free electron like according to (23) with an effective mass of unity and that the phonon dispersion relation is as simple as in (16) with a common Debye energy $\hbar\omega_D \simeq 18$ meV. These two characteristics therefore unify the 9 compounds.

Before proceeding let us inspect three compounds with a somewhat unusual behavior that may be traced to different characteristics of the phonon and electron subsystems of the one-dimensional conductor than the majority of compounds in Table I. Two compounds of interest in this respect are Pb/KCP(Cl) and GCP(Br), which in Table VI and Figure 14 are characterized by unusually low Δ and λ, and a third conductor LiPt(mnt) which clearly deviates from the salts considered up to now.

The compound Pb/KCP(Cl) [30] is isostructural with KCP with an intrachain separation of $d_{\|} = 2.90$ Å. Its electronic band filling as measured by diffuse X-ray scattering corresponds to $k_F = 0.885\ \pi/d_{\|}$, or a DPO = 0.23, i.e., lower than expected from the usual relation (15). The room temperature conductivity is as expected (cf. Table IV) and $\sigma_{\|}(T)$ has the same qualitative behavior as demonstrated in Figure 3. However, the low temperature activation energy is markedly lower than expected, namely $\Delta = 27$ meV. This anomaly is Pb/KCP(Cl) with respect to the usual behavior of $Pt(CN)_4$-compounds may be explained as follows. There seems to be no reason to assume that the free electron model is not adequate for this compound, so that (22) allows a determination of $\lambda = 0.23$ from the given values of $d_{\|}$ and Δ. This should be compared to $\lambda = 0.28$ expected from the other compounds. Analyzing λ in terms of (20) suggests that ω_0 is unusually high, which is indeed to be expected from the inclusion the divalent cation Pb^{++} in the lattice. Divalent ions are well known to form much stronger bonds than monovalent ions. Hence the small gap Δ of the compound Pb/KCP(Cl) within the class of Pt-chain conductors that form simple charge density waves may be understood directly from a change in λ caused by the increased lattice strength from the divalent Pb^{++}.

Also the compound GCP(Br) [32] has a low activation energy Δ and therefore a small electron phonon coupling constant λ compared to expected values for its $d_{\parallel}$ = 2.908 Å. This may be understood in a similar way as in Pb/KCP(Cl), namely as a result of anomalously strong bondings giving rise to large elastic energies $\hbar\omega_D$. In GCP(Br) the increased bond strength may be attributed to an extensive hydrogen bonding network along and between the chains [32].

Finally, LiPt(mnt) is an example of a simple charge density wave model compound [66]. Despite the fact that the electronic overlap is significantly diminished as a result of the large separation $d_{\parallel}$ = 3.64 Å so that this compound might behave more like an organic conductor than a Pt-conductor, LiPt(mnt) resembles the compounds of Table I as far as its physical properties are concerned. The low temperature activation energy is Δ = 36 meV. Infrared reflectivity studies and thermopower measurements show that the high temperature metallic phase has a tight binding band structure with a rather small band width $W_{\parallel}$ = 0.4 eV. In order to derive the electron–phonon coupling constant λ we use the specific expression for (3):

$$\Delta = 2W_{\parallel}[1 - \cos^2(k_F d)]\, e^{-1/\lambda}, \tag{38}$$

according to the tight binding case [7]. This yields λ = 0.34 for LiPt(mnt). The magnitude of λ is similar to those of Table VI but its origin is different here. Whereas in the simple Pt-chain conductors λ is dominated by the coupling between the electrons and the acoustic phonon, this contribution is relatively small in organic conductors [93, 122, 123]. Therefore λ becomes dominated by the coupling to other modes both external and internal molecular vibrations:

$$\begin{aligned}\lambda &= \sum_i \lambda_i \\ &= N(\epsilon_F) \sum_i \frac{g_i^2}{\hbar\omega_i(2k_F)}\end{aligned} \tag{39}$$

Hence LiPt(mnt) is an example of a conductor with a charge density wave ground state, where both the electronic characteristics and the lattice dynamics are quite different from the regularly behaving $[Pt(CN)_4]$-conductors.

4.6. Transport Mechanisms

The study of low temperature transport in platinum chain compounds have not been carried out in great detail, but the recent developments in the understanding of transport in related compounds seems to justify a renewed interest in this aspect of their behavior. The purpose of reviewing transport studies here is therefore in part to provoke new interest in the subject; but it seems also relevant to place the results in the context of the rather different ideas that have been proposed over the years for transport in quasi-one-dimensional conductors.

The thermopower along the chain direction $S_{\|}(T)$ below the phase transition contains direct information about the relative mobilities of electrons μ_e and holes μ_h since $S_{\|}(T)$ may be expressed by [124]:

$$S_{\|}(T) = -\frac{k_B}{e}\,\frac{\mu_e - \mu_h}{\mu_e + \mu_h}\left(\frac{\Delta}{k_B T} + \text{constant}\right). \tag{40}$$

As pointed out by Conwell and Bannik $\mu_{e,h} \sim (m^*_{e,h})^{-3/2}$, where $m^*_{e,h}$ are the band masses of the Peierls insulator [125]. These may be calculated from the distorted band structure according to the Fröhlich Hamiltonian (1):

$$\begin{aligned} E_k &= \tfrac{1}{2}[\epsilon_k + \epsilon_{k-2k_F}] \pm \sqrt{\tfrac{1}{4}[\epsilon_k - \epsilon_{k-2k_F}]^2 + \Delta^2} \\ &\approx \epsilon_F \pm \left(\Delta + \frac{\hbar^2 (k - k_F)^2}{2m^*_{e,h}}\right) \quad \text{for} \quad |k - k_F| \ll k_F, \end{aligned} \tag{41}$$

where + or − apply to electrons and holes, respectively. Assuming the free-electron approximation for the metallic band structure ϵ_k we get $m^*_{e,h} = m(\Delta/2\epsilon_F)(1 \pm \Delta/2\epsilon_F)$. This in turn gives:

$$S_{\|}(T) = -\frac{k_B}{e}\left(\frac{3}{4}\cdot\frac{\Delta}{\epsilon_F}\,\frac{\Delta}{k_B T} + \text{constant}\right). \tag{42}$$

Inserting numbers of Δ and ϵ_F for KCP as given above (42) predicts that the slope of $eS_{\|}/k_B$ *vs* $1/T$ should correspond to $0.02\Delta = 1.4$ meV. Experimentally the value is 3–5 meV. This reasonable agreement between the theory and the experiment indicates that the transport in KCP is rather close to that predicted for an *intrinsic* Peierls semiconductor. In particular this means that very high mobilities can be achieved as a consequence of the light masses of both electrons and holes. Specifically (41) leads to carrier masses of $0.01m$ for KCP. The temperature dependence of the mobility may in principle be determined from conductivity. As shown in Figure 13 the low temperature conductivity $\sigma_{\|}(T)$ of tetracyanoplatinates is dominated by an activated behavior which was used above to deduce the low temperature gap Δ. The mobility $\mu(T)$ is often assumed to follow a power law behavior in temperature so that the conductivity becomes:

$$\begin{aligned} \sigma_{\|}(T) &= e\mu_{\|}(T)n(T) \\ &\sim T^{-\beta} e^{-\Delta(T)/T}. \end{aligned} \tag{43}$$

However, the practical applicability of (43) to find β is limited since the temperature dependence of the mobility contributes relatively little to the total temperature variation of the conductivity. An interesting way of isolating the mobility was proposed by Soda *et al.* [126] who suggested to analyze the conductivity anisotropy $\sigma_{\|}(T)/\sigma_{\perp}(T)$. Under the assumption that the conductivity is coherent along the chains but diffusive perpendicular to them these authors suggested that the temperature dependence of $\sigma_{\|}/\sigma_{\perp}$ would be dominated by the product of the in chain scattering time $\tau_{\|}$ which varies as $T^{-\beta}$ and the interchain 'escape time' $\tau'_{\perp}$ [126, 127]:

$$\sigma_{\|}(T)/\sigma_{\perp}(T) \propto \tau'_{\perp}\tau_{\|}. \tag{44}$$

In most organic conductors it has been found that $\sigma_{\parallel}/\sigma_{\perp}$ is either rather independent of temperature or decreases as the temperature is increased, indicating that $\tau'_{\perp}$ (assumed to increase with increasing temperature) varies less than or similarly to $\tau_{\parallel}$. The experimental result for KCP, taken from Zeller and Bech [128] and shown in Figure 19 is therefore

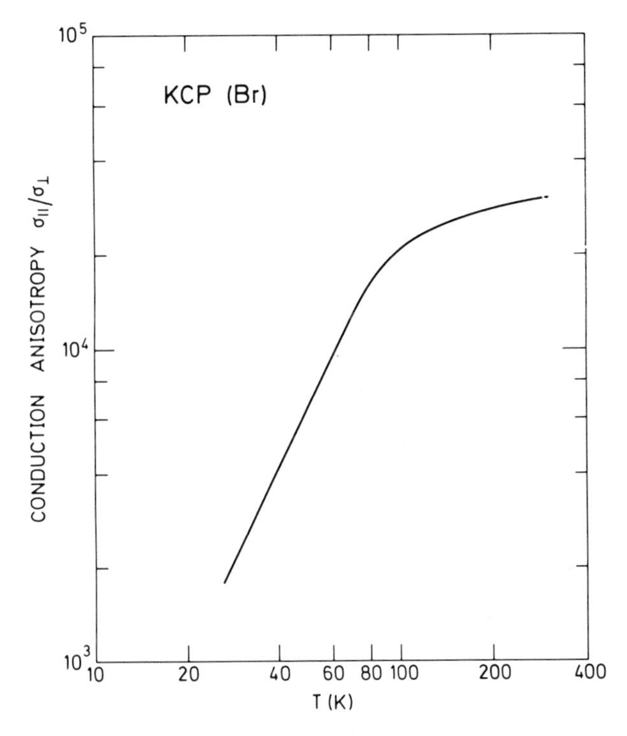

Fig. 19. Temperature dependence of the conductivity anisotropy in KCP.

astonishing since at low temperatures $\sigma_{\parallel}/\sigma_{\perp}$ increases as the temperature is raised in contrast to the usual behavior of organic conductors. Therefore in KCP when the temperature is increased the increase in the transverse escape time $\tau'_{\perp}$ appears to dominate over the decrease in the scattering time $\tau_{\parallel}$. Hence, the latter is not well determined from the conductivity anisotropy, and this anomalous behavior of KCP(Br) seems to warrant more consideration.

Ever since the concept of collective transport by nonlinear excitations in one-dimensional systems emerged, Pt-conductors have been considered potential model compounds. Recently Thomas has proposed that the conductivity $\sigma_{\parallel}(T)$ reflects solitary waves (solitons) in KCP(Br) and KCP(Cl) when $T \lesssim 50$ K [129]. The experimental observation is a decreased activation energy in the conductivity, and the interpretation is that nonlinear excitations of the phason modes contributes to the conduction as proposed theoretically by Rice *et al.* [130]. However, the stability of phase solitons (or Φ-particles) in incommensurate systems is at present not entirely clear from theoretical arguments, whereas the existence of phase solitons in commensurately distorted systems seems well documented. It is therefore possible that a field, commensurate with $2k_F$, is provided by the random Br^--ions, and that this field is the origin of the current carrying nonlinear excitations.

An alternative proposal for the existence of static solitons, similar to what has been proposed for 2H-$TaSe_2$ [131] has been made by Mehring *et al.* [132]. If the charge density wave in KCP were commensurate with the crystallographic unit cell and therefore the excess charge (1.70–5/3 electrons per Pt) were accumulated in a soliton lattice, then this lattice would cause the NMR-signal to change with respect to that from the incommensurate charge density wave. Evidence for such solitons stems from line shape analysis of high resolution NMR spectra [133]. However, NMR-experiments do not distinguish between a static soliton lattice which does not carry current and the thermally excited, mobile solitons proposed by Thomas. Hence both experiments may be reconciled by adopting Thomas' picture, but as just mentioned the theoretical picture is not totally clear.

Nonohmic behavior, i.e., frequency ω and field E dependent conductivities [134] have been reported from several conducting platinum chain complexes. In KCP, Schegolev *et al.* [135] has measured $\sigma_{\parallel}(T, \omega)$ at microwave frequencies (10 and 35 GHz), and CoOP was investigated at 35 GHz [136]. Carneiro *et al.* showed that a significant frequency dependence were observable in ACP(Cl) even at 0.1–10 MHz [58]. ACP(Cl) has a commensurate band filling as shown in Table IV, so the charge density wave is here exposed to a commensurate field from the crystallographic lattice which makes it an obvious candidate for transport by solitons.

A subject which has barely been dealt with at all in Pt-conductors is the depinning of the charge density wave, similar to what occurs in $NbSe_3$ [137] and in the spin density wave system $(TMTSF)_2PF_6$ [138]. Assuming that the CDW is pinned by a periodic potential commensurate with $2k_F$, the depinning field strength may be calculated to be [139]:

$$E_0 = \frac{1}{2k_F} \frac{M^*}{e} \omega_\phi^2(2k_F). \tag{45}$$

In the case of KCP this yields $E_0 \approx 3 \times 10^7$ V/cm, using parameters above, i.e., out of experimental range. However, the low experimental $E_0 = 120$ V/cm found by Stucky *et al.* [32] in GCP suggests a quite different $M^*\omega_\phi^2(2k_F)$ for this compound indicative of a great variation in pinning strengths among the series of Pt-chain conductors. Critical fields are $E_0 \lesssim 1$ V/cm for $NbSe_3$ and $E_0 \approx 10$ mV/cm for $(TMTSF)_2PF_6$. However, it should be kept in mind that the light mass of the carriers may lead to pronounced 'hot carrier' effects, which also give nonlinear conduction [140, 141].

In summary, transport studies of Pt-chain conductors display features which are related to both single particle transport (electrons and holes of the Peierls semiconductor) as well as collective excitations (phase solitons). It seems an interesting task of the future to investigate these features more quantitatively making use of the systematic variations of the band filling that the series of salts in Table I provides.

4.7. Summary

The charge density wave ground state of the one-dimensional conductor with prevailing electron–phonon coupling has been described. As originally predicted by Peierls it is characterized by a gap Δ in the electronic spectrum and a concomitant periodic lattice distortion of amplitude α and period $2k_F$. These two basic characteristics are associated with a wealth of interesting features which have been very well illuminated. Knowing the electronic structure ϵ_k, the band filling k_F, the lattice dynamics $\omega_0(q_\parallel)$, and the electron–phonon coupling g (or λ) of the metallic state one can in principle predict all properties at low temperatures. But the lack of precise knowledge about g prevents us from doing this, and instead we have built a consistent picture of the tetracyanoplatinates by studying (i) the variation of parameters upon the change of k_F and (ii) the detailed behavior of KCP(Br).

This picture explains why the low temperature activation energy Δ varies dramatically from 18 to 125 meV in the 9 regularly-behaving $[Pt(CN)_4]$ conductors as a result of a quite small lattice expansion, giving a platinum separation $d_\parallel$ ranging from 2.798 to 2.92 Å. It also provides insight into the behavior of the anomalously behaving compounds GCP(Br), Pb/KCP(Cl) and PtLi(mnt). In these three compounds $\omega_0(q_\parallel)$ has been modified and in the latter also ϵ_k.

The prototype conductor KCP(Br) may be understood in great detail. With ϵ_k being free electron like, $k_F = 0.85\ \pi/d_\parallel$, $\hbar\omega_0(2k_F) = 8$ meV and $g \simeq 0.15$ eV ($\lambda \simeq 0.30$) we can quantitatively explain: dc-conductivity and thermopower, the lattice distortion, the optical conductivity, together with the characteristics of the amplitude and phase modes. Many aspects are conveniently discussed by introducing an effective mass M^* of the charge density wave. In KCP one finds $M^* \simeq 1000m$.

Despite the fact that both the ground state and its microscopic excitations are so well understood, the transport mechanisms responsible for the electronic transport in the $[Pt(CN)_4]$-conductors have not been treated systematically. But there is a fair amount of experimental data available indicating that also here is something to be learned from this class of compounds.

5. The Critical Temperature Region

Above we have described the two limiting states of the one-dimensional conductor as illuminated by the conducting $[Pt(CN)_4]$-salts of Table I and LiPt(mnt) of Table III. These are the high-temperature metal and the low-temperature charge density wave semiconductor. It should be noted that both states have been characterized strictly from a one-dimensional point of view without any considerations regarding the three-dimensional nature of real compounds. Hence, in this respect the two outer regions of the phase-diagram in Figure 7 are determined purely by the one-dimensional Peierls instability

We now turn to the intermediate region of Figure 7, *the critical region*, where the metal is distorted by one-dimensional fluctuations when $T \approx T_P^{MF}$ and by three-dimensional fluctuations when $T \simeq T_P$. In this region the behavior of the quasi-one-dimensional conductor is determined in a composite way by the one-dimensional instability together with the interchain interactions. The former sets the scale temperature T_P^{MF} and the latter sets the real transition temperature T_P.

According to the theory of second-order phase transitions [142, 143] below the critical temperature T_C the symmetry of the high temperature phase is broken as characterized by an order parameter with long-range order, which we may take as either the temperature dependent lattice distortion $\alpha(T)$ or the electronic gap $\Delta(T)$ (they are related by (32)). The critical temperature region is defined by the temperature region in which the order parameter locally differs significantly from the mean value (which is finite when $T < T_C$ and zero when $T \geqslant T_C$). These local fluctuations enhance both the susceptibility of the system and the specific heat capacity $C_P(T)$. They both diverge at T_C. In an isotropic system the critical region is narrow and T_C is close to T_C^{MF}, but in strongly anisotropic systems there is a broad one dimensional critical region around T_P^{MF} and a narrow three-dimensional critical region at T_C.

As has been mentioned already, long range order does not develop in KCP, from this point of view the only studied Pt-chain compound. Hence, there is no phase transition. However, as we have demonstrated above the physical properties are so different at high and low temperatures that it is illuminating to discuss their changes in terms of a transition. This is further justified from Figure 20 where the measurements associated with

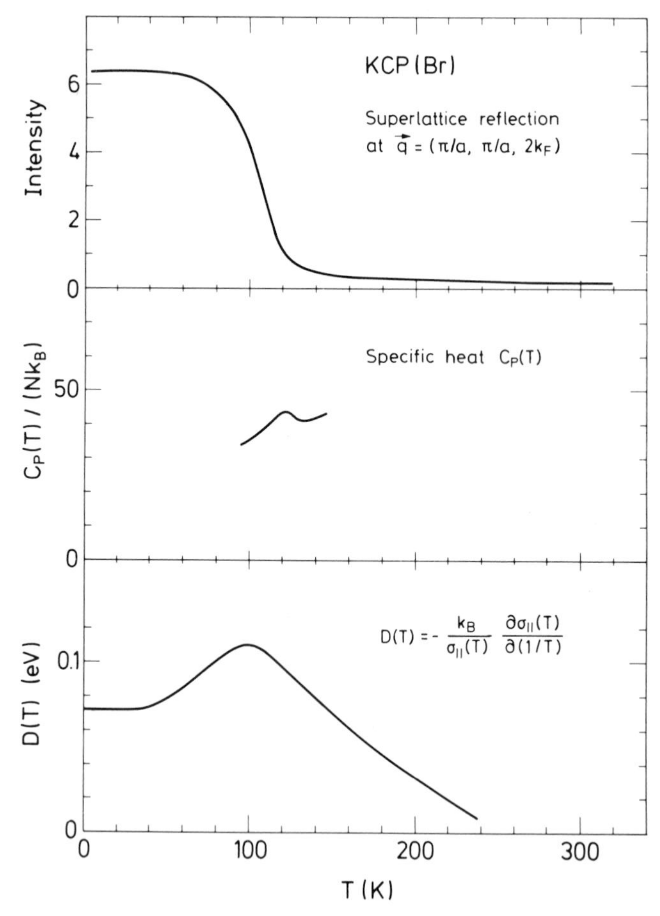

Fig. 20. Experimental evidence for the phase transition in KCP from neutron scattering, specific heat and conductivity.

$\alpha(T)$, $C_P(T)$ and $\Delta(T)$ are shown. The neutron diffraction intensities [89] measured at $\mathbf{q} = (\pi/a, \pi/a, 2k_F)$ identifies the symmetry breaking wave vector corresponding to an antiferroelastic distortion in the **ab** plane and from the temperature variation one would identify $T_P \approx 120$ K. One also notices the extent of the distortion to high temperatures, consistent with the high $T_P^{MF} = 474$ K from (4). The specific heat $C_P(T)$ shows an anomaly at T_P riding on a slowly-varying background [144]. This is as expected for the quasi-one-dimensional system and has been observed in several quasi-one-dimensional magnets [145]. The 'background' is associated with the one-dimensional fluctuations and the 'anomaly' is related to the three-dimensional ordering. It is interesting that the measured anomaly corresponds to an entropy change of approximately $\Delta S = 0.6k_B$, i.e., very close to the theoretical $\Delta S = k_B \ln 2$ for the superlattice formation determined by diffraction. The specific heat identifies $T_P = 117 \pm 5$ K for KCP. Finally Figure 20 also shows the experimental logarithmic derivative $D(T)$ of the conductivity in KCP [56, 146]. If $\sigma_\parallel(T)$ behaves according to (43) $D(T)$ may be written:

$$\begin{aligned} D(T) &= -\partial[\ln \sigma_\parallel(T)]/\partial[1/T] \\ &= \Delta(T) - T\partial\Delta(T)/\partial T - \beta T. \end{aligned} \tag{46}$$

Hence $D(T)$ should peak at T_P because of the divergence of the second term in (46). This offers a simple way of deriving an approximate value for T_P from the conductivity [146]. From this we get $T_P \approx 100$ K in KCP, somewhat lower than the above given values.

Having established that for practical purposes the Peierls transition at T_P is a real transition we discuss below some of its characteristics. We also discuss how recent pressure work has illuminated the interesting interplay between the one-dimensional instability and the interchain coupling.

5.1. 'QUASI-1D' AND 'QUASI-3D' THEORIES

Theoretically, one may approach the critical phenomena of coupled chain conductors from two extreme starting points [10–16, 49, 147–150]. Firstly one may start from a strictly one-dimensional model which of course gives $T_P = 0$ and then build in the interchain coupling η. This will be a good approximation for small η's, i.e., to the left in Figure 7, and we will call such models 'quasi-one-dimensional' (Q1D). It turns out to be very important to distinguish between the two cases (i) where the order parameter is complex (or has two components, corresponding to both amplitude α and phase ϕ fluctuating) or (ii) where the order parameter is real (or has one component, corresponding to α fluctuating but ϕ being fixed), because the thermodynamic behavior is different in the two cases. Case (ii) corresponds to a half filled band conductor ($k_F = \frac{1}{2}\pi/d_\parallel$) which we do not encounter in Table IV, so we should be concerned with theories treating the complex order parameter only.

A second theoretical approach is to start with a mean field theory, neglecting fluctuations, and then build in approximately the relevant fluctuations. This corresponds to relatively high interchain coupling, i.e., to the right in Figure 7. We call these models 'quasi-three-dimensional' (Q3D).

The microscopic origin of the interchain coupling in conducting Pt-chain compounds have been proposed to be either due to Coulombic repulsion between charge density waves on neighboring chains [147, 148] or due to interchain electronic overlap giving a finite bandwidth $W_\perp$ [14] as indicated in Figure 1. But its relation to the theoretical interchain parameters is not always clear since these are often phenomenological in nature or chosen to be analogous to interactions in anisotropic magnetic systems such as the quasi-one-dimensional Ising and Heisenberg models.

5.2. THE TRANSITION TEMPERATURE AND THE INTERCHAIN COUPLING

In order to demonstrate how the above mentioned theories relate to experiments we analyze the observed transition temperature T_P of the 'well behaved' $[Pt(CN)_4]$-conductors and LiPt(mnt). Their T_P are listed in Table VII, and the task is now to use these temperatures to understand the interchain coupling, to which we have so far loosely referred to as η. As an example of a quasi-one-dimensional theory as just defined we consider the results of the Ginsburg–Landau theory for weakly coupled chains as reviewed by Dieterich [49]. Here one considers the free energy $\mathscr{F}\{\phi_i(z)\}$ as a functional of the complex order parameter $\phi_i(z)$ on the ith chain:

$$\mathscr{F}\{\phi_i(z)\} = \int \mathrm{d}z \left[a|\phi_i|^2 + b|\phi_i|^4 + c|\mathrm{d}\phi_i/\mathrm{d}z|^2 + \tfrac{1}{2}\phi_i d \sum_j \phi_j^* \right], \tag{47}$$

where the first three terms describe the usual Ginsburg–Landau expansion of the order parameter on one chain with $a = a'(T - T_P^{MF})$ and the last term describes the interaction between chains. If $d = 0$ the problem may be solved exactly to find the expected result for the transition temperature and the correlation length [151]:

$$T_P^{1\mathrm{D}} = 0 \quad \text{and} \quad \xi_\parallel^{1\mathrm{D}} = \frac{\hbar v_F}{2\pi} \frac{1}{k_B T} = d_\parallel \frac{1}{\pi k_F \cdot d_\parallel} \frac{\epsilon_F}{k_B T}, \tag{48}$$

where we have used this free electron relation $\epsilon_F = \frac{1}{2}\hbar v_F k_F$, and this may be used to relate the parameters a, b, and c to the parameters of materials. Introducing now a finite d one finds:

$$T_P^{\mathrm{Q1D}} = \frac{\Delta(0)}{\pi_{kB}} \sqrt{2\eta^{\mathrm{Q1D}}} \approx T_P^{MF} \sqrt{\eta^{\mathrm{Q1D}}}, \quad \text{when} \quad \eta^{\mathrm{Q1D}} \ll 1 \tag{49}$$

where we have defined $\eta^{\mathrm{Q1D}} = |d|/(4|a|)$. Using the measured activation energies and transition temperatures the interchain coupling constant of this theory is derived from (49). The results are shown in Table VII.

As an example of a quasi-three-dimensional theory we consider the results of Horowitz, Gutfreund and Weger [14]. They specifically considered transverse electronic overlap

TABLE VII

Interchain coupling η derived from the observed transition temperatures T_P (derived from conductivity results via (46)) for compounds which have a simple Peierls transition. The results are shown using both a 'quasi-one-dimensional' theory (Q1D) and a 'quasi-three-dimensional' theory (Q3D) as defined in the text.

Compound	T_P(K)	T_P^{MF}(K)	η^{Q1D}	η^{Q3D}
RbCP(FHF)	80	118	0.73	0.020[b]
RbCP(DSH)	80	283	0.13	0.015[b]
CsCP(FHF)	80	178	0.32	0.016[b]
CsCP(Cl)	90	198	0.33	0.016[b]
KCP(Cl)	95	461	0.07	0.020[b]
$CsCP(N_3)$	110	270	0.26	0.019[b]
KCP(Br)	100	474	0.07	0.020[b]
Pb/KCP(Cl)	85	178	0.37	0.015[b]
RbCP(Cl)	110	494	0.08	0.022[b]
GCP(Br)	100	296	0.18	0.015[b]
ACP(Cl)	195	824	0.09	0.039[c]
LiPt(mnt)	216[a]	237	–	0.95[c,d]

[a] Taken from diffuse X-ray results.
[b] Strongly fluctuating.
[c] Weakly fluctuating.
[d] Here (51) has been modified to account for the actual band structure.

so that the band structure looks as follows, assuming free electron dispersions along the chains:

$$\epsilon_{\mathbf{k}} = \frac{\hbar^2 k_\parallel^2}{2m} - \eta^{Q3D}\epsilon_F[\cos(d_\perp k_x) + \cos(d_\perp k_y)]. \tag{50}$$

The transverse band width (cf. Figure 1) becomes $W_\perp = 2\eta^{Q3D}\epsilon_F$. First they derived an upper limit for η for the occurrence of a phase transition as a result of the Peierls instability $\eta < \eta_C = 3\sqrt{\tau}$, $\tau = k_B T_P^{MF}/\epsilon_F$, quantifying the argument that if η becomes too large the one-dimensional argument leading to the Peierls transition loses its meaning. They then modified the mean field theory result by considering fluctuations, using the Ornstein–Zernike criterion to find T_P. This criterion states that the size of the fluctuations of the order parameter at T_P^{MF} is equal to the size of the mean field order parameter at T_P. Their results for the transition temperature are as follows:

(i) There exists a region of interchain coupling, where fluctuations are small and the transition temperature is given by

$$T_P = T_P^{MF} \exp\left[-\left(\frac{2\tau}{\eta^{Q3D}}\right)^2\right], \qquad 4\tau \lesssim \eta^{Q3D} \lesssim 3\sqrt{\tau}. \tag{51}$$

(ii) For smaller η's fluctuations are more important and an approximate value of the transition temperature is given by:

$$T_P \simeq T_P^{TF} \exp\left[-\left(\frac{2.5\tau}{\eta^{Q3D}}\right)\right], \qquad \eta^{Q3D} \leqslant 4\tau. \tag{52}$$

Hence one may use (51) and (52) to characterize the compounds in 'weakly' and 'strongly' fluctuating materials and to calculate the according interchain coupling. The results are shown in Table VII and shown together with the results for η^{Q1D} in Figure 21.

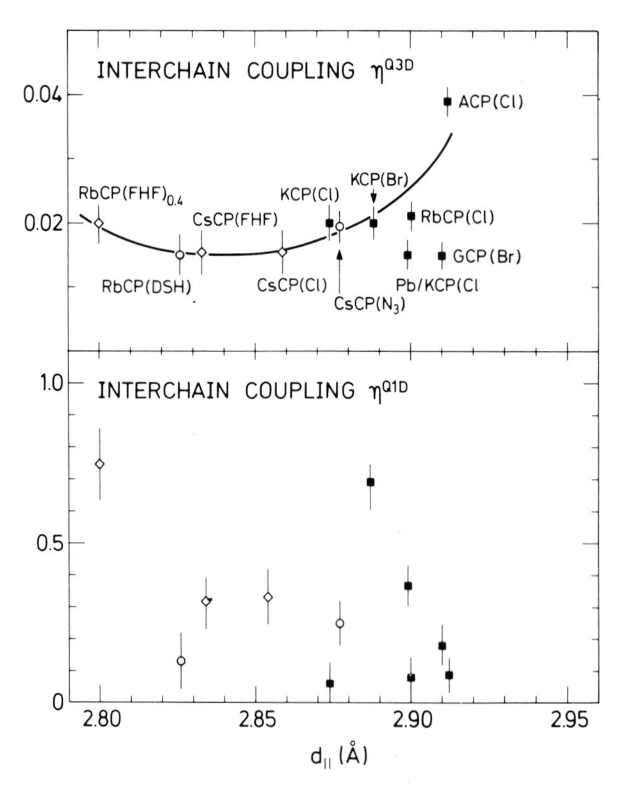

Fig. 21. Interchain coupling parameters from Table VII for conducting $[Pt(CN)_4]$-salts. The line is a guide to the eye.

Several lessons may be learned from Table VII and Figure 21. Firstly, there is no reason why η^{Q1D} and η^{Q3D} should be the same since they are defined quite differently, and indeed they are not. But if we believe that the interchain coupling varies regularly over the series of $[Pt(CN)_4]$-conductors they are not to be treated with in quasi-one-dimensional theories since η^{Q1D} varies quite irregularly, whereas η^{Q3D} shows a more smooth behavior. This difference may be traced to the fact that in the Q1D-theory the ratio T_P/T_P^{MF} is determined by η only, whereas in Q3D theory T_P/T_P^{MF} depends on both η and $\tau = k_B T_P^{MF}/\epsilon_F$. Hence, the regular behavior of η^{Q3D} indicates that the ratio τ, which is inherent in the mean field picture of the Peierls instability and apparently absent in the Ginsburg–Landau treatment, is relevant for the compounds considered. On the other hand all materials but ACP(Cl) and LiPt(mnt) are classified as strongly fluctuating since they have $\eta < 4\tau$, and therefore the quasi-three-dimensional theory is not very accurate either. All compounds, even LiPt(mnt) have $\eta < 3\sqrt{\tau}$, so the observed Peierls transitions are theoretically meaningful.

Secondly, since η^{Q3D} has a specific meaning from (50) it may be quantitatively discussed. From the typical value of $\eta^{Q3D} = 0.02$ we get a transverse bandwidth $W_\perp \approx$ 0.1 eV, which is of similar magnitude as $2\Delta(0)$ from Table VI. If $W_\perp > 2\Delta(0)$ then it appears from Figure 1 that the low temperature state is a semi-metal and not a semiconductor, and the results therefore suggest that η^{Q3D} only in part stems from actual electronic overlap but also from Coulomb repulsion between chains, which does not give rise to electrons hopping from chain to chain. This composite picture is consistent with the fact that if one estimates the transverse bandwidth from the almost temperature independent conductivity anisotropy at high temperatures, using the result by Soda *et al.* [126]:

$$W_\perp = W_\| \frac{d_\|}{d_\perp} \sqrt{\frac{\sigma_\perp}{\sigma_\|}} \tag{53}$$

which gives $W_\perp = 0.01$ eV for KCP(Br). Regarding the variation of η^{Q3D} from compound to compound, Figure 21 indicates a slight increase at low chain separations $d_\perp$ ($d_\perp$ follows $d_\|$). This is expected, but the increase for larger separations is at first sight astonishing. However, when considering that bulky ions which can naturally mediate electronic coupling are used to separate the chains, it is only natural that these do not become more effectively isolated as they are geometrically more separated. Finally one may notice that the unusual compounds Pb/KCP(Cl) and GCP(Br) from the point of view of the one dimensional electron–phonon coupling, are also somewhat off the regular behavior sketched for η^{Q3D}. This indicates that their special chemical bonding has an effect both along and perpendicular to the chains.

5.3. THE CORRELATION LENGTHS

The correlation lengths in the critical temperature region has been a subject of appreciable theoretical interest, in particular $\xi_\|(T)$ along chains. This is related to the fact that in strictly one dimension $\xi_\|^{1D}(T)$ for the complex order parameter as given by (48) is very much different from the case of a real order parameter. For finite interchain couplings the theoretical results are as follows:

$$\xi_\|^{Q1D}(T) = \xi_\|^{1D}(T)[1 - (T_P/T)^2]^{-1/2}$$

and

$$\xi_\|^{Q3D}(T) = \xi_\|^{1D}(T)[\tfrac{1}{2}\ln(T/T_P)]^{-1/2}, \tag{54}$$

where $\xi_\|^{1D}(T)$ is given by (48). It is interesting to notice that despite the fact that the interchain coupling leading to a finite T_P seems to have no resemblance in the two theories, the correlation lengths are quite similar when expressed in terms of the resulting T_P. In

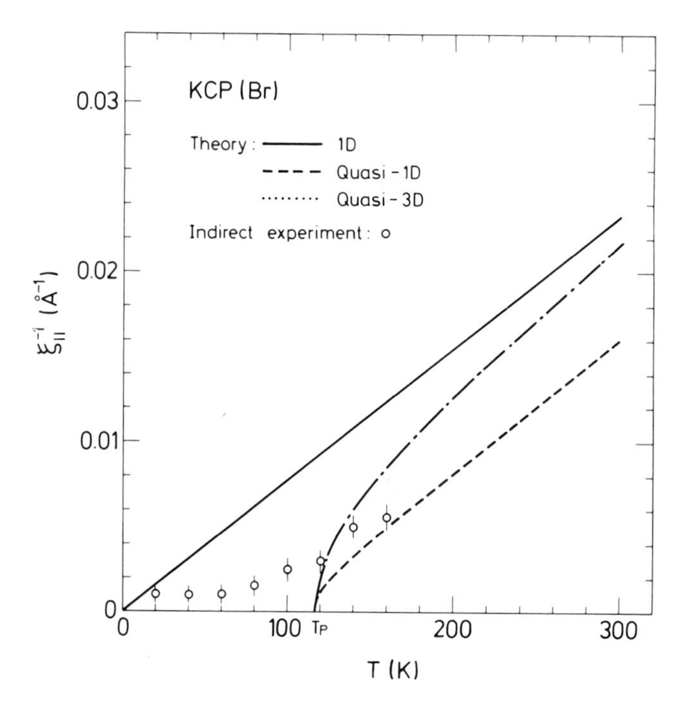

Fig. 22. The inverse of the longitudinal correlation length $\xi_{\parallel}$ in KCP. The transition temperature T_P has been taken from specific heat data. There are no adjustable parameters.

Figure 22 we show the inverse correlation length $\xi_{\parallel}^{-1}(T)$ with parameters appropriate to KCP. T_P has here been taken from the specific heat data to be 117 K. Only an indirect attempt has been made to estimate the longitudinal correlation length experimentally. The difficulty in direct determination, e.g., as the halfwidth measured in the chain direction of the $2k_F$ diffuse X-ray scattering or elastic neutron scattering is that at low temperatures where the scattering is intense $\xi_{\parallel}$ exceeds experimental resolution of conventional diffractometers, and at high temperatures where the linewidth is greater the intensity is very low. However, one may associate $\xi_{\parallel}^{-1}$ with the half width $\Delta Q/2$ of the inelastic neutron scattering linewidth at the pinned phason frequency $\omega_{\phi}(2k_F)$. ΔQ is indicated in Figure 17 and was estimated by Carneiro *et al.* below $T = 160$ K [91]. The results are shown in Figure 22 to demonstrate that in this respect present experiments are only suggestive and do not discriminate between the different theoretical approaches.

The transverse correlation length $\xi_{\perp}(T)$ as measured by the width of the $2k_F$ diffuse scattering perpendicular to the chains has been studied accurately in KCP(Br) and in Figure 23 we show the results by Lynn *et al.* [90]. These are compared to the theoretical results:

$$\xi_{\perp}^{\mathrm{Q1D}}(T) = \frac{d_{\perp}}{2}\frac{T_P}{T}\left[1 - (T_P/T)^2\right]^{-1/2}$$

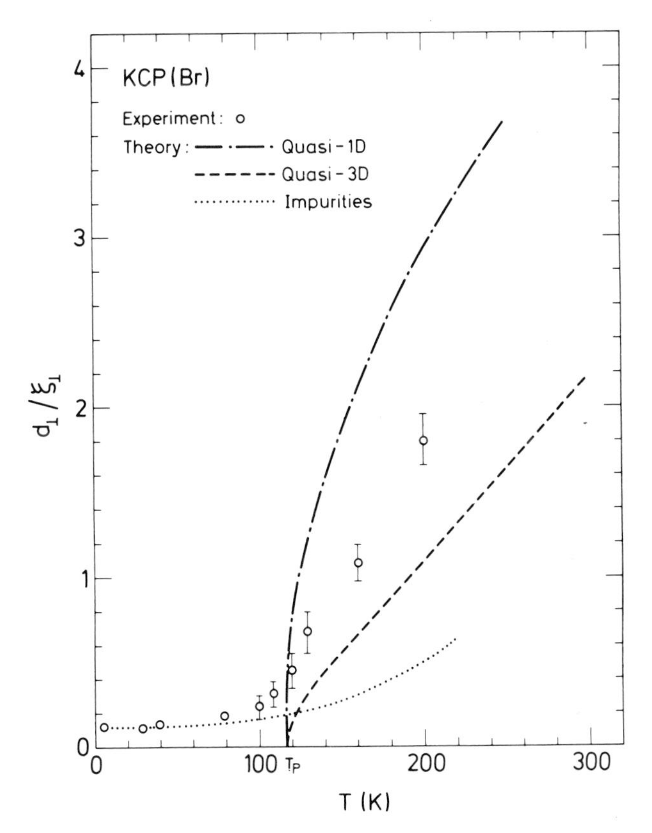

Fig. 23. The inverse of the transverse correlation length $\xi_\perp$ in KCP. $\xi_\perp$ is measured in units of the perpendicular distance between chairs $d_\perp$, which is 9.91 Å.

and

$$\xi_\perp^{Q3D}(T) = \frac{d_\perp}{2} \frac{\eta^{Q3D}}{\pi} \frac{\epsilon_F}{k_B T} [\ln(T/T_P)]^{-1/2}, \tag{55}$$

where η^{Q3D} is determined by (52) so that in both cases there are no adjustable parameters. As for $\xi_\parallel$ the two theoretical results for $\xi_\perp$ are similar and differ mainly in magnitude. From Figure 23 one finds that the experimental results for KCP lie salomonically between the two theoretical results demonstrating that these materials maintain their right to have parameters that are not within easy reach of theory!

Figure 23 also shows the theoretical result for $\xi_\perp^{-1}(T)$ obtained by Sham and Patton [16]. It shows that if one attributes the finite correlation length below T_P to pinning by random impurities only, it is not possible to reproduce its temperature dependence which $T > T_P$. This problem is probably related to the difficulty in explaining the pinning frequency $\omega_\phi(2k_F)$ by random impurities. It is in this respect interesting to notice that whereas $\omega_\phi(2k_F)$ has always been reported to be approximately 2 meV independent of sample, considerable range for low temperature correlation length has been reported.

Assuming that the random impurity concentration may vary from sample to sample whereas a periodic potential does not change, these results indicate a different effect on $\xi_\perp$ and ω_ϕ by random and periodic potentials.

5.4. The soft phonons

As originally pointed out by Rice and Strässler [6] the Peierls transition in the mean field theory is associated with a soft phonon mode, and one might therefore expect a dramatic (or 'critical') temperature dependence of both the amplitude and phase modes in the critical temperature region in close analogy with ferroelectrics. The first inelastic neutron scattering results from KCP were in fact interpreted this way [114], but later work [91] together with Raman scattering [116] and far infrared reflection studies [118] demonstrate that both $\omega_\alpha(2k_F)$ and $\omega_\phi(2k_F)$ vary remarkably little with temperature. The experiment results are shown in Figure 24.

The lack of critical dependence of the lattice dynamics in KCP may be qualitatively understood from the work of Dieterich [49] although the detailed calculations are fairly complicated. In mean field theory both the amplitude mode and the soft phonon in the metallic region vary with the square root of the Landau parameter $|\mathbf{a}|$ in (47)

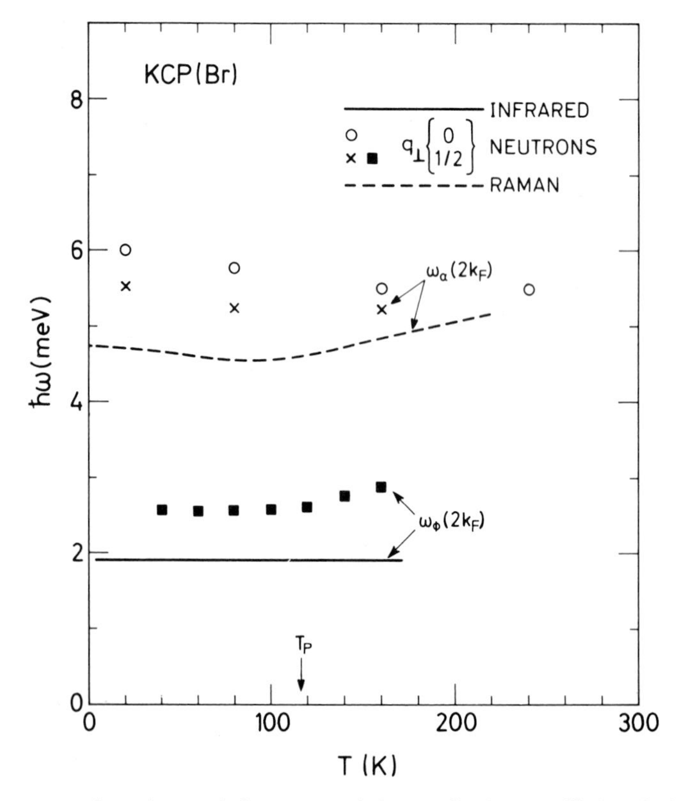

Fig. 24. Temperature dependence of the energy of the amplitudon $\omega_\alpha(2k_F)$ and phason $\omega_\phi(2k_F)$ by different experiments. Energies are given meV (cf. Figure 11).

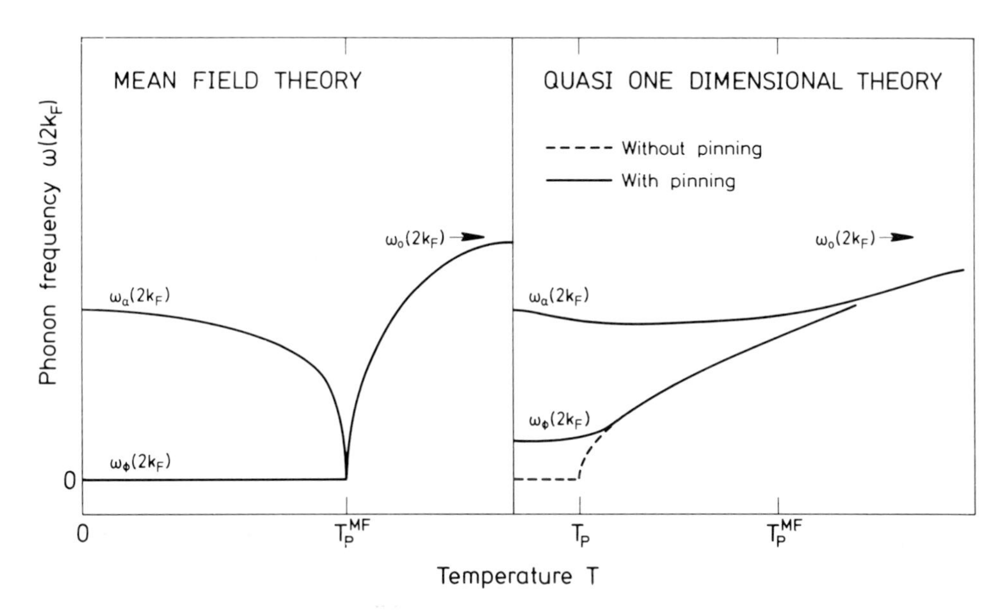

Fig. 25. Temperature-dependent phonon frequencies of a quasi-one-dimensional conductor. ω_0 is the phonon of the unperturbed metallic phase, ω_α and ω_ϕ denote amplitudons and phasons, respectively.

leading to a square root dependence $\sqrt{|T - T_P^{MF}|}$ as illustrated in Figure 25 for ω_ϕ and ω_α. However, when fluctuations suppress the critical temperature to T_P the Landau parameter $|a|$ varies only little around T_P and so do ω_ϕ and ω_α. The fluctuations that diverge at T_P do then manifest themselves in the low energy phason, and $\omega_\phi(2k_F)$ would then go to zero if pinning did not require that it be nonzero. In this situation fluctuations are then taken up by still lower lying excitations which in KCP(Br) appear to be of very low frequency. These modes are not identified but from NMR-experiments their frequency $\omega/2\pi$ is estimated to be of the order of 10 kHz corresponding to $10^{-8} \times \omega_\phi(2k_F)$ [133]. This situation, that as the critical temperature is approached soft modes do not really soften is also seen in ferroelectrics [152] but in quasi-one-dimensional conductors the large critical region magnifies these effects. Finally we notice that it is not astonishing that the temperature dependence of $\omega_\alpha(2k_F)$ does not reflect the calculations of Kurihara [111], since it is determined by quite complex mechanisms relating to the critical behavior, which he disregarded entirely.

Although their temperature dependence agree the absolute values of the amplitude and phase frequencies in Figure 24 differ considerably when the results of neutron scattering are compared with those of optical experiments. In order to investigate this problem we compare in detail the measured lineshape for $\sigma(\omega)$ at $\omega_0(2k_F)$, the Raman intensity $a_L^2(\omega)$ at $\omega \simeq \omega_\alpha(2k_F)$ and the neutron intensities in Figure 26. The neutron intensities are shown both for $q_\perp = 0$ and $q_\perp$ = '1/2' ($\mathbf{q}_\perp = \frac{1}{2}(\mathbf{a}^* + \mathbf{b}^*)$). The 'accidentally' crossing transverse acoustic mode (cf. Figure 17) is also shown. From Figure 26 it is clear that the differences in measured frequencies are not an artifact of differences in data

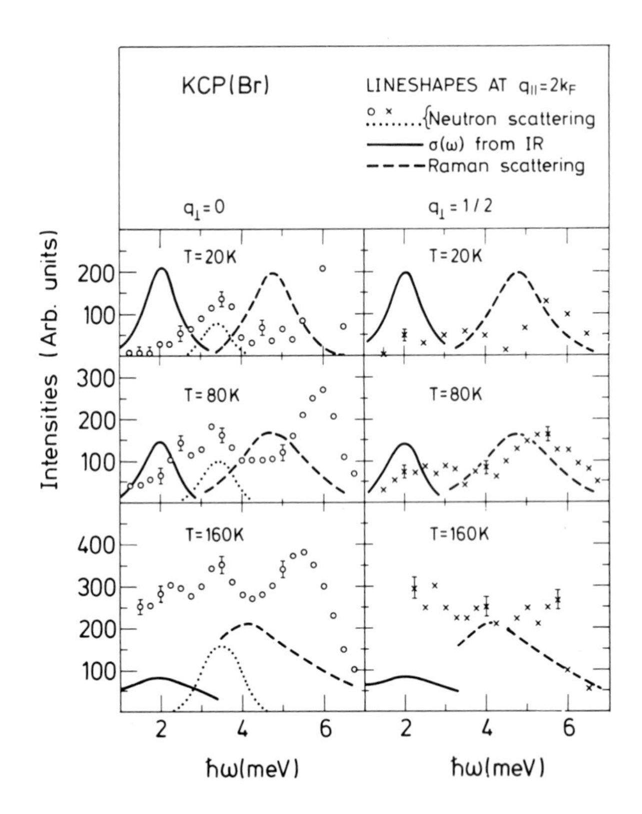

Fig. 26. Different measured lineshapes relating to the phase (ω_{ϕ}) and amplitude (ω_{α}) modes in KCP(Br). The absolute intensities of different measurements are not related, but the temperature dependence of each is shown as measured. In principle neutron scattering should reflect the sum of the IR-lineshape, the Raman lineshape, and at $q_{\perp} = 0$ also the crossing transverse acoustic mode. Energies are given in meV (cf. Figure 11).

analysis, etc., but appear to be a real effect. The reason for this discrepancy is not clear; but once again we learn that if we are asking too much, the answers become more and more equivocal.

5.5. The Order Parameter

Throughout this section we have tacitly assumed that an order parameter could be defined that goes continuously to zero at $T = T_P$. Above we have argued that the amplitude of the lattice distortion $\alpha(T)$ and the gap in the electronic spectrum $\Delta(T)$ could both be taken to be the order parameter. However, to measure the critical temperature dependence of these quantities is not feasible. This is illustrated in Figure 27 where $\alpha(T)$ and $\Delta(T)$ are shown as deduced from the intensity of diffuse elastic neutron scattering and from conductivity results, respectively. Both vary very smoothly through T_P a fact which may at least in part be related to that fluctuations show up in the measured-

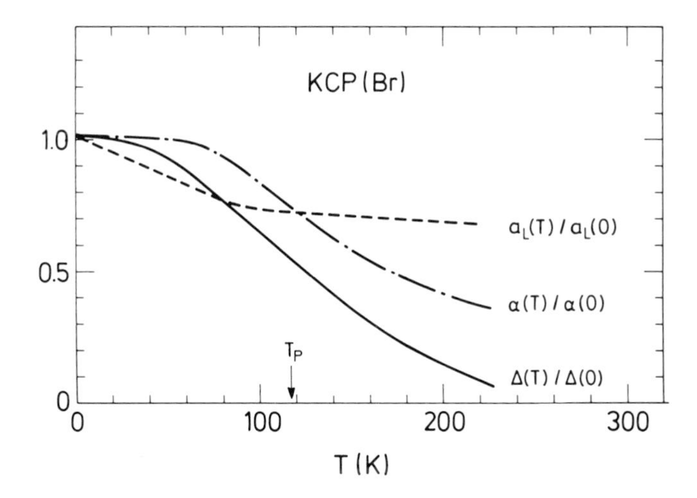

Fig. 27. Measurements of the order parameter in KCP(Br). Δ comes from conductivity measurements, α from neutron scattering and a_L from Raman scattering.

order parameter. But also the influence of the different kinds of pinning on Δ and α may be of importance.

The role of fluctuations on the measured-order parameter is clearly demonstrated when one considers the Raman oscillator strength a_L for KCP also shown in Figure 27. a_L may be taken as an order parameter because it is only because of the lattice distortion that $q_{\parallel} = 2k_F$ becomes equivalent to $q = 0$ giving a finite oscillator strength. However, if the size of the domains above T_P is comparable to the wave length of the light used in the experiment, the situation is effectively static and cannot be distinguished from $T < T_P$. Hence the lack of temperature dependence of a_L is a further evidence for a long correlation length $\xi_{\parallel}$ in KCP(Br).

5.6. Pressure Effects

Given the remarkable effects of pressure in organic conductors one might hope that pressure in Pt-conductors would reveal similarly exciting results. However, experiments on KCP(Br) show that pressure has a less dramatic effect; but it is nevertheless interesting to analyze the pressure effects within the framework developed above.

Thielemann *et al.* applied pressure to KCP(Br) and measured $\sigma_{\parallel}(T)$ for $P \leqslant 32$ kbar [153]. They found that T_P increased to 171 K whereas the low temperature activation energy decreased to 26 meV suggesting a decrease in electron phonon coupling λ and an increase in interchain coupling η with increasing pressure. Shortly before Roth *et al.* [154] had measured the two elastic constants C_{33} and C_{44} both at $P = 0$ varying the temperature as well as varying the pressure at room temperature. According to (20) the main reason for a change in λ is the ω^2 term related to C_{33} so from this a direct relation between the ultrasonic measurements and $\sigma_{\parallel}(T)$ may be made, provided that k_F is known as a function of pressure as well as the lattice parameters a and c. It is therefore interesting that these have recently been reported.

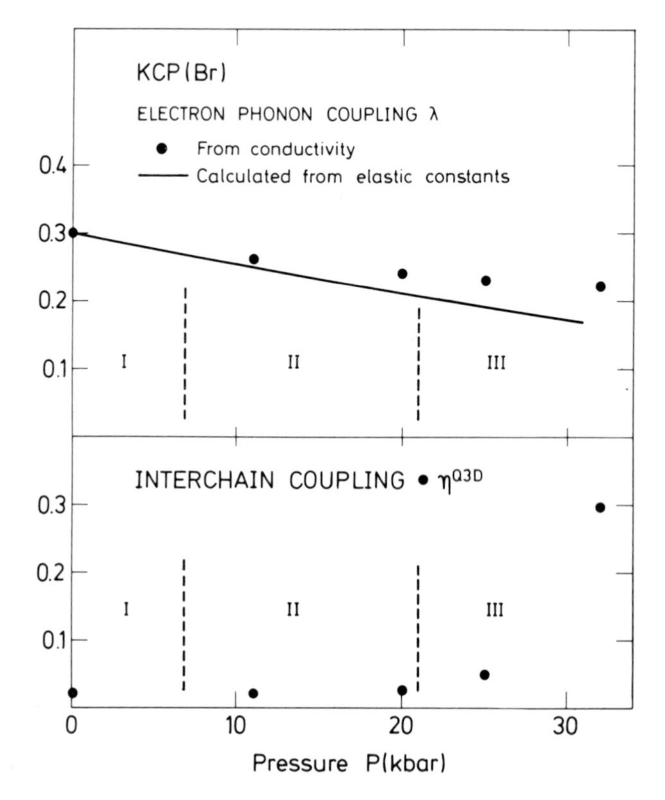

Fig. 28. Electron–phonon coupling constant λ and interchain coupling η, deduced from conductivity. I, II, III indicate the structural transitions reflecting the change in lattice parameter c at 7 kbar and in a at 21 kbar.

Kobayashi *et al.* [155] have determined the lattice parameters for $P < 30$ kbar and found that two phase transitions occur. At $P \simeq 7$ kbar there is an abrupt decrease in c and at $P \simeq 21$ kbar a similar change occurs in a. And recently Renker *et al.* have found that $k_F d_{\|}$ is unchanged at least up to $P = 11$ kbar [156]. (22) therefore enables us to calculate λ from the measured activation energies, and the result is shown in Figure 28. Assuming that (20) is correct apart from a constant factor we may calculate λ independently in the following way:

$$\frac{\lambda(P)}{\lambda(P=0)} = \left[\frac{\epsilon_F(P)}{\epsilon_F(0)}\right]^{1/2} \left[\frac{d^*(P)}{d^*(0)}\right]^3 \left[\frac{C_{33}(P)}{C_{33}(0)}\right]^{-1}, \tag{55}$$

and this result reproduces the results from the activation energy analysis very well.

Analyzing the interchain coupling from the measured transition temperature as shown in Figure 28 indicates that η^{Q3D} is fairly constant until after the second transition found by Kobayashi *et al.* at $P \simeq 25$ kbar. This is different from the conclusion by Renker *et al.* that there is a change in dimensionality already between 1 and 11 kbar. To investigate this further we plot in Figure 29 the measured pressure dependent correlation lengths $\xi_\perp$ and

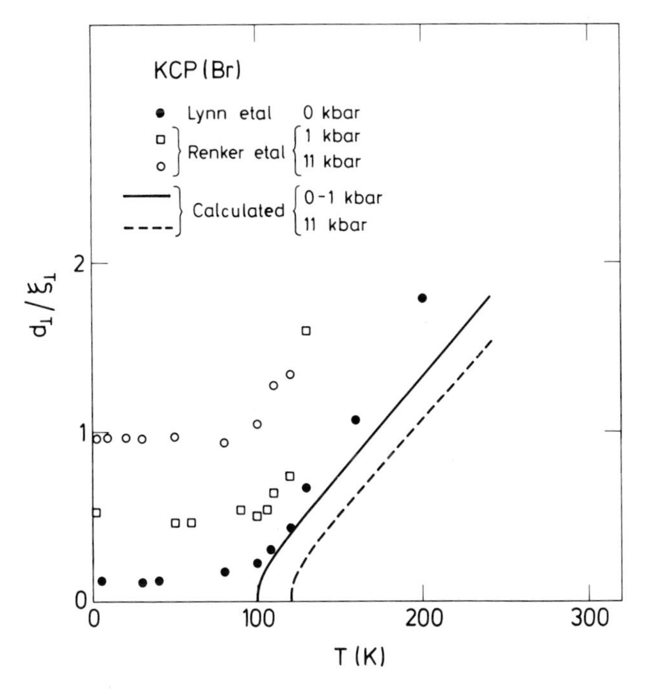

Fig. 29. Temperature dependence of the perpendicular correlation length $\xi_\perp$ as measured by elastic neutron scattering, compared to the theoretically calculated $\xi_\perp$ using T_P from conductivity and the quasi-three-dimensional theory discussed in the text.

compare them to the ones calculated from (5) using the pressure dependence of T_P from conductivity. From this it appears that the major change in the measured $\xi_\perp$ is in its value below T_P. The change in $\xi_\perp$ $(T = 0)$ between 1 and 11 kbar is actually similar to the difference between the low pressure values for $\xi_\perp$ $(T = 0)$ obtained by Renker *et al.* [89] and obtained by Lynn *et al.* [90]. This suggests different sample perfections in crystals from different groups as well as a change in perfection with pressure since below T_P the finite correlation length is attributed to defects. In view of the above discusion one has to be appreciably above T_P before η has an appreciable effect on $\xi_\perp$. Therefore, there is no real discrepancy between the elastic neutron scattering results and the conductivity data, within the framework developed here, since the former measurements do not reflect in a clear way the interchain coupling.

5.7. SUMMARY

The study of quasi-one-dimensional compounds gives new insight into critical phenomena since the critical temperature region, where fluctuations are important, is very large. However, it is unfortunate that the interchain coupling η in the platinum conductors is such that it is hard to treat theoretically both when elaborating an exact one-dimensional theory as well as when starting from isotropic models. In cases where these two approaches give qualitative different results, e.g., for the magnitude of η itself experiments suggest that the 'quasi-three-dimensional' models give the most consistent answer. This is true both when analyzing the series of compounds in Table I and when study KCP under pressure. Most illuminating is the study of the correlation lengths, the soft phonons and the order parameter. The perpendicular correlation length $\xi_\perp(T)$ is well measured and the experimental results are in semiquantitative accordance with theory. Estimates of $\xi_\parallel(T)$ suggest that the same is true here. The lack of critical behavior of both the amplitudon and the phason as well as of the measured-order parameter demonstrate the paramount influence of fluctuations in quasi-one-dimensional systems.

6. Other Instabilities

The argument by Peierls as quantified by the analysis of the Fröhlich model for the electron–phonon coupling leads to a charge density wave ground state for a one dimensional conductor. Above we have described how in platinum chain compounds one may derive the properties of both the high temperature metal and the low temperature insulating (or semiconducting state). Importantly one may build a bridge between the two states since the electron–phonon coupling λ may be calculated via (20) using the following properties of the high-temperature state: (i) the electrons are free, (ii) the phonons are simple and (iii) the Fermi vector is known. A consistent picture is then obtained which explains:

(a) The detailed behavior of KCP(Br), where $\lambda = 0.3$.
(b) The pressure dependence of the conductivity in KCP(Br).
(c) Why λ varies from 0.2 to 0.4 in the series of compounds in Table I.

Second, phase transitions in strongly anisotropic systems give rise to unusual features because of the large temperature region in which strong fluctuations occur. From studies of KCP we learn that:

(a) The specific heat and the transverse correlation length show evidence of critical behavior although the mean field transition temperature is depressed by a factor of about 4,
(b) The temperature dependence of both the order parameter and the soft phonons are difficult to characterize quantitatively.

Hence the platinum conductors have given their share to our understanding of quasi-one dimensional systems. But in order to illustrate that they can teach us even more we review below two subjects which cannot be explained in terms of the picture developed above. They are related to the elastic properties and to the occurrence of molecular ordering, respectively. Whereas the former has shown to be important in the past, the latter is likely to contribute to future developments in solid state physics.

6.1. Elastic Properties of KCP. The Possibility of Superconductivity

Whereas electron–phonon coupling in a linear conductor invariably leads to a metal–insulator transition, as first pointed out by Peierls [1], electron–electron coupling may lead to superconductivity [157]. Whereas the important interaction occurs at $q = 2k_F$ for the Peierls instability, $q \simeq 0$ is the relevant wave vector for superconductivity. It is therefore natural to look for small wave vector anomalies in the phonon spectra.

Based on the comparison between the longitudinal sound velocity along **c** as measured by ultrasonic measurements and inelastic neutron scattering Roth *et al.* [154] attributed the difference to such a $q = 0$ coupling, but unfortunately the neutron data were not very accurate and later high-resolution neutron scattering indicate that this effect, if present at all, is very small. This is demonstrated in Table VIII where the results of both neutron [158] and ultrasonic measurements [154, 159] are shown for the six elastic constants of the tetragonal crystal symmetry. The agreement between the different experiments is within experimental accuracy which for the neutron results primarily is determined by the distortion of the measured velocities by resolution effects [160]. On the other hand only neutron scattering have been used to derive the off-diagonal constants C_{12} and C_{23}, which offers a possibility to check the so-called Cauchy relations. Assuming isotropic forces between Bravais lattice points these relations predict that $C_{21} = C_{44}$ and that $C_{32} = C_{66}$. From these results in Table VIII one finds that $C_{11} \approx C_{44}$ whereas the latter relation is far from obeyed.

TABLE VIII

Elastic constants of protonated KCP(Br) measured by inelastic neutron scattering (zero sound) and ultrasonic measurements (First sound). The unit is 10^{10} dyne/cm^2.

	$T = 290$ K			$T = 90$ K	
Elastic constant	Zero sound [158]	First sound [159]	First sound [154]	Zero sound [158]	First sound [159]
C_{11}	36	35.2	–	–	39.5
C_{33}	33	36.2	33.8	47.2	43.6
C_{44}	6.4	6.0[a]	5.4	7.5	6.75
C_{66}	6.8	5.5[a]	–	9.3	<2[a]
C_{21}	9.6	–	–	–	–
C_{32}	20	–	–	33	–

[a] Extrapolated values.

The reason why the elastic constants derived from ultrasonics should be different from those derived from neutron scattering is that the wavelength of sound propagation is dramatically different in the two cases. It corresponds to adiabatic (or first sound) propagation in the former case and to collisionless (or zero sound) propagation in the latter. If electron–phonon coupling modifies the adiabatic velocity v_1 one finds

$$v_1^2 = (1 - \lambda')v_0^2 \tag{56}$$

where λ' is the forward scattering ($q = 0$) electron–phonon coupling [161]. This effect has recently been demonstrated in the organic conductor $(TMTSF)_2PF_6$ [162], but from Table VIII, the closeness of the reported C_{33}'s indicate that λ' is negligible in KCP.

A pronounced decrease in the elastic constant C_{44} and C_{66} was seen by ultrasonics around $T = 60$ K by Doi *et al.* [159], whereas nothing unusual was seen at this temperature with neutron scattering. The temperature dependences of C_{44} and C_{66} are shown in Figure 30. These anomalies have been explained in terms of the coupling between strain and rotation of water molecules [163] where the latter effect has been studied independently by NMR [164]. Hence the anomalous behaviour of C_{44} and C_{66} gives no evidence of the importance of forward scattering in KCP. One might still hope to reach a superconducting

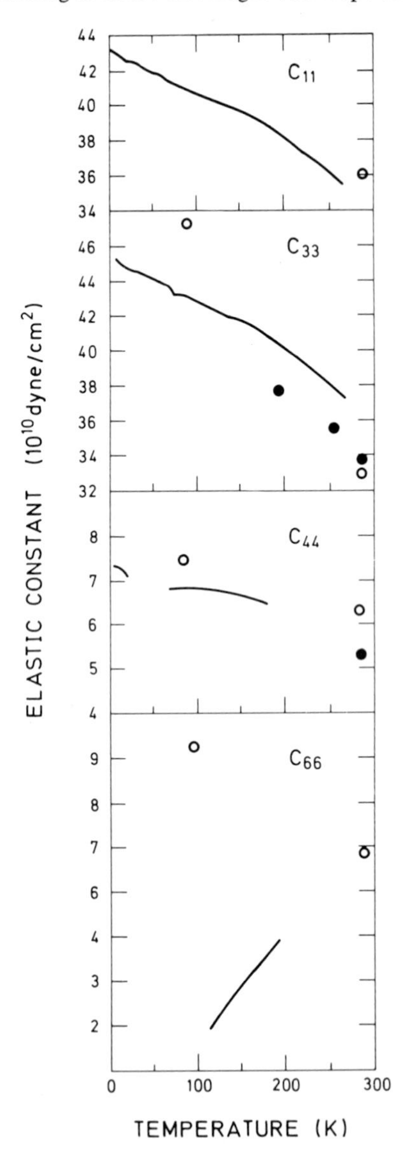

Fig. 30. Temperature dependence of the diagonal elastic constants in KCP. Lines [159] and dots [154] are the results of ultrasonic measurements. Open circles stem from neutron inelastic scattering.

state in the $[Pt(CN)_4]$-conductor if the Peierls instability could be suppressed. Unfortunately pressure increases T_P rather than decreases it, but the decrease in both T_P and Δ for the compounds with the shortest $d_{\parallel}$'s suggests that the Peierls transition might be suppressed if $d_{\parallel}$ could be made small enough. However, in the competition between superconductivity and charge density waves, the large bandwidths of $[Pt(CN)_4]$-conductors (or rather the small ratio $\omega_0(2k_F)/W_{\parallel}$) leaves very little theoretical room for superconductivity [165]. If this state is therefore ever found in Pt-chain conductors it is most likely to occur in compounds like those of Table III where the bandwidths are relatively small.

6.2. Competing Instabilities in M-OP's and in K(def)TCP

The existence of phase transitions in Pt-conductors, which cannot be related to the Peierls instability in a simple way was first reported by Braude *et al.* [166] who found superlattice reflections displaced k_F from the main Bragg reflections. These reflections are spot-like indicating a three-dimensional nature of the instability and they occur at a temperature which is higher than the T_P as estimated from conductivity. Their results are shown in Figure 31. At T_P a second set of spots occur, now at $2k_F$, i.e., on the diffuse lines stemming from the one-dimensional Peierls instability. Hence from a structural point of view there are two transitions at T_1 and T_2 ($T_1 > T_2$) and the lower one at T_2 has both the symmetry, as well as the conduction characteristics, of the Peierls instability although the metal–insulator transition is usually sharp. The compounds NiOP, ZnOP and CoOP appear to be very similar in this respect with slightly different T_1 and T_2. In ZnOP the detailed symmetry below T_2 has been reported by Bertinotti [167], and in NiOP Kobayashi *et al.* [168] identified the 'k_F' spots with the ordering of the cation column. The two temperatures T_1 and T_2 are given in Table IX [169].

The compounds MnOP and MgOP differ from the above-mentioned. In MgOP the cations order at T_1 = 285 K with a simultaneous small increase in this conductivity [169], as opposed to the marked decrease at T_1 in ZnOP. This difference in conduction behaviour is associated with a difference in the symmetry of the superstructure pattern. In fact, the cation superlattice seems to prevent the Peierls transition in MgOP and in consequence no second transition is observed. MnOP appears to be very similar to MgOP except for the fact that T_1 is greater than room temperature [169].

Both in ZnOP [170] and in MgOP [171] there is evidence of a strong competition between different instabilities, evidenced by very long thermal relaxation times near the transitions as well as by other strange conduction behaviours. Left alone the cations would crystallize in a Wigner lattice of period k_F along the Pt-chains as a result of their mutual Coulomb repulsions. However, the crystal lattice exerts an incommensurable potential of period $d_{\parallel}^*$ on the cations and this gives rise to a competition between ordering at k_F *or* ordering commensurate with $d_{\parallel}^*$. In ZnOP the first possibility is chosen and in this case the scene is naturally set for the Peierls transition which occurs at T_2. In MgOP the second possibility is chosen which makes it symmetry-forbidden to undergo a simple Peierls transition at a lower temperature. Hence the absence of T_2. The fact that different M-OP's with close to identical crystallographic structures demonstrate such a competition between instabilities of almost equal strength makes them promising candidates for studies of chaotic states.

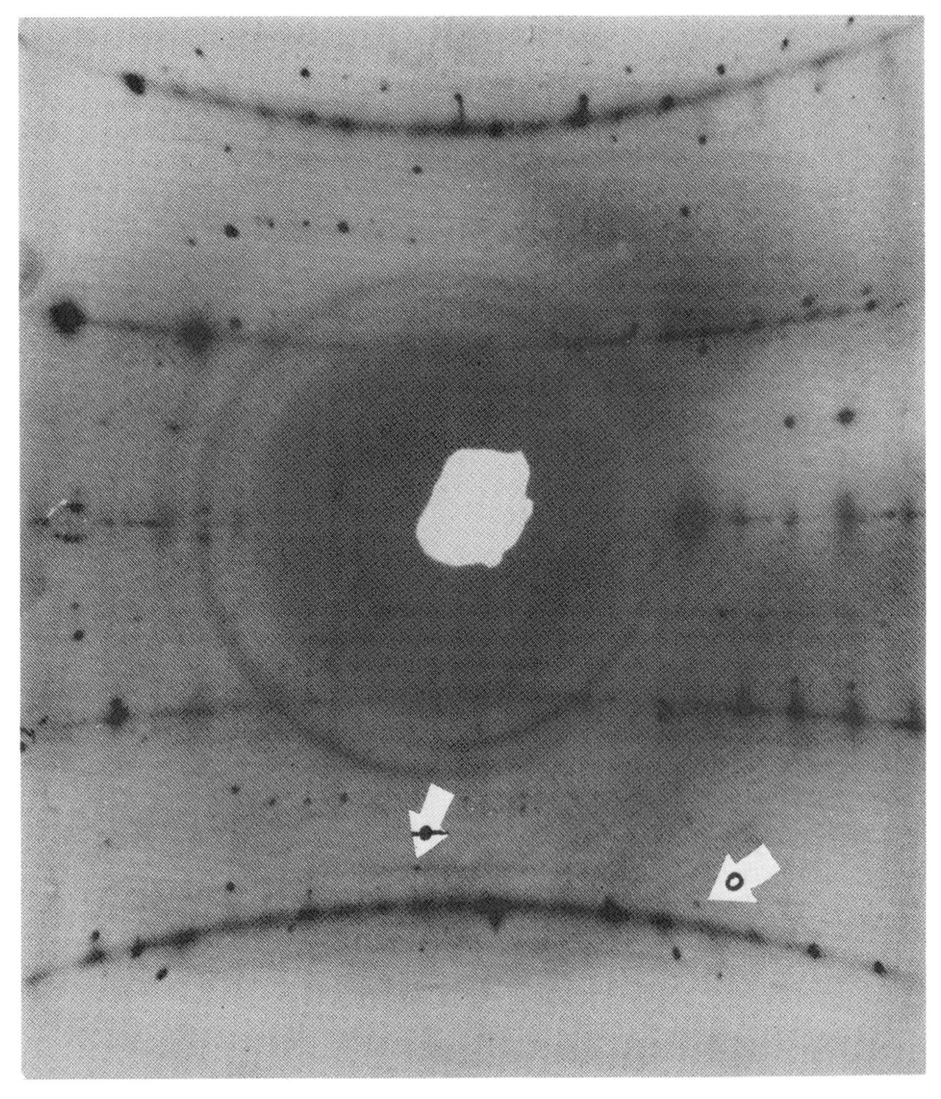

Fig. 31. Diffuse X-ray scattering from CoOP. One arrow (○) indicates the 'k_F'-superlattice reflection which occur at a higher temperature than the $2k_F$-spots (-●-).

It seems useful to discuss K(def)TCP within the same context as for the M-OP's. This compound has two transitions at T_1 = 296 and $T_2 \simeq 100$, both demonstrated in conductivity [59, 60] and X-ray scattering [172] but in neither transition has the symmetry of the lower temperature state been fully characterized. This makes it difficult to identify the Peierls transition temperature T_P. On the whole it has been hard to get a consistent picture of K(def)TCP. If for instance one identifies T_1 with T_P, K(def)TCP must be characterized as almost mean-field-like. This is consistent with the observation that the linewidth of the amplitude mode is narrower than predicted by Kurihara's

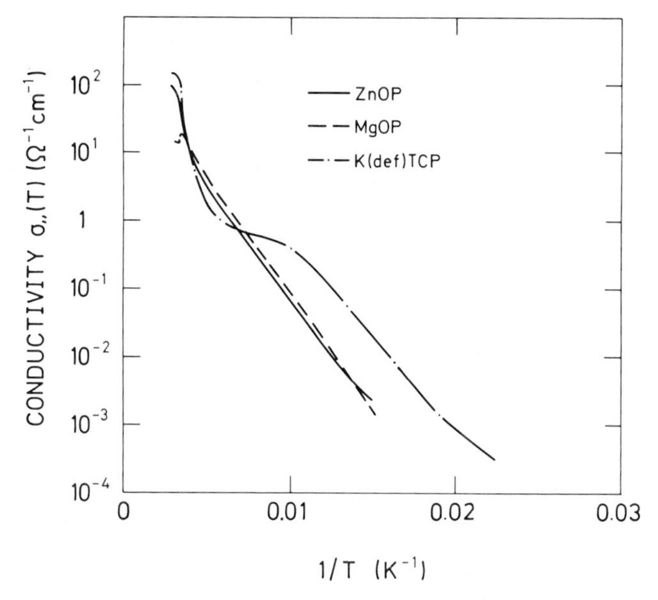

Fig. 32. Temperature-dependent conductivities of ZnOP, MgOP and K(def)TCP.

TABLE IX

Critical temperature T_1 and T_2 as determined by occurrence of superlattice reflections in M-OP's and K(def)TCP. [172]

Compound	T_1 (K)	T_2 (K)
MnOP	>310	–[b]
ZnOP	303	278[a]
CoOP	297	280[a]
NiOP	315	301[a]
MgOP	≈285[b]	–[b]
K(def)TCP	296[c]	≈60[c]

[a] The Peierls transition is identified with T_2, both from conductivity and diffuse X-ray scattering.
[b] No Peierls transition occurs.
[c] Both T_1 and T_2 have been associated with the Peierls transition.

one-dimensional theory; but it is inconsistent with the fact that a large gap is seen at room temperature in the optical absorption. Of course one then has the reverse problem if K(def)TCP is very one-dimensional, i.e., $T_P = T_2$.

Perhaps it is educational to conclude this review with a compound which does not fit into any simple picture. The properties of conducting platinum chain compounds are **in general well** explained by concepts developed for almost one-dimensional conductors, and they have played an important role in developing those concepts. But still a lot has to be understood.

Acknowledgments

This paper was written in part during my stay with IBM Research Laboratory at San Jose, California, supported by IBM World Trade. It describes work which has been done together with many collaborators. In particular, I would like to thank D. Baeriswyl, A. Braude, L. K. Hansen, A. Lindegaard-Andersen, G. A. McKenzie, B. Mogensen, J. B. Nielsen, A. S. Petersen, G. Shirane, E. Steigmeier, J. M. Williams and A. E. Underhill for collaboration and fruitful discussions. The careful reading of the manuscript by H. Højgaard Jensen resulting in several improvements is also gratefully acknowledged.

References and Notes

1. R. E. Peierls, *Quantum Theory of Solids* (Oxford University Press, 1954), p. 108.
2. J. W. Döbereiner, *Pogg. Ann* **28**, 180 (1833).
3. K. Krogman and H. D. Hausen, *Z. Anorg. Allg. Chem.* **358**, 67–81 (1968).
4. K. Krogman, *Z. Anorg. Allg. Chem.* **358**, 97–110 (1968).
5. H. Frölich, *Proc. R. Soc. (London)* **A 223**, 296–305 (1954).
6. M. J. Rice and S. Strässler, *Solid State Commun.* **13**, 125–128 (1973).
7. J. B. Nielsen and K. Carneiro, *Solid State Commun.* **33**, 1097–1101 (1980).
8. C. G. Kuper, *Proc. R. Soc. (London)* **A 227**, 214–228 (1955).
9. L. D. Landau and E. M. Lifschitz, *Statistical Physics* (Pergamon Press, 1980).
10. D. J. Scalapino, M. Sears and R. L. Ferrell, *Phys. Rev. B* **6**, 3409 (1972).
11. P. A. Lee, T. M. Rice and P. W. Anderson, *Phys. Rev. Lett.* **31**, 462–465 (1973).
12. W. Dieterich, *Z. Phys.* **270**, 239 (1974).
13. D. J. Scalapino, Y. Imry and P. Pincus, *Phys. Rev. B* **11**, 2042 (1975).
14. B. Horowitz, H. Gutfreund and M. Weger, *Phys. Rev. B* **12**, 3174–3185 (1975).
15. P. Bak, *Electron–phonon Interactions and Phase Transitions*, T. Riste, ed. (Plenum Press, 1977), pp. 66–87.
16. L. J. Sham and B. R. Patton, *Phys. Rev. Lett.* **36**, 733 (1976).
17. J. M. Williams and A. R. Schultz, *Molecular Metals*, W. E. Hatfield, ed. (Plenum Press, 1979), pp. 337–368.
18. J. W. Williams, A. J. Schultz, A. E. Underhill and K. Carneiro, *Extended Linear Chain Compounds*, J. S. Miller, ed. (Plenum Press, 1977), pp. 73–118.
19. A. E. Underhill, D. M. Watkins, J. M. Williams and K. Carneiro, *Extended Linear Chain Compounds*, J. S. Miller, ed. (Plenum Press, 1977), pp. 119–156.
20. W. Knop, *Annalen* **43**, 111 (1842).
21. A. J. Schultz, C. C. Coffey, G. C. Lee and J. M. Williams, *Inorg. Chem* **16**, 2129–2131 (1977).
22. G. S. V. Coles, A. Lindegaard-Andersen, J. M. Williams, A. J. Schutz, R. K. Brown, R. E. Besinger, J. R. Ferraro, A. E. Underhill and D. M. Watkins, *Physica Scripta* **25**, 873–878 (1982).
23. A. J. Schultz, D. P. Gerrity and J. M. Williams, *Acta Cryst.* **B 34**, 1673 (1978).
24. R. K. Brown and J. M. Williams, *Inorg. Chem.* **18**, 1922 (1979).
25. J. M. Williams, M. Iwata, S. W. Peterson, K. A. Leslie and H. J. Guggenheim, *Phys. Rev. Lett.* **34**, 1653 (1975).
26. R. K. Brown, D. A. Vidusek and J. M. Williams, *Inorg. Chem.* **17**, 2607–2609 (1978).
27. J. M. Williams, J. L. Petersen, H. M. Gerdes and S. W. Peterson, *Phys. Rev. Lett.* **33**, 1079 (1974).
28. G. Heger, H. J. Deiseroth and H. Schultz, *Acta Cryst.* **B 34**, 725 (1978).
29. C. Peters and C. F. Eagen, *Inorg. Chem.* **15**, 782 (1976).
30. G. S. V. Coles, A. E. Underhill and Kim Carneiro, *J. Chem. Soc. (Dalton Trans.)*, 1411–1415 (1983).

31. J. M. Williams, P. L. Johnson, A. J. Schultz and C. Coffey, *Inorg. Chem.* **17**, 834 (1978).
32. G. D. Stucky, C. Putnik, J. Kelber, M. J. Schaffman, M. B. Salamon, G. Pasquali, A. J. Schultz, J. M. Williams, T. F. Cornish, D. M. Washecheck and P. L. Johnson, *Ann. N. Y. Acad. Sci.* **313**, 525 (1978).
33. P. L. Johnson, A. J. Schultz, A. E. Underhill, D. M. Watkins, D. W. Wood, and J. M. Williams, *Inorg. Chem.* **17**, 839 (1978).
34. J. M. Williams, K. D. Keefer, D. M. Washecheck and N. P. Enright, *Inorg. Chem.* **15**, 2446 (1976).
35. A. E. Underhill, D. J. Wood and K. Carneiro, *Synth. Metals* **1**, 395 (1979–80).
36. K. Carneiro, *Mol. Cryst. Liq. Cryst.* **81**, 163–181 (1982).
37. D. M. Watkins, A. E. Underhill and C. S. Jacobsen, *J. Phys. Chem. Solids* **43**, 183–187 (1982).
38. A. E. Underhill, D. M. Watkins and C. S. Jacobsen, *Solid State Commun.* **36**, 477–480 (1980).
39. A. J. Schultz, A. E. Underhill and J. M. Williams, *Inorg. Chem.* **17**, 1313–1315 (1978).
40. A. Kobayashi, H. Kondo, Y. Sasaki, H. Kobayashi, A. E. Underhill and D. M. Watkins, *Bull. Chem. Soc. Japan* **55**, 2074–2078 (1981).
41. H. Toftlund, *J. C. S. Chem. Comm.* 837–838 (1979).
42. H. Toftlund, Personal Communication.
43. A. E. Underhill and M. N. Ahmad, *J. C. S. Chem. Comm.* 67–68 (1981).
44. A. Kobayashi, Y. Sasaki, H. Kobayashi, A. E. Underhill and M. M. Ahmad *J. Chem. Soc. Chem. Commun.* 390–391 (1981).
45. P. Brenni, D. Brinkmann, H. Huber, M. Mali, J. Roos and H. Arena, *Solid State Commun.* **47**, 415–418 (1983).
46. H. Kobayashi and A. Kobayashi, *Extended Linear Chain Compounds*, J. S. Miller, ed. (Plenum Press, 1982), pp. 259–300.
47. K. Krogmann and D. Stefan, *Z. Anorg. Chem.* **362**, 290 (1968).
48. M. Stock and H. Yersin, *Solid State Commun.* **27**, 1305 (1978).
49. W. Dieterich, *Adv. Phys.* **25**, 615–655 (1976).
50. H. J. Schultz, *Mol. Cryst. Liq. Cryst.* 79, 199–212 (1982).
51. D. M. Washecheck, S. W. Peterson, A. H. Reis and J. M. Williams, *Inorg. Chem.* **15**, 74 (1976).
52. *Handbook of Chemistry and Physics*. CRC Press.
53. D. J. Wood, A. E. Underhill, A. J. Schultz and J. M. Williams, *Solid State Commun.* **30**, 501 (1979).
54. D. J. Wood, A. E. Underhill and J. M. Williams, *Solid State Commun.* **31**, 219 (1979).
55. A. E. Underhill, G. S. V. Coles, J. M. Williams and K. Carneiro, *Phys. Rev. Lett.* **47**, 1221–1223 (1981).
56. D. Kuse and H. R. Zeller, *Phys. Rev. Lett.* **27**, 1060–1063 (1971).
57. A. E. Underhill, D. M. Watkins and D. J. Wood, *J. C. S. Chem. Commun.* 809 (1976).
58. K. Carneiro, A. S. Petersen, A. E. Underhill, D. J. Wood and D. M. Watkins, *Phys. Rev. B* **19**, 6279–6288 (1979).
59. A. J. Epstein and J. S. Miller, *Solid State Commun.* **29**, 627 (1979).
60. K. Carneiro, C. S. Jacobsen, and J. M. Williams, *Solid State Commun.* **31**, 345 (1979).
61. J. H. O'Neill, A. E. Underhill and G. A. Toombs, *Solid State Commun.* **29**, 557–560 (1979).
62. A. E. Underhill and D. J. Wood, *Molecular Metals*, W. E. Hatfield, ed. (Plenum Press, 1979), p. 377.
63. H. R. Zeller and P. Brüesch, *Phys. Stat. Sol. B* **65**, 537–542 (1974).
64. R. Musselman and J. M. Williams, *J. C. S. Chem. Comm.* 186–188 (1977).
65. L. H. Greene, D. B. Tanner, A. J. Epstein and J. S. Miller, *Phys. Rev. B* **25**, 1331–1338 (1982).
66. M. M. Ahmad, D. J. Turner, A. E. Underhill, C. S. Jacobsen, K. Mortensen and K. Carneiro, *Phys. Rev. B* **29**, 4796–4799 (1984).
67. E. F. Steigmeier, D. Baeriswyl, H. Auderset and J. M. Williams, *Lecture Notes in Physics* **96**, 229 (1979).
68. E. F. Steigmeier, D. Baeriswyl, H. Auderset and J. M. Williams, Personal Communication.
69. D. Kuse and H. R. Zeller, *Solid State Commun.* **11**, 355–359 (1972).
70. D. M. Watkins, C. S. Jacobsen and K. Carneiro, *Chemica Scripta* **17**, 193–194 (1981).
71. L. Fritsche and M. Rafat-Mehr, *Lecture Notes in Physics* **34**, 97–107 (1975).

72. R. P. Messmer and D. R. Salahub, *Phys. Rev. Lett.* **35**, 533–536 (1975).
73. M.-H. Whangbo and R. Hoffmann, *J. Amer. Chem. Soc.*, 6093–6098 (1978).
74. J. S. Miller *Inorg. Chem.* **15**, 2357 (1976).
75. A. H. Reis Jr. and S. W. Peterson, *Ann. N. Y. Acad. Sci.* **313**, 560 (1978).
76. N. V. Smith, G. K. Wertheim, S. Hüfner and M. N. Traum, *Phys. Rev. B* **10**, 3197–3206 (1974).
77. J. H. Weaver, *Phys. Rev. B* **11**, 1416–1425 (1975).
78. N. V. Smith, *Phys. Rev. B* **9**, 1365 (1975).
79. H. Yersin, G. Gliemann and U. Rossler, *Solid State Commun.* **21**, 915–918 (1977).
80. R. C. Jacklevic and R. B. Saillant, *Solid State Commun.* **15**, 307–311 (1974).
81. H. Nagasawa, *Phys. Stat. Sol. B* **109**, 749–759 (1982).
82. J. M. Williams, *Inorg. Nucl. Chem. Lett.* **12**, 651–656 (1976).
83. L. Pauling, *The Nature of the Chemical Bond* (Cornell University Press, 1960), p. 368.
84. L. Pauling, *J. Amer. Chem. Soc.* **69**, 542 (1947).
85. A. Braude, A. Lindegaard-Andersen, K. Carneiro and A. S. Petersen, *Solid State Commun.* **33**, 365–369 (1980).
86. R. Comès, M. Lambert, M. Launois and H. R. Zeller, *Phys. Rev. B* **8**, 571–575 (1973).
87. A. J. Schultz, G. D. Stucky, J. M. Williams, T. R. Koch and R. L. Maffey, *Solid State Commun.* **21**, 197 (1977).
88. A. E. Underhill, Personal Communication.
89. B. Renker, L. Pintschovius, W. Gläser, H. Rietschel, R. Comès, L. Liebert and W. Drexel, *Phys. Rev. Lett.* **32**, 836 (1974).
90. J. W. Lynn, M. Iizumi, G. Shirane, S. A. Werner and R. B. Saillant, *Phys. Rev. B* **12**, 1154 (1975).
91. K. Carneiro, G. Shirane, S. A. Werner and S. Kaiser, *Phys. Rev. B* **13**, 4258–4253 (1976).
92. K. Carneiro, J. Eckert, G. Shirane and J. M. Williams, *Solid State Commun.* **20**, 333–336 (1976).
93. E. M. Conwell, *Phys. Rev. B* **22**, 1761–1780 (1980).
94. L. Friedman, *Phys. Rev.* **140**, 1649 (1965).
95. The result given here differs from that given in [36], but is believed to be more correct.
96. W. L. McMillan, *Phys. Rev.* **167**, 331–344 (1968).
97. F. Mehran and B. A. Scott, *Phys. Rev. Lett.* **31**, 1347 (1973).
98. T. Takanashi, H. Akagawa, H. Doi and H. Nagasawa, *Solid State Commun.* **23**, 809–814 (1977).
99. P. A. Lee, T. M. Rice and P. W. Anderson, *Solid State Commun.* **14**, 703–709 (1974).
100. P. Brüesch, S. Strässler and H. R. Zeller, *Phys. Rev. B* **12**, 219–225 (1975).
101. D. K. Campbell and A. R. Bishop, *Phys. Rev. B* **24**, 4859–4862 (1981).
102. R. Comès, M. Lambert and H. R. Zeller, *Phys. Stat. Solidi (b)* **58**, 587–592 (1973).
103. C. F. Eagan, S. A. Werner and R. B. Saillant, *Phys. Rev.* **B12**, 2036–2041 (1975).
104. M. J. Rice, S. Strässler and W. R. Schneider in *One-Dimensional Conductors*, ed. T. Schuster (*Lecture Notes in Physics*, Springer, 34, 1975), pp. 282–334.
105. S. Barisic, A. Bjelis and K. Saub, *Solid State Commun.* **13**, 1119–1124 (1973).
106. B. Horovitz, M. Weger and H. Gutfreund, *Phys. Rev. B* **9**, 1246–1260 (1974).
107. P. Bak and S. A. Brazovsky, *Phys. Rev. B* **17**, 3154–3164 (1978).
108. G. Guilliani and E. Tosatti, *Il Nuevo Cimento* **47B**, 135–148 (1978).
109. H. J. Schultz, *Phys. Rev. B* **18**, 5756–5767 (1978).
110. K. Käfer in *Quasi One-Dimensional Conductors II*, ed. S. Barisic (*Lecture Notes in Physics* **96**, Springer, 1979), pp. 219–223.
111. S. Kurihara, *J. Phys. Soc. Japan* **48**, 1821–1828 (1980).
112. M. Apostol and F. Baldea, *J. Phys. C: Solid State Phys.* **15**, 3319–3331 (1982).
113. There is an error in [99] regarding the expression for the amplitude mode. Equation (33) corresponds to later work by several authors. I would like to thank P. Eisenrigler for pointing this out to me.
114. B. Renker, H. Rietschel, L. Pinschovius, W. Gläser, P. Brüesch, D. Kuse and M. J. Rice, *Phys. Rev. Lett.* **30**, 1144–1146 (1973).

115. R. Comès, B. Renker, L. Pinschovius, R. Currat, W. Gläser and G. Scheiber, *Phys. Stat. Sol. (b)* **71**, 171–178 (1975).
116. E. F. Steigmeier, R. Loudon, G. Harbeke, H. Auderset, and G. Scheiber, *Solid State Commun.* **17**, 1447–1452 (1975).
117. E. F. Steigmeier, D. Baeriswyl, G. Harbeke, H. Auderset and G. Scheiber, *Solid State Commun.* **20**, 661–666 (1976).
118. P. Brüesch and H. R. Zeller, *Solid State Commun.* **14**, 1037–1040 (1974).
119. E. F. Steigmeier, H. Auderset, D. Baeriswyl, A. E. Underhill and K. Carneiro, *Mol. Cryst. Liq. Cryst.* **81**, 205–216 (1982).
120. H. Fukuyama, *J. Phys. Soc. Japan* **45**, 1266–1275 (1978).
121. H. Fukuyama, *J. Phys. Soc. Japan* **45**, 1474–1481 (1978).
122. A. J. Berlinsky, *Solid State Commun.* **19**, 1165–1168 (1976).
123. M. J. Rice, L. Pietronero and P. Brüesch, *Solid State Commun.* **21**, 757–760 (1977).
124. H. Fritzsche, *Solid State Commun.* **9**, 1813 (1971).
125. E. M. Conwell and N. C. Bannik, *Solid State Commun.* **39**, 411–413 (1981).
126. G. Soda, D. Jérome, M. Weger, J. Alizon, J. Gallice, H. Robert and J. M. Fabre, *J. de Physique* 38, 931–948 (1977).
127. J. R. Cooper, D. Jérome, S. Etemad and E. M. Engler, *Solid State Commun.* **22**, 257–263 (1977).
128. H. R. Zeller and A. Beck, *J. Phys. Chem. Solids* **35**, 77–80 (1974).
129. J. F. Thomas, *Solid State Commun.* **42**, 567–570 (1982).
130. M. J. Rice, A. R. Bishop, J. A. Krumhansl and S. E. Tullinger, *Phys. Rev. Lett.* 36, 4321–4325 (1976).
131. W. L. McMillan, *Phys. Rev. B* **14**, 1496–1502 (1976).
132. M. Mehring, U. Deininghaus, H. Fischer, H. Seibel and H. Weber, Proceedings of the Ampère summer school Portoroz, Yugoslavia (1982).
133. M. Mali, O. Kanert, M. Mehring and D. Brinkmann, *Phys. Rev. B* **20**, 4442–4446 (1979).
134. For a recent review see: G. Grüner, *Chemica Scripta* **17**, 207–213 (1981).
135. F. F. Shchegolev, *Phys. Stat. Sol. (a)* **12**, 9–45 (1972).
136. H. J. Pedersen and A. E. Underhill, *Solid State Commun.* **33**, 289–292 (1980).
137. P. Monceau, J. Richard and M. Renard, *Phys. Rev. B* **25**, 931–947 (1982).
138. W. M. Walsh, F. Wudl, G. A. Thomas, D. Nalewajek, J. J. Hauser, P. A. Lee and T. Poechler, *Phys. Rev. Lett.* **45**, 829–832 (1980).
139. G. Grüner, A. Zawadowski and P. M. Chaikin, *Phys. Rev. Lett.* **46**, 511–514 (1981).
140. N. C. Bannik, E. M. Conwell and C. S. Jacobsen, *Solid State Commun.* 38, 267–270 (1981).
141. E. M. Conwell and N. C. Bannik, *Phys. Rev. B* **24**, 4883–4885 (1981).
141. H. E. Stanley, *Introduction to Phase Transitions and Critical Phenomena* (Clarendon, 1971).
143. S.-K. Ma, *Modern Theory of Critical Phenomena* (Frontiers in Physics **46**, Benjamin, 1976).
144. K. Franulovic and D. Djurek, *Phys. Lett.* **51A**, 91–92 (1975).
145. M. Steiner, J. Villain, G. G. Windsor, *Adv. Phys.* **25**, 87–209 (1976).
146. Kim Carneiro, *Molecular Metals*, W. E. Hatfield, ed. (Plenum, 1979), pp. 369–376.
147. S. Barisic, *Phys. Rev. B* **5**, 941 (1972).
148. S. Barisic, *Ann. Phys. (France)* **7**, 23 (1972).
149. M. J. Rice and S. Strässler, *Solid State Commun.* **13**, 1389–1392 (1973).
150. W. Dieterich, *Solid State Commun.* **17**, 445–449 (1975).
151. S. Brazovsky and I. E. Dzyaloshinsky, *Sov. Phys. JETP* **44**, 1233 (1976).
152. See for instance *Anharmonic Lattices, Structural Transitions and Melting*, T. Riste, ed. (Plenum Press, 1974).
153. M. Thielemann, R. Deltour, D. Jérome and J. R. Cooper, *Solid State Commun.* **19**, 21–27 (1976).
154. S. Roth, R. Ranvaud, A. Waintal and W. Drexel, *Solid State Commun.* **15**, 625–627 (1974).
155. H. Kobayashi, A. Kobayashi, K. Asanumi and S. Minomura, *Solid State Commun.* **35**, 293–296 (1980).

156. B. Renker, L. Bernard, C. Vettier, R. Comès and B. P. Schweiss, *Solid State Commun.* **41**, 935–937 (1982).
157. Y. Byschkov, L. P. Gorkov and Z. E. Dzyaloshinskii, *Sov. Phys. JETP* **23**, 489– (1966).
158. K. Carneiro, G. A. McKenzie and J. M. Williams, unpublished results. We are grateful to R. Pynn for adapting his resolution program to the tetragonal case.
159. H. Doi, H. Nagasawa, T. Ishiguro, S. Kagoshima, *Solid State Commun.* **24**, 729–731 (1977).
160. R. Pynn and S. A. Werner, *J. Appl. Phys.* **42**, 4736–4749 (1971).
161. M. J. Rice, *Low Dimensional Cooperative Phenomena* (ed. H. J. Keller, Plenum, 1975), p. 23.
162. P. M. Chaikin, T. Tiedje and A. N. Bloch, *Solid State Commun.* **41**, 739–742 (1982).
163. S. Kurihara, H. Fukuyama and S. Nakajima, *J. Phys. Soc. Japan* **47**, 1403–1410 (1979).
164. H. Niki, H. Doi, M. Nagasawa, *J. Phys. Soc. Japan* **51**, 2470–2477 (1982).
165. B. Horovitz and A. Birnboim, *Solid State Commun.* **19**, 91–95 (1976).
166. A. Braude, A. Lindegaard-Andersen, K. Carneiro, A. E. Underhill *Synth. Met.* **1**, 35–42 (1979).
167. A. Bertinotti and D. Luzet, *J. de Physique* **44**, C3/1551–1554 (1983).
168. H. Kobayashi, Y. Hano, T. Damno, A. Kobayashi and Y. Sasaki, *Chem. Lett.*, 177–178 (1980).
169. A. Braude, K. Carneiro, C. S. Jacobsen, K. Mortensen, D. J. Turner, A. E. Underhill to be published.
170. K. Carneiro and A. E. Underhill, *J. Physique* **44** (1983). C3-1007–C3-1010.
171. M. Mizuno, A. E. Underhill and K. Carneiro, *J. Phys. C* **16**, 2105–2113 (1983).
172. A. Kobayashi, Personal Communication.

PHYSICAL PROPERTIES OF $(SN)_x$ AND $(SNBr_y)_x$

K. KANETO, K. YOSHINO AND Y. INUISHI

Faculty of Engineering, Osaka University,
Yamada-Kami, Suita, Osaka, Japan

1. Introduction

During the last ten years our understanding of polymeric sulfur nitride $(SN)_x$ has achieved remarkable progress thanks to many research workers throughout the world. $(SN)_x$ is the first synthetic polymeric material which shows metallic and superconductivity with strong anisotropy (quasi-one-dimensional, 1-D) properties between parallel and perpendicular direction with respect to the chain (b) axis.

After successful growth of crystalline $(SN)_x$ with good quality by Labes [1] and by MacDiarmid [2] at the beginning of 1970, its metallic nature was first confirmed by the positive temperature dependence of resistivity and the existence of a plasma edge in the optical reflectance. One prominent feature of $(SN)_x$ in comparison with the other polymeric conductors such as polyacetylene is its metallic behaviour down to cryogenic temperatures and its superconductivity, discovered by Greene *et al.* [3], without a Peierls transition, despite its quasi-1-D properties.

In order to elucidate the metallic nature down to the onset of superconductivity of crystalline $(SN)_x$, the energy band calculation was done extensively by a large number of workers, using different approaches. Two distinguishable band models, i.e., quasi-one-dimensional metal with overlapping band and strongly anisotropic three-dimensional (3-D) semimetal were proposed by Kamimura [4] and by Rudge and Grant [5], respectively. This difference is attributable to the fact that the former considered $(SN)_x$ as the ionic structure of S^+N^- with smaller s–p orbital interaction and the latter regarded it as the result of covalent bonding of natural S and N. Afterwards, using a self-consistent LCAO technique, recalculation by Oshiyama and Kamimura [6] resulted in a similar structure to that of Rudge and Grant. The idea of the stronger interchain interaction giving rise to a highly anisotropic semimetal is widely accepted to account for the suppression of Peierls instability in such a linear chain structure as $(SN)_x$.

Various physical properties of crystalline and thin film $(SN)_x$ such as crystal structure, electrical conductivity, galvanomagnetic properties and optical properties etc. have been extensively investigated experimentally utilizing well-developed and up-to-date techniques.

Early dc measurements on the conductivity anisotropy ratio ($10^2 \sim 10^3$) parallel to perpendicular with respect to the chain (b) axis suggested a quasi-1-D metal, which is ascribed to the inherent fibril structure of $(SN)_x$ crystals. The experimental evidence, such as plasmon dispersion, optical reflectance and microwave conductivity etc., revealed a smaller anisotropy ratio of several units to several tens.

From the temperature dependence of resistivity in $(SN)_x$, the following components were distinguished; (i) residual resistivity due to defects and/or impurities, (ii) resistivity due to carrier-carrier scattering ($\rho \propto T^2$), (iii) electron–phonon interaction ($\rho \propto T$). The

P. Monceau (ed.), Electronic Properties of Inorganic Quasi-One-Dimensional Materials, II, 69–109.

behaviour of the case (i), observed predominantly at lower temperature for low quality samples, is ascribed mainly to defects, i.e., chain breaks, which should present a significant barrier for carriers in the systems consisting of a linear chain conducting path. The experimental facts such as negative magnetoresistance, Curie–Weiss-type susceptibility in γ-ray irradiated $(SN)_x$ and the Kondo effect type behaviour will be discussed in terms of spin dependent scattering of carriers at defects having a magnetic moment.

The quadratic temperature dependence of resistivity is also a characteristic behaviour of $(SN)_x$, which is firstly explained by the carrier–carrier (electron–electron Umklapp) scattering. Oshiyama *et al.* [6] have shown that the electron–hole scattering plays the most important role in the semimetallic material for the mechanism (ii) by a comparison of theoretical calculation and experimental results obtained upon bromination of $(SN)_x$ by Kaneto *et al.* [7].

The discovery of superconductivity of $(SN)_x$ at around 0.3 K by Greene *et al.* [3] has attracted much interest, since it is the first polymeric superconductor composed of V–VI elements in the Periodic Table with low-dimensional properties.

The dependence of superconductivity of $(SN)_x$ on magnetic field, pressure, Br_2 doping, sample quality and pretransitional behaviour has been studied in detail. The superconductivity of $(SN)_x$ is considered to be a highly anisotropic type II superconductor, being influenced strongly by the weakly coupled fibrous structure of the $(SN)_x$ crystal.

The second important feature of $(SN)_x$ is the possibility of intercalation, to large extent without destroying the original chain [8]. For example, the bromine intercalation decreases both (i) residual resistivity and (ii) resistivity due to the electron–hole scattering. The existence of bromine at chain break and interchain sites enhances carrier hopping to the neighbouring chains, giving rise to the decrease of residual resistance and also enhancement of the 3-D character which increases the transition temperature of superconduction with suppression of low-dimensional fluctuations. The bromine takes off an electron from the host $(SN)_x$, which causes the expansion of the hole pocket and shrinking of the electron pocket, resulting in a decrease of the electron–hole scattering probability and revealing the carrier–phonon scattering ($\rho \propto T$) [6].

Effects of intercalation on the physical properties, such as normal conductivity, superconductivity, magnetoresistance, optical properties etc. will be summarized briefly.

Although $(SN)_x$ does not show typical low-dimensional metallic characteristics, the possibility of preparing relatively high quality crystalline samples may still provide an attractive way to understand the basic physics of polymer metal and superconductors.

2. Preparation of $(SN)_x$ and $(SNBr_y)_x$

2.1. CRYSTALLINE $(SN)_x$

The first report on the synthesis of $(SN)_x$ appeared about seventy years ago [9], however, it was during the last decade, after the discovery of metallic behaviour [1, 2, 10–11], that the preparation technique of analytically pure $(SN)_x$ has been established. The currently adopted procedures [1, 2, 11–15] for obtaining $(SN)_x$ crystals are essentially similar. Tetrasulfur tetranitrile, S_4N_4 as a starting material, is synthesized by passing dry ammonia gas through disulfur dichloride diluted in chloroform [13, 14, 16], benzene [16, 17] or carbon tetrachloride [18] at the volume ratio of *ca.* 1:10. The mixture of sulfur and S_4N_4 is extracted from the ammonium chloride formed. Then sulfur and S_4N_4 are separated by the recrystallization from the chloroform solution. Purification of S_4N_4 is performed by recrystallization from solution of chlorofrom [2, 13], benzene [11], dioxane [18] or toluene [18], and by vacuum sublimation [2, 11, 12] at *ca.* 80 °C. Crystalline S_4N_4 has a thermochromic orange colour [12] and a melting point of 175–190 °C [2, 11, 17].

Although several types of apparatus for the synthesis of $(SN)_x$ have been described [2, 11, 12, 19], they are basically similar, consisting of three sections, i.e., (i) splitting of S_4N_4 into S_2N_2, (ii) vapour phase crystallization of S_2N_2 and (iii) solid-state polymerization of S_2N_2 to $(SN)_x$.

Figure 1 shows the apparatus used in the author's laboratory for the synthesis of $(SN)_x$ from S_4N_4, in which ground-glass joints and stopcock are completely dispensed

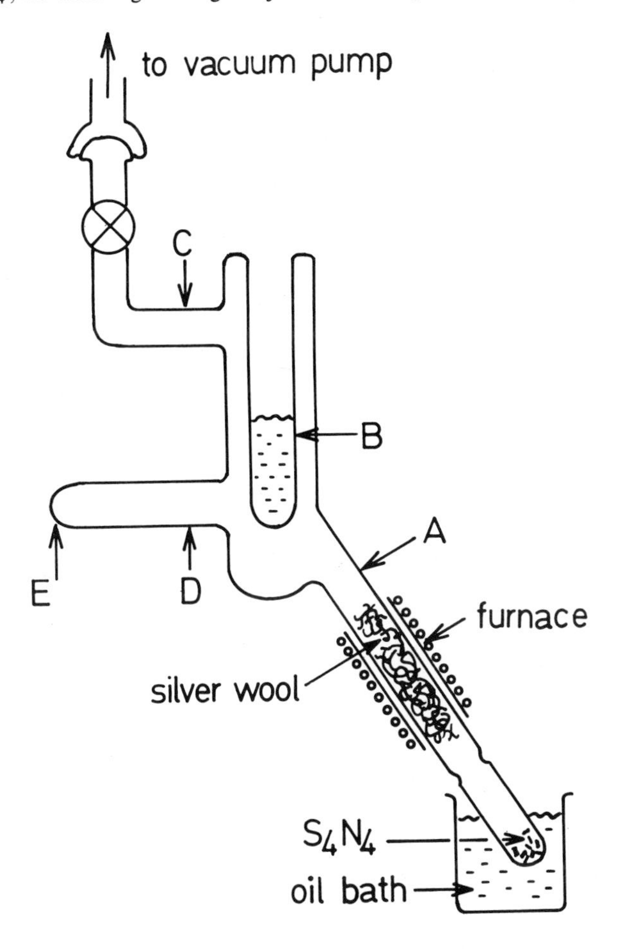

Fig. 1. Apparatus for the polymerization of $(SN)_x$ single crystal from S_4N_4.

with after process (ii). Usually, the reaction times of (i), (ii) and (iii) are several hours, a few days, and several weeks, respectively.

After placing 0.5 g S_4N_4 and 1.0 g silver wool into the glass tubing (~1.5 cm diameter, 20 cm length; shown below the part A), this was connected to the trap apparatus at A by a blown glass joint, as shown in Figure 1. The silver wool was heated up to *ca.* 300 °C by the furnace, then the oil bath was warmed up to *ca.* 90 °C to sublimate S_4N_4. The vapour phase S_4N_4 was split into S_2N_2 on the catalytic silver wool, and amorphous S_2N_2 having a dark brown colour was collected at the trap B, which was cooled by liquid nitrogen.

After sublimation of S_4N_4, the temperature of trap B was raised to *ca.* 0 °C, while pumping for several minutes to remove some of the volatile impurities, such as residual S_4N_2 etc., followed by sealing off at C. At the temperature of *ca.* 0 °C the amorphous S_2N_2 at B was dark grey. The apparatus was then turned around to place the part E into an ice bath at 0~5 °C, so that amorphous S_2N_2 transferred from the trap B to the colder trap E. Transparent crystals of S_2N_2 slowly grew at E for several days. The temperature of whole apparatus was cooled down to *ca.* 10 °C to slow down the crystallization speed of S_2N_2 in order to obtain a higher-quality sample. This should be also useful to slow down any harmful crystallization at B. After several days, the part D was sealed off, and S_2N_2 crystals in the portion of E was allowed to undergo solid-state polymerization by keeping the apparatus at room temperature for several weeks.

The $(SN)_x$ thus obtained was shiny gold in colour with the largest dimensional of *ca.* $2 \times 2 \times 5$ mm^3, and consisted of fibrous bundles of several tenths μm diameter, as shown in the scanning electron micrograph of Figure 2.

The fibrous bundle structure shown in Figure 2 is inherent in $(SN)_x$ crystals obtained

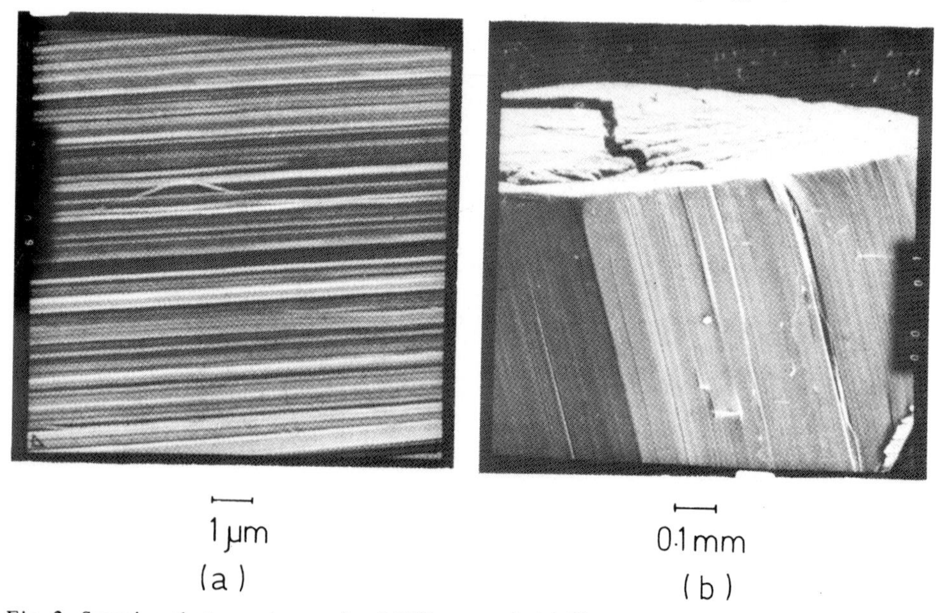

Fig. 2. Scanning electron micrograph of $(SN)_x$ crystal, (a) fibrous structure of surface, (b) an end of crystal.

by the solid-state polymerization of crystalline S_2N_2, since the lattice constant of $(SN)_x$ is significantly different from that of S_2N_2 [2, 14].

The electrical conductivity measured by the usual four probes method was $1{\sim}4 \times 10^3$ S/cm along the fiber direction, and the conductivity ratio at 4.2 K to that at room temperature ($\sigma_{4.2}/\sigma_R$) ranged between several tens and a thousand, depending remarkably on the sample quality [1, 2, 20, 21].

The elemental analysis of the $(SN)_x$ crystal has been performed by many workers [2, 11–14] and showed considerably good agreement with the calculated value.

The crystal structure of β-phase $(SN)_x$ was determined using the electron diffraction technique by Boudeulle [22] and the X-ray method by Cohen *et al.* [14]. Figure 3

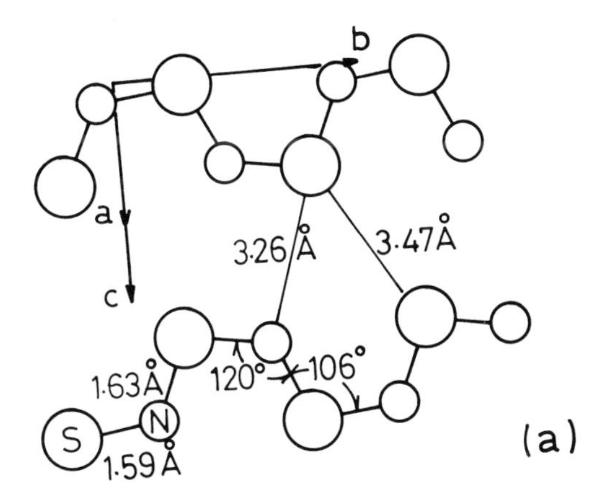

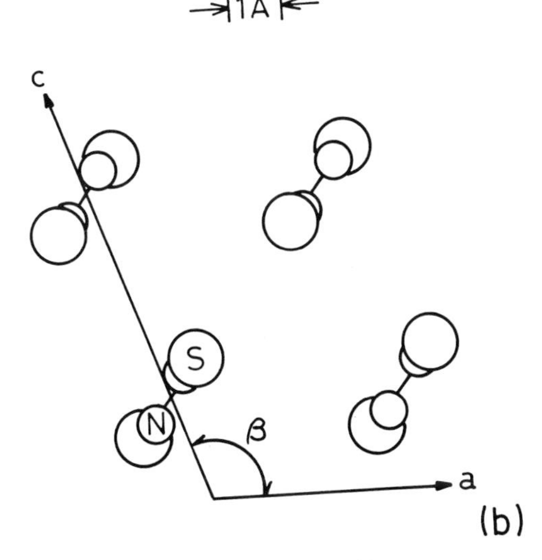

Fig. 3. Crystal structure of $(SN)_x$ [14], (a) projection onto $(\bar{1}02)$ plane (b) downward view of the *ac* plane.

shows the crystal structure reported by Cohen *et al.*: the monoclinic space group of $P2_{1/c}$ with lattice constant of a = 4.153, b = 4.439 and c = 7.637 Å with β = 109.7°. The crystal structure obtained by electron diffraction is slightly different from Figure 3, in respect of the SN bond length and bond angles of SNS and NSN. The linear chain structure with the interchain distance slightly shorter than the Van der Waals distance [20] gives rise to the highly anisotropic or quasi-1-D nature.

2.2. THIN FILM $(SN)_x$

Thin film $(SN)_x$ can be prepared either by (i) sublimation of $(SN)_x$ onto substrates [23–26] or (ii) polymerization of thin film S_2N_2 on substrate [27–29]. The vacuum sublimation of $(SN)_x$ crystal, i.e., method (i), was performed under pressures of less than 10^{-5} torr at a temperature of *ca.* 150 °C. The evaporated $(SN)_x$ cannot have a long chain and the size of x was estimated to be 2~4 [25, 26] by mass spectroscopy. In the case (ii), thermally split S_2N_2 through the heated silver or quartz wool catalyst from the vapour phase S_4N_4 was used as source substance similar to the procedure for obtaining amorphous S_2N_2 outlined in Section 2.1. The vapour phase $(SN)_x$ or S_2N_2 was deposited onto the temperature-controlled substrates, such as glass [23, 24, 29], Teflon [23], Mylar [23], polyethylene [23], fused silica [28], sapphire [29] and alkali halide [27, 29].

The polymerization of deposited S_2N_2 and sublimated $(SN)_x$ were significantly influenced by substrate materials. Sublimation onto streched polymer films [23] or fused silica [28] resulted in orientated film, i.e., the $(SN)_x$ chain aligned along the streched direction. These thin films, even though carefully prepared on the high temperature substrate, did not show high electrical conductivity like crystalline $(SN)_x$ [29, 30].

2.3. BROMINATION OF $(SN)_x$

Exposure of crystalline or thin film $(SN)_x$ to bromine vapour yields bromine-intercalated $(SN)_x$ [7, 8, 31–33] with an increased conductivity as high as $\sim 10^5$ S/cm. The procedure to obtain brominated $(SN)_x$ is the following. By the introduction of bromine gas into a well-dried and evacuated chamber containing an $(SN)_x$ sample at room temperature for *ca.* 1 hour, $(SN)_x$ is brominated. The colour of $(SN)_x$ changed from a golden luster to shiny copper or blueish-purple depending on the bromine content [31]. Bromine concentration was determined by the weight change of $(SN)_x$ or by elemental analysis. The composition $(SNBr_y)_x$ of fully brominated stable $(SN)_x$ obtained by *ca.* 1 hour exposure to bromine gas at room temperature followed by pumping for several tens of minutes was revealed to be $0.38 < y < 0.42$. Heavy doping by long time exposure to bromine vapour léads the crystal to burst and become brittle.

Upon bromination, $(SN)_x$ crystal swells 46~50% in volume [8, 31] with no measurable change in the length of the b axis direction, the density of $(SN)_x$ of 2.3 g/cm^3 increases to 2.65~2.7 g/cm^3 at the dopant concentration of $y \simeq 0.4$.

The electron diffraction pattern [8] revealed that the crystallographic lattice did not change upon bromination, indicating that the bromine does not enter between $(SN)_x$ chains. The Raman scattering in $(SNBr_y)_x$ crystal studied by Iqbal *et al.* [34] and

Temkin *et al.* [35] suggested the model, in which bromines intercalated as Br_3^- form around the surface of fibrils with diameter of 20~30 Å. This model is also supported by the result of the electron micrograph study of $(SNBr_y)_x$ [8]. The possibility of insertion of Br into the $(SN)_x$ lattice, i.e., between chains, is also claimed by Iqbal *et al.* [34] from the pressure dependence of the Raman scattering spectra. Street and Gill [36] proposed two models of the bromine intercalation i.e., (i) insertion of bromine between $(SN)_x$ chains and filling up chain breaks in the bulk crystal, (ii) adsorption at the surface of fibers or crystal. In fact, the change of temperature dependence of resistivity and increased dimensionality of $(SN)_x$ upon bromination should be mainly due to (i).

Intercalation of electron acceptors such as iodine [24], ICl [34], and the electron donor Na [37] were also performed. In iodine vapour at room temperature, the electrical conductivity of $(SN)_x$ crystal did not change, contrary to the case of the thin film [24], indicating that the fairly large iodine molecular cannot diffuse into the bulk of the $(SN)_x$ crystal.

3. Electronic Band Structure

The metallic properties of $(SN)_x$ down to the onset of superconductivity have stimulated a large number of theoretical discussions [4–6, 38–48] on the origin of suppression of the Peierls transition [49] in such a linear chain structure. The idea that three dimensional interchain interaction suppresses Peierls instability is widely accepted, contrary to the case of polyacetylene [50] in which the proton beside the chain may keep chains apart from the interchain interactions.

Six electrons from $3s$–$3p$ orbitals of sulfur and five electrons from $2s$–$2p$ orbitals of nitrogen contribute to form the $(SN)_x$ chain as shown in Figure 4(a) and (b), where solid and open circles indicate the electrons and holes, respectively, in the orbitals. The s orbitals of sulfur and nitrogen atoms do not contribute predominantly to the bonding (lone pairs).

Each two electrons from the $3p$ and $2p$ orbitals of sulfur and nitrogen atoms are exchanged to form a covalent σ bond (large solid circles), filling up the bonding σ orbitals as shown in Figure 4(c). These σ bonds constitute the backbone of the $(SN)_x$ chain. The rest of two and one electrons of sulfur and nitrogen, respectively, contribute to form π orbitals as indicated by circles of broken line as shown in Figure 4(a) and (b). Two of them fill up the bonding π orbital and the other fills a half of the antibonding π^* orbital as shown in Figure 4(c). One electron at this band provided from the SN unit does delocalize on the extended chain and contributes to the electronic transport. Figure 4(a) and (b) show a couple of possible orbitals located at different SN sites. The electron at π^* orbital derived from sulfur distributes on the nitrogen atom by the ratio of 0.4~0.5 [38].

In 1-D systems, however, the lattice distortion due to the Peierls instability [49] causes the dimerization of the lattice, resulting in a metal–insulator transition or gap formation in the π^* band at a half of the reciprocal lattice ($k = \pi/b$; b is the lattice constant along the b axis) as shown in Figure 5(a). This can be associated with the fact that the states of (a) and (b) in Figure 4 are not in equilibrium and slight bond alternation as shown in Figure 3(a) does exist. Because of the screw axis symmetry [10, 14] of two

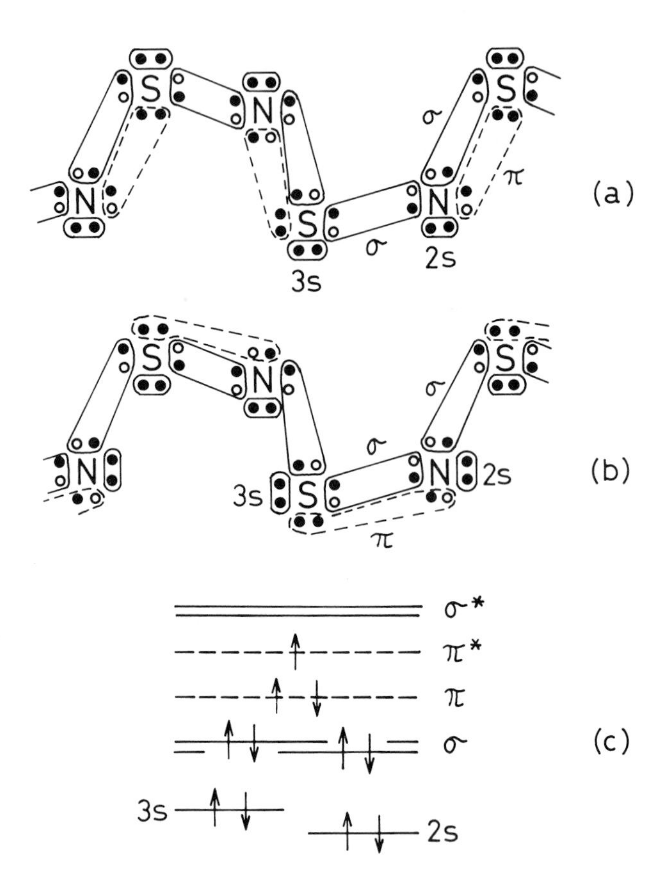

Fig. 4. Schematics of electronic bonding and bands in $(SN)_x$, (a) and (b) are a couple of π orbitals at different site, (c) energy levels of s, σ and π orbitals.

SN units along the b axis, this π^* band bends over half way along the reciprocal lattice and sticks together at the Brillouin zone boundary [38, 39] as shown in Figure 5(b). It has been pointed out [4, 40] that this band structure is not stable to Peierls distortion. In the real $(SN)_x$, two SN chains per unit cell and the interchain interaction should give rise to the splitting of this band [38, 41, 42], as shown in Figure 5(c). Hence, this band structure is stable to the Peierls distortion and displays a metallic ground state. Detailed band calculations for one- or three-dimensional models of $(SN)_x$ based on the Extended Hückel method by Parry and Thomas [39], Friesen *et al.* [40], Bright and Soven [41], and OPW calculations by Rudge and Grant [5] for the crystal structures determined by the electron [10] and/or X-ray diffraction methods [14] gave essentially the similar concept as stated above. Self consistent field (SCF) calculations by Salahub and Messmer [42, 43] and Mihich [44] also gave nearly the same conclusion.

The band structure proposed by Rudge and Grant [5] is a semimetal, i.e., electron and hole pockets locate at the Brillouin zone boundary with a Fermi surface consisting of anisotropic tubings [38].

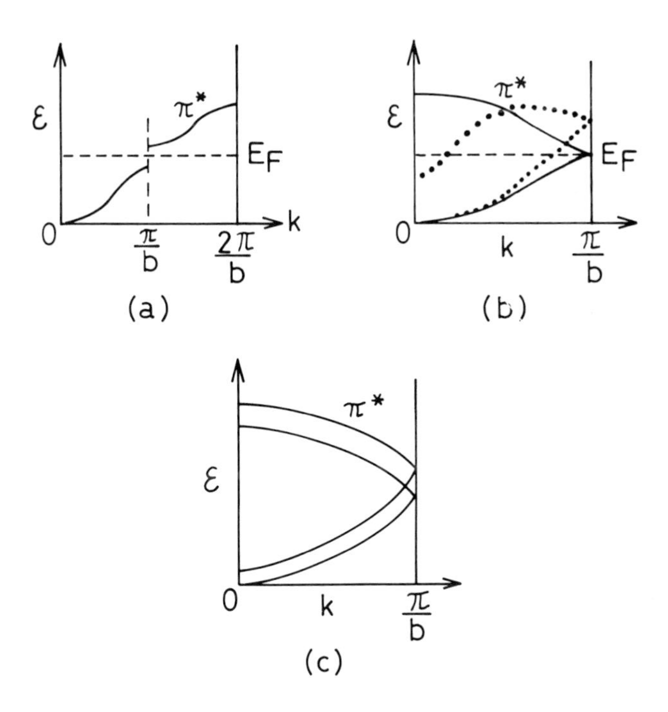

Fig. 5. Energy band of π^* band along k_b direction, (a) formation of Peierls gap (b) stick together at zone boundary due to the screw symmetry, dotted lines shows the overlapping band [4], (c) splitting of π^* band by the interchain coupling.

Contrary to these band structures deduced from the strong s–p interaction, a non-empirical calculation using the tight-binding approximation by Kamimura *et al.* [4, 51] based on the electron diffraction result [10] predicted the overlapping band structure as illustrated in Figure 5(b) by dotted lines. In this band model, the Fermi level crosses two conduction bands with one-dimensional characteristics [51]. Taking the interchain interaction into account, they obtained a biconcave Fermi surface [4]. The similar result of overlapping band model was also suggested by Rajan and Falicov [45] using LCAO, and by Schlüter *et al.* [46] using the empirical pseudopotential scheme.

Afterward, Oshiyama and Kamimura [6] recalculated the band structure of $(SN)_x$ using SCF–LCAO method and gave their final band structure, which showed a Fermi surface having a considerable similarity to the Rudge and Grant result [5]. Oshiyama *et al.* suggest that if the ratio of electron transfered from sulfur to nitrogen *vs.* total electron exceeds 0.8, the conduction band crosses the Fermi level and the overlapping band is plausible. However, X-ray photoelectron spectroscopy measurement by Mengel *et al.* [52] gave the result of electron transfer of 0.3~0.42 to nitrogen from sulfur atoms. This result indicates that the overlapping band model is not favoured.

Figure 6 shows the band structure proposed by Oshiyama *et al.* [6]. The band width along the conducting axis (ΓZ, BD, YC and AE) are relatively large compared to those of perpendicular directions as shown in Figure 7, indicating highly anisotropic features.

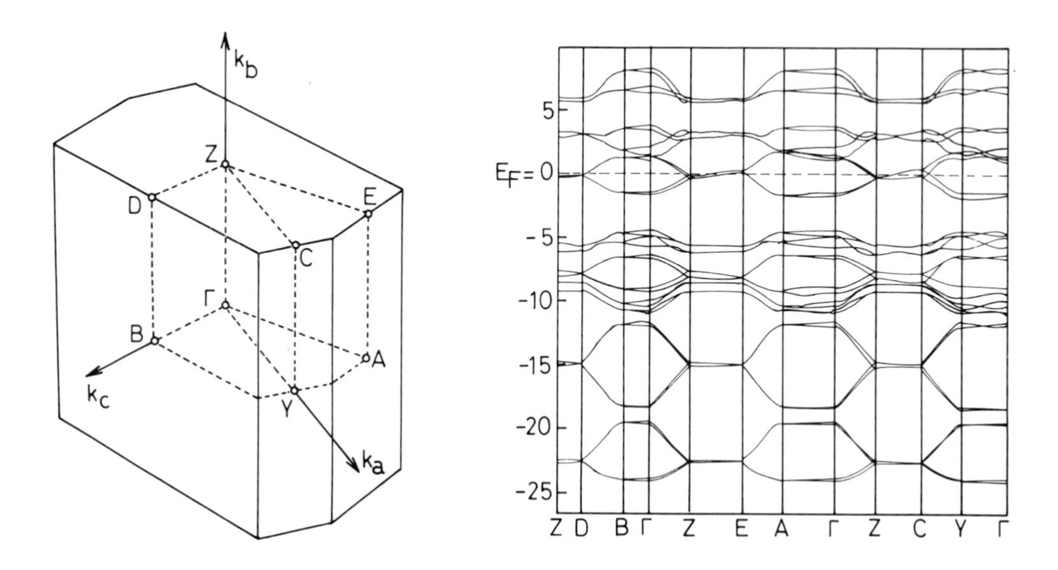

Fig. 6. Energy structure of $(SN)_x$ [6], Fermi level is shown by broken line. The left is Brillouin zone of $(SN)_x$ crystal.

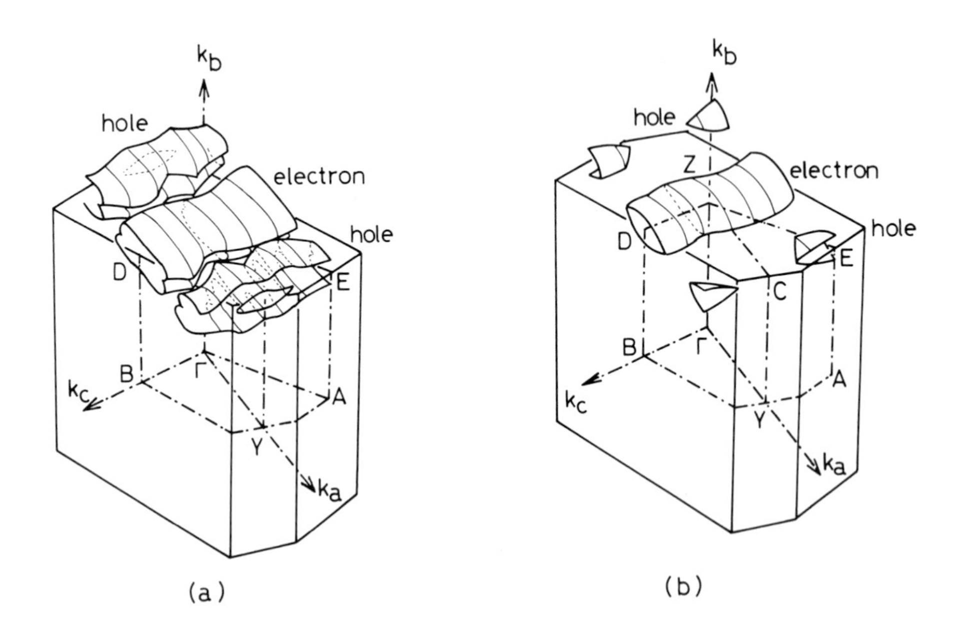

Fig. 7. Fermi surface of $(SN)_x$ obtained from LCAO–SCF band calculation by Oshiyama *et al.* [6], (a) the outermost surfaces, (b) innermost surfaces.

The conduction band also crosses the Fermi level in the k_a direction due to the interchain interaction, resulting in the closed Fermi surface for the k_a and k_b directions [6]. Fermi surfaces consisting of four conduction bands are indicated in Figure 7(a) and (b) for the outermost and innermost surfaces, respectively. The density of states at the Fermi level $D(E_F)$ = 0.14 states/(eV · spin · molecule), from this band model seems to be fairly consistent with the experimental values from XPS by Mengel [52] and by Ley [53] and from the specific heat data [54] and with that of OPW calculation [5].

The Fermi surfaces of brominated $(SN)_x$ were also discussed by Oshiyama *et al.* [6, 47] using the rigid band model of $(SN)_x$, i.e., an idea that the acceptor of bromine lowers only the Fermi level in the unchanged band structure. According to this simple model, the electron pocket shrinks and the hole pocket expands with increasing bromine contents, leading to disappearance of the electron pocket at the charge transfer ratio of *ca.* 0.1 from SN to bromine. The temperature dependence of electrical conductivity in $(SNBr_y)_x$ was calculated [47] using this model and compared with the experimental result [7, 55] as will be discussed later.

4. Electrical Properties of $(SN)_x$ and $(SNBr_y)_x$

4.1. ELECTRICAL CONDUCTIVITY IN $(SN)_x$

Metallic conductivity of $(SN)_x$ was first reported by Walatka *et al.* [1] contrary to the previous observation of semiconductive characteristics [56]. The temperature (T) dependence of electrical resistivity (ρ) was reexamined by several workers [11, 21, 57–59] and revealed the relation of $\rho \propto T^n$ (n = 1.9 2.3).

Figure 8 shows the temperature dependence of electrical resistivity normalized to the room temperature value along the b axis ($\rho_{\parallel}$) with different ratio of the conductivity at 4.2 K to 300 K, $\sigma_{4.2}/\sigma_R$. This quadratic temperature dependence of the resistivity has been interpreted in terms of carrier–carrier (electrons or holes) Umklapp scattering [21]. Oshiyama and Kamimura [47, 48] discussed this problem taking their band model of $(SN)_x$ into account. They pointed out that in the system with quasi-1-D Fermi surface or in the semimetal with strong electron–hole interaction, the normal electron–hole scattering can be dominant to carrier–carrier Umklapp and carrier–phonon scattering, resulting in a T^2 dependence of resistivity.

The negative thermoelectric power in $(SN)_x$ [57, 59, 61] as shown in Figure 9 indicates that the contribution of electrons to the electrical transport should be larger than that of holes in the semimetal model [6, 38].

The sample with smaller value of $\sigma_{4.2}/\sigma_R$ (<50) shows a remarkable resistance minimum for temperature as shown in Figure 8. Deformation by the mechanical stress [57], heat treatment at above 60 °C [13] or γ-ray irradiation [58] gives rise to the resistivity minimum. These results indicate that the resistance minimum is mainly due to crystal imperfections.

In highly anisotropic conductors, mobile carriers have to hop or tunnel to the next conducting channel at the chain break or at the termination of the percolation path, probably giving rise to the significant residual resistance in imperfect samples. The mechanism of increase of resistivity below 30 K can be attributed to (i) the carrier

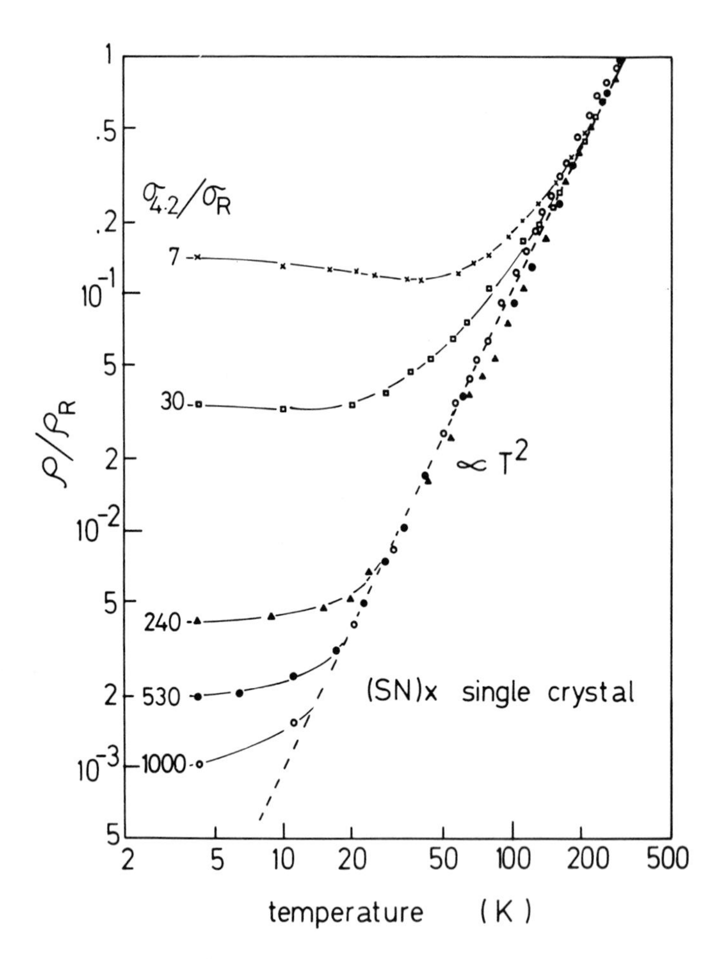

Fig. 8. Temperature dependence of the electrical resistivity ρ/ρ_R along the b axis for various samples with different $\sigma_{4.2}/\sigma_R$ values.

scattering by either (i–a) charged defects and/or impurity [21, 58] or (i–b) localized magnetic spins [2, 58] (Kondo effect [62]) or to (ii) the decrease of hopping probability at chain breaks at lower temperature. The observation of negative transverse magneto-resistance [38, 57, 63, 64] at low temperature and the paramagnetic behaviour of the magnetic susceptibility in the γ-ray irradiated sample [58] seems to suggest the existence of localized magnetic spins.

Exact measurement of the electrical conductivity perpendicular to the b axis ($\sigma_\perp$) is difficult because of the fibrous structure of $(SN)_x$ crystal. $\sigma_\perp$ measured by dc methods had sample-dependent values of 1∼5 [20, 57, 65] and 70 [59] S/cm which do not seem to be the intrinsic value. For example, $\sigma_\perp$ obtained by the dc method shows peculiar behaviour [38, 57, 59, 65] of almost temperature independence, as shown in Figure 10.

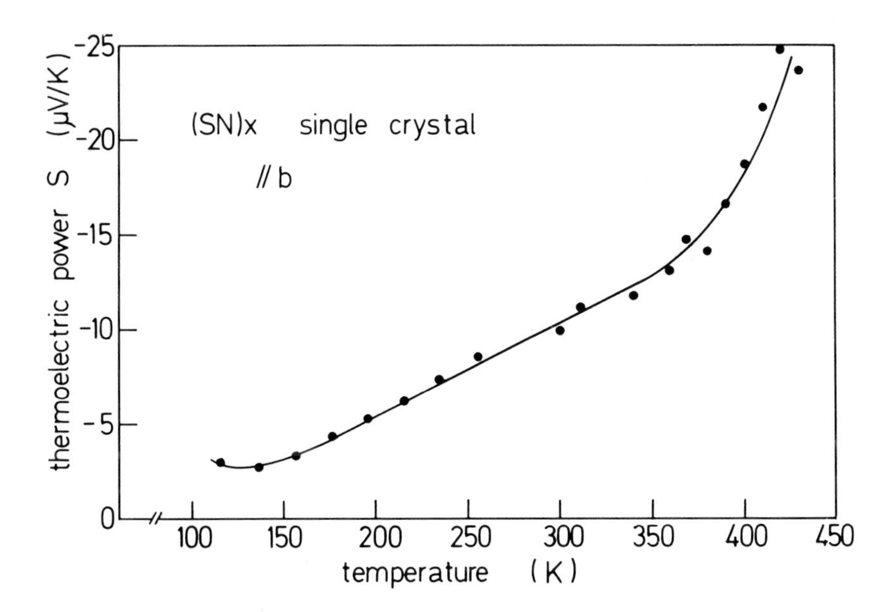

Fig. 9. Temperature dependence of the thermoelectric power of $(SN)_x$ crystal along the b axis.

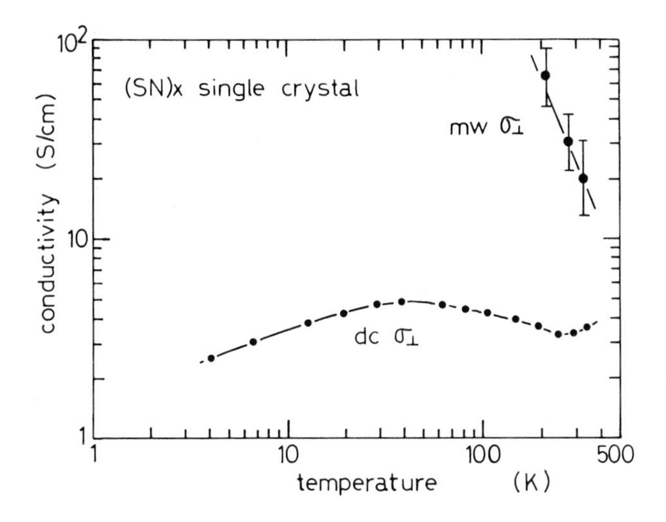

Fig. 10. Temperature dependence of $\sigma_\perp$ obtained by dc and microwave measurements.

To exclude the boundary effect of the interfiber resistance, the conductivity measurement at high frequency (9GHz) was carried out by Kaneto *et al.* [65]. The microwave conductivity $\sigma_\perp$ obtained at room temperature was *ca.* 40 S/cm, being greater than dc $\sigma_\perp$ by several times, and increased with decreasing temperature as shown in Figure 10. This result shows that the anisotropy factor of $\sigma_\parallel/\sigma_\perp < 100$ at room temperature is considerably smaller than the previously reported value of $\sim 10^3$ from dc measurement, indicating the remarkable contribution from interfiber resistance to dc $\sigma_\perp$.

From the calculated Fermi surface such as Figure 7, however, it should be noted that there is also anisotropy within the *ac* plane, suggesting considerably higher conductivity along the *a* axis than the *c* axis. The large discrepancy of the dc $\sigma_\perp$ obtained by different workers as stated previously may be partly related to the anisotropy within the *ac* plane.

4.2. Other Electrical Properties of $(SN)_x$

The temperature dependence of specific heat in $(SN)_x$ crystal was measured by Greene *et al.* [54] and Harper *et al.* [66]. They found that the specific heat C obeys fairly well, below 4 K, the usual relation of

$$C/T = \gamma + \beta T^2 \tag{4-1}$$

where γ is the term which arises from free carriers, and the second term from lattice specific heat. The estimated value of γ = 0.83 ± 0.09 mJ/mol · K gives the density of states at Fermi level as 0.14 states/(eV · spin · mol). From $\beta = 0.14 \pm 0.02$ mJ/mol · K^3, the Debye temperature was estimated to be Θ_D = 148 K [54, 66]. Deviation of the lattice specific heat from the cubic temperature dependence was observed above 4 K, i.e., intermediate region of $T^{2.7}$ dependence, followed by a linear term at the higher temperature of 20 K. As the cubic temperature dependent term is considered to be derived from the 3–D phonon mode, Harper *et al.* [66] interpreted the Debye temperature obtained as the characteristic energy due to the interchain coupling. Moreover, they attributed the deviation of C from T^3 dependence to the highly anisotropic nature of the linear chain $(SN)_x$.

The $(SN)_x$ crystal with no observable resistivity minimum showed the temperature independent Pauli paramagnetism [58] of $\chi_P = (0.2 \pm 0.1) \times 10^{-6}$ emu/g. Although the observed value may include some diamagnetism, this χ_P gives the lowermost density of state at the Fermi level, in good agreement with the specific heat result and band calculations [5, 6].

Detailed thermopower measurements by Azevedo *et al.* [67] have revealed that the phonon drag dominates the electron transport above 1 K from the cubic temperature dependence of thermopower. They also suggested that the slight increase of thermopower below 1 K can be explained in terms of the Kondo effect.

Tunneling spectroscopies in $(SN)_x$ crystal with the junction of $(SN)_x$/Al at room temperature [68], $(SN)_x$/Pb at around superconduction region [69] and $(SN)_x/(SN)_x$ [70] which will be stated in Section 5 were also studied.

4.3. Electrical Properties of $(SNBr_y)_x$

The increase of the electrical conductivity and the change of the thermoelectric power from negative to positive [31] upon bromination of the $(SN)_x$ are interesting in terms of mechanisms of the electron transfer from host material to acceptor and of the carrier scattering.

The electrical resistivity ρ of $(SN)_x$ along the *b* axis decreased immediately upon exposure to bromine vapour, being followed by continuous decrease of the resistivity for about one day even after the replacement of bromine gas by inert gas (Ar) [7]. The slow decrease of resistivity after removal of bromine vapour indicates that the diffusion of bromine from the surface into the interior of fibrous bundles takes fairly long time (several hours).

Figure 11 [7] shows the resitivity decrease expressed by $\ln[(\rho - \rho_c)/(\rho_i - \rho_c)]$ as a function of time *t* after the removal of bromine gas ($t = 0$), where ρ_i and ρ_c are resistivities at $t = 0$ and at $t \rightarrow \infty$, respectively. In this experiment, the dopant concentration was estimated to be $y \simeq 0.25$ from the weight change. If one assumes that the decrease of resistivity is attributed to more homogeneous dopant distribution in the bulk, the diffusion time constant τ_D may be estimated to be *ca.* 12 hours from the straight line region as shown in Figure 11. Taking the cylindrical model of diffusion process [71] into account, the diffusion coefficient *D* is evaluated to be about 2×10^{-12} cm^2/sec from the relation of $D = r^2/(2.4)^2\tau_D$ and the fiber radius *r* of *ca.* 70 μm. Rapid initial decrease of the resistivity for $t < \tau_D$ compared to logarithmic decay should be consistent with the characteristic of the cylindrical diffusion model [71].

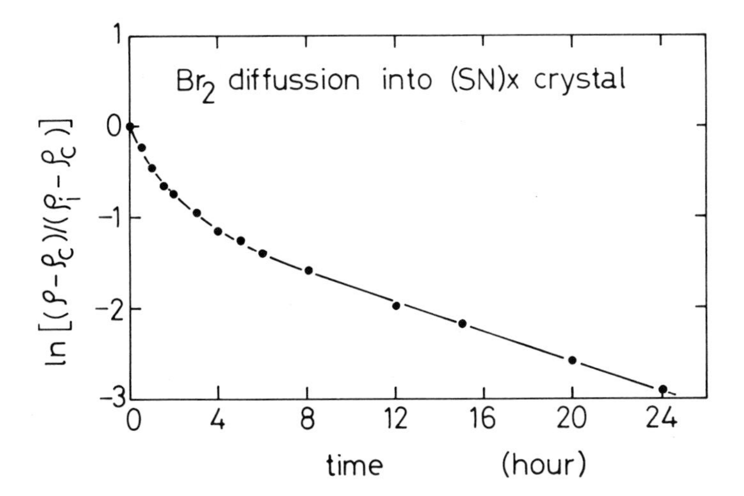

Fig. 11. Time dependence of electrical resistivity along the *b* axis by the diffusion of bromine into $(SN)_x$ crystal.

Figure 12 shows the resistivity of $(SNBr_y)_x$ crystal (ρ_{Br}) normalized to that of the pristine $(SN)_x$ (ρ_{prist}) *vs.* various bromine contents *y*, immediately after the exposure to bromine gas and diffusion equilibrium. Downward arrows from solid to open circles indicate the decrease of the resistivity from the values immediately after bromination to those of quasi-diffusion equilibrium under the inert gas atmosphere. The smallest value of ρ_{Br}/ρ_{prist} along the *b* axis was found to be *ca.* 0.06 for $y \simeq 0.4$ at room temperature, indicating σ_{Br} was *ca.* 4×10^4 S/cm.

Figure 13 shows the temperature dependence of $\rho_{Br}/\rho_{prist \cdot R}$ for parallel ($\rho_\parallel$) and perpendicular ($\rho_\perp$) directions to the *b* axis in a sample with successive doping (increase

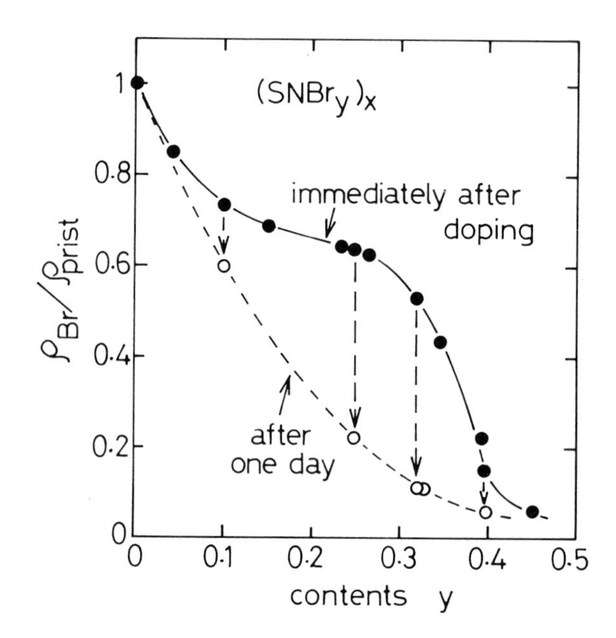

Fig. 12. Electrical resistivity of $(SNBr_y)_x$ normalized to the pristine $(SN)_x$ *vs.* bromine content y for the immediately after doping and quasi-equilibrium state.

y) [7]. The residual resistivity below 30 K observed in the pristine sample decreases remarkably upon bromination, indicating that the dopant bromine assists the carrier movement at the chain breaks or termination of conducting path.

The temperature dependence of $\rho_{\|}$ above 30 K becomes weaker than the quadratic dependence ($\rho_{\|} \propto T^2$) with increasing bromine contents as seen in Figure 13. The resistivity $\rho_{\|}$ of brominated $(SN)_x$ can be expressed by $\rho_{\|} = \rho_{res} + AT + BT^2$, where ρ_{res}, A and B are the residual resistivity at low temperature, and constants related to phonon scattering and carrier–carrier scattering, respectively. The coefficient B obtained experimentally for pristine $(SN)_x$ is 5.6×10^{-9} cm/SK^2 [7] which agrees fairly well with the calculated value of 3.16×10^{-9} cm/SK^2 by Oshiyama *et al.* [47]. The coefficients A and B are estimated to be 4.2×10^{-8} cm/SK and 2.2×10^{-10} cm/SK^2, respectively, at the dopant concentration of $y \simeq 0.4$ [7]. The value of B decreases remarkably by bromination from that of pristine $(SN)_x$ by about 25 times.

Oshiyama *et al.* [47] developed the model of electron–hole scattering offered by Kukkonen *et al.* [72] to calculate the temperature dependence of resistivity in brominated $(SN)_x$. They considered that the total current is divided into two terms, i.e., J_c: center-of-mass motion; and J_r: relative motion of electrons and holes as defined by following equations [48, 72],

$$J = J_c + J_r, \tag{4-2}$$

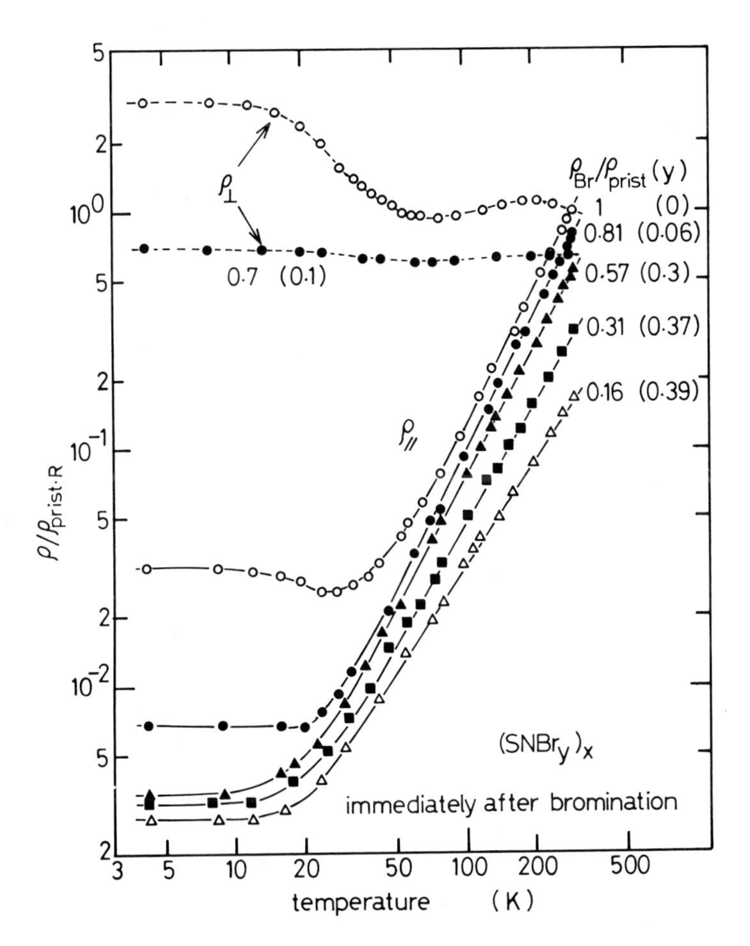

Fig. 13. Temperature dependence of $\rho_{\parallel}$ and $\rho_{\perp}$ of $(SNBr_y)_x$ normalized to those of pristine $(SN)_x$ at room temperature for various y, pristine $\sigma_{\parallel} = 2 \times 10^3$, $\sigma_{\perp} = 5$ S/cm at room temperature.

$$J_c = \frac{-e(n_e - n_h)}{n_e m_e + n_h m_h} \left(\sum_i^{n_e} P_i + \sum_j^{n_h} P_j \right), \tag{4-3}$$

$$J_r = -e \left(\frac{1}{m_e} + \frac{1}{m_h} \right) \left(\frac{1}{n_e m_h} + \frac{1}{n_h m_e} \right) \times$$

$$\times \left(\frac{1}{n_e m_e} \sum_i^{n_e} P_i - \frac{1}{n_h m_h} \sum_j^{n_h} P_j \right), \tag{4-4}$$

where n_e and n_h, m_e and m_h are electron and hole concentrations and masses, respectively, and P_i and P_j are momenta at electron i and hole j bands, respectively, in their band model [6]. The first term J_c is attenuated only by carrier–phonon scattering not by electron–hole scattering, although J_r depends on electron–hole scattering and the other scattering mechanisms [48].

For the case of pristine $(SN)_x$, i.e., $n_e \simeq n_h$, the current should be dominantly determined by J_r and follow T^2 dependence due to the electron–hole scattering [48, 72]. Upon bromination, however, n_e decreases and n_h increases with bromine contents ($n_e < n_h$), resulting in the increase of the contribution of J_c to current. Furthermore the inverse relaxation time of electron–hole scattering τ_{eh}^{-1} decreases with increasing Fermi vector of hole [48], which comes from the expansion of the hole pocket and shrinkage of the electron pocket of the Fermi surface by the bromination.

Oshiyama *et al.* [48] found that the rate of charge transfer $\Delta\tilde{n}$ from $(SN)_x$ to bromine is proportional to $n_h - n_e$ for their band model, and calculated the temperature dependence of resistivity using this parameter. Figure 14 shows the comparison of their calculation and our experimental data [7, 55], indicating quantitatively good agreement. The smaller value of $\Delta\tilde{n}$ compared to the bromine contents y should coincide with the idea of existence of Br_2 and Br_3^- as suggested by Raman scattering results [34, 35].

The $\sigma_\perp$ also increases by bromination, as shown in Figure 13, and becomes almost temperature-independent at the dopant level of $y \simeq 0.1$. Somewhat metallic behaviour of $\sigma_\perp$ in $(SNBr_y)_x$ was observed [8] at $y \simeq 0.4$ with the room temperature conductivity

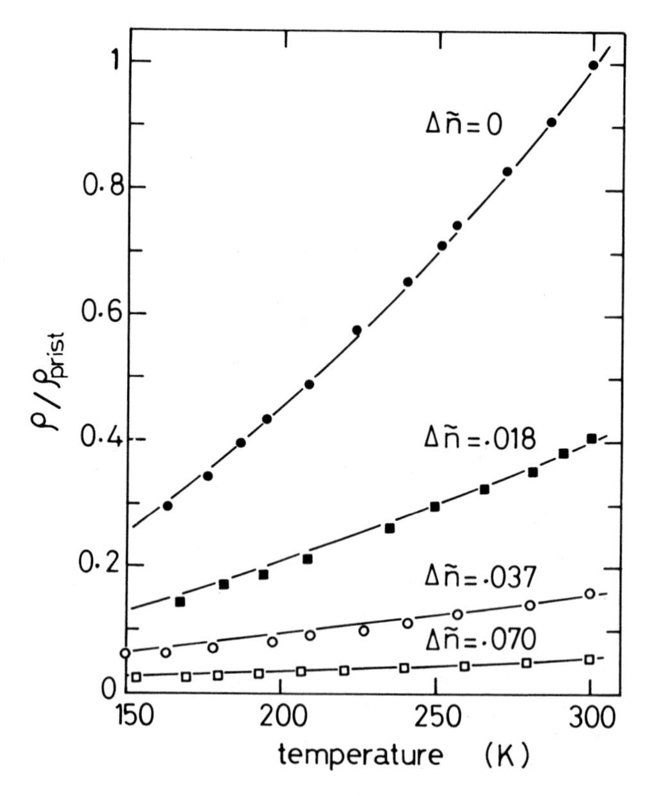

Fig. 14. Temperature dependence of calculated resistivity [47] along the b axis in $(SNBr_y)_x$ for various $\Delta\tilde{n}$ value, • (pristine) ▪ ($y \simeq 0.15$), ○ ($y \simeq 0.28$) and □ ($y \simeq 0.4$) are experimental data.

of *ca.* 30 S/cm. These results indicate that the intercalated bromine influences on the electronic states near the Fermi level and assists the carrier transport perpendicular to the b axis, namely, inter- and intra-fibrils. The latter effect will also decrease residual resistivity along the b axis caused by chain breaks.

5. Superconductivity

Superconductivity of $(SN)_x$ was discovered by Greene *et al.* [3] and has attracted much interest, since it is the first polymeric superconductor composed of V–VI elements in the Periodic Table and the first superconductivity found in low dimensional materials. They pointed out that $(SN)_x$ becomes superconducting at 0.2~0.3 K and that the transition temperature T_c depends strongly on the sample quality, i.e., $\sigma_{4.2}/\sigma_R$. The width of transition temperature ΔT_c indicated in Figure 15 is also one of the characteristics of

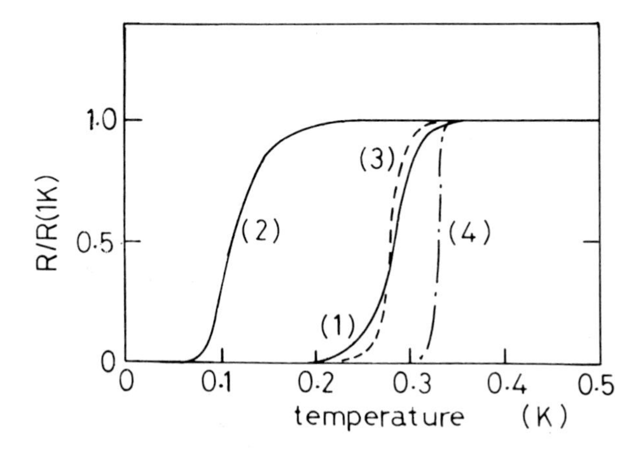

Fig. 15. Superconducting transition temperature in $(SN)_x$ [3] and $(SNBr_y)_x$ [33], (1) applied magnetic field $H = 0$ G, (2) $H = 335$ G, (3) $H = 0$ G (our sample) (4) $(SNBr_{0.4})_x$.

$(SN)_x$, being strongly dependent on the quality of sample. Low quality sample shows lower T_c and wider ΔT_c.

The transition temperature T_c decreases remarkably with applied magnetic field. Especially applied magnetic field perpendicular to the fiber axis $H_\perp$ is found to be more effective than that parallel to the fiber. This magnetic field dependence of T_c is considered to be a conclusive evidence of the BCS type superconductivity in $(SN)_x$. Greene *et al.* [3] suggested that $(SN)_x$ should be a filamentary superconductor.

Critical magnetic field for superconducting $(SN)_x$ and its anisotropy were studied by Azevedo *et al.* [73] in detail. The upper critical fields $H_{c2\parallel}$ and $H_{c2\perp}$ parallel and perpendicular to the b axis, respectively, were evaluated to be 8.1 ± 0.4 KOe and 870 ± 80 Oe. This extremely large anisotropy of the upper critical field is one of the characteristics of $(SN)_x$. $H_{c2\parallel}$ is much larger and exceeds that of the paramagnetic limit ($H_p = 18.4T_c$ = 5.4 KOe). Contrary to this large anisotropy with respect to the b axis, the anisotropy in the basal plane perpendicular to the b axis was estimated to be much smaller, indicating a quasi-1-D nature of the material.

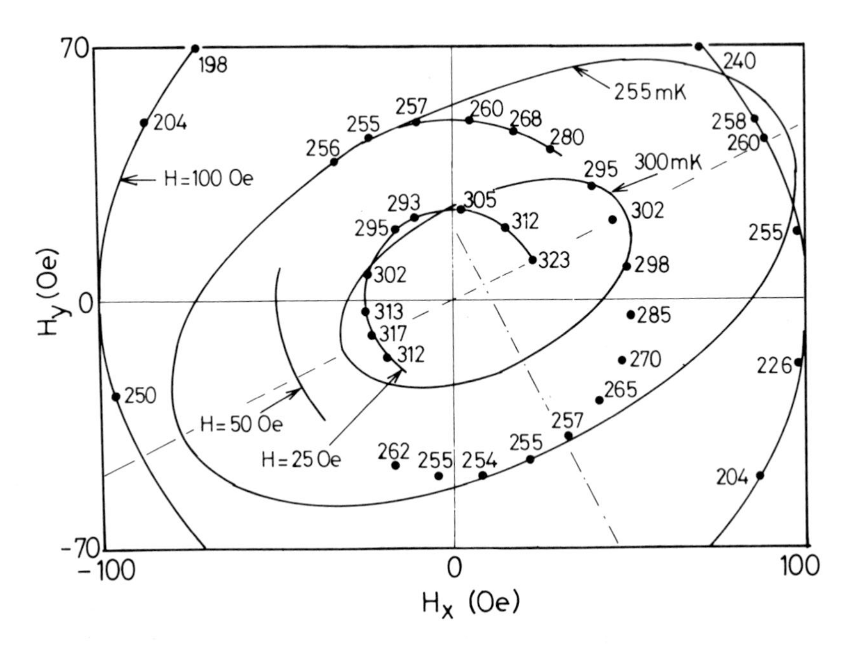

Fig. 16. Anisotropy of $H_{c2\perp}$ in basal plane [74].

Figure 16 shows the anisotropy of the upper critical field in the basal plane [74]. Each point in this figure indicates magnitude and orientation of the magnetic field in *ac* plane. At a given temperature, the major axis indicating the direction with highest $H_{c2\perp}$, coincides with the *c* direction. On the other hand, the minor axis corresponding to the direction of lowest $H_{c2\perp}$ coincided with a^* direction (perpendicular to both *c* and *b* axis).

Azevedo *et al.* [73] tried to explain these characteristics in terms of either intrinsic anisotropy of Fermi surface or fibrous nature of morphology. The coherence length perpendicular to fiber $\xi_\perp(T)$ and parallel to it $\xi_\parallel(T)$ are evaluated to be $\xi_\perp(0) = 135 \pm 15$ Å and $\xi_\parallel(0) = 3500 \pm 1500$ Å, respectively. Temperature dependence of $\epsilon^{-1} = H_{c2\parallel}/H_{c2\perp}$ and angular dependence of H_{c2} are explained by taking the fibrous structure into consideration. Namely, they pointed out that $(SN)_x$ is an anisotropic 3-D superconductor influenced strongly by the fiber size, but not an intrinsic 1-D superconductor.

They also found anomalous upward curvature in the temperature dependence of H_{c2} near T_c as shown in Figure 17 [33, 74]. This upward curvature in the temperature dependence of the critical field was explained by the decreased coupling of the filaments from Ginzburg–Landau theory of filamentary superconductors by Turkevich and Klemm [75].

Magnetic properties are also studied by Dee *et al.* [76, 77]. They first measured complex ac magnetic susceptibility χ at 0.2 K. The positive value of observed χ and its decrease with increasing magnetic field are pointed out to be the evidence of type II superconductivity. They concluded that $(SN)_x$ crystal behaves as bundles of fibers which

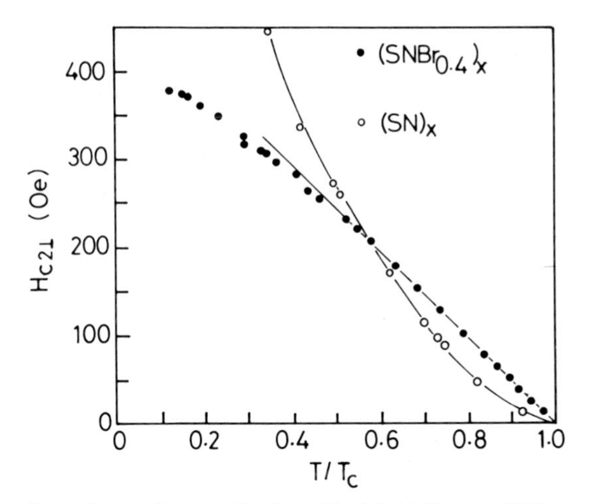

Fig. 17. Temperature dependence of perpendicular critical field $H_{c2\perp}$ in $(SN)_x$ and $(SNBr_{0.4})_x$ [33].

are weakly coupled at best with the coherence length larger than diameter of fibers, in accordance with the conclusion of Azevedo *et al.* [73, 74].

The Meissner effect was also confirmed by Dee *et al.* [77, 78] by the observation of the change of magnetization. In their second report [78] they found full diamagnetic response of $-1/4\pi$ in the case of perpendicular magnetic field and slightly smaller value of $0.6 \times (-1/4\pi)$ for the case of parallel field, as indicated in Figure 18.

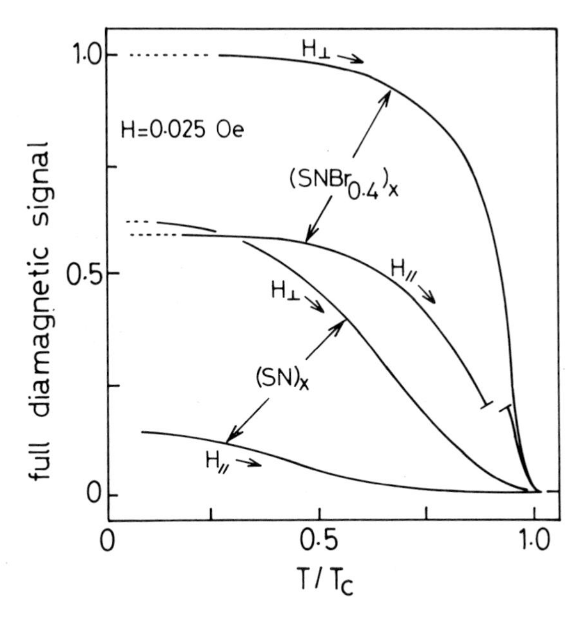

Fig. 18. Magnetization of $(SN)_x$ and $(SNBr_{0.4})_x$, $H_\perp$ magnetic field applied perpendicular to the b axis, $H_\parallel$ magnetic field applied parallel to the b axis [78].

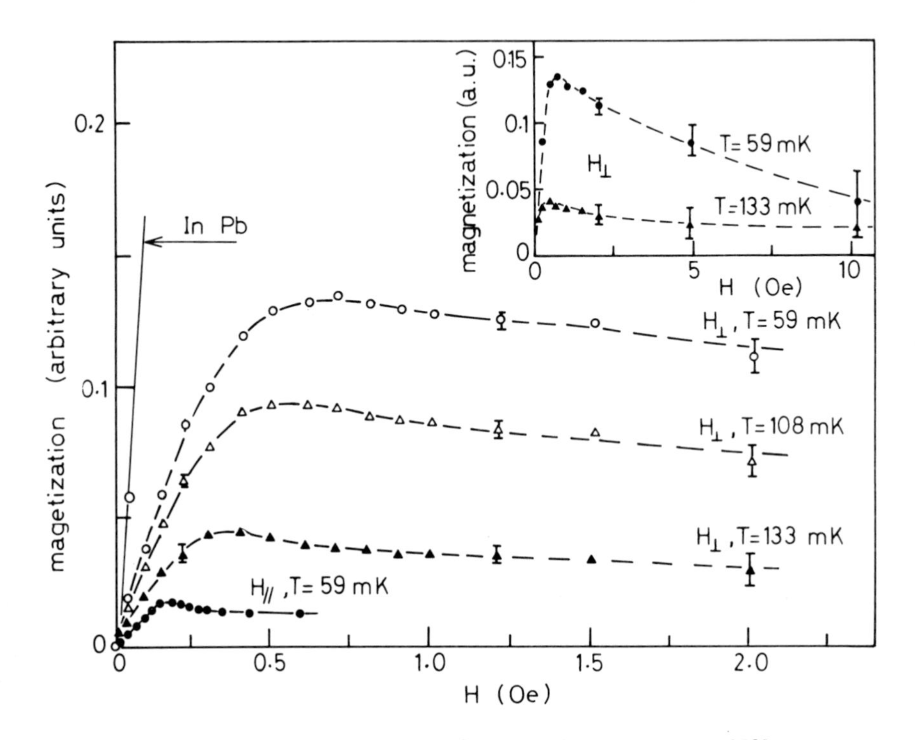

Fig. 19. Magnetization curves for $(SN)_x$ at various temperatures [78].

Figure 19 shows an example of magnetization curve in $(SN)_x$. The lower critical field H_{c1} as a type II superconductor can be evaluated from the field H_p at which the peak magnetization is observed by the relation of $H_{c1} = H_p/(1 - n_s)$, where n_s is a demagnetization factor. Using observed H_p [78] and n_s of 0.2~0.4, $H_{c1\perp}$ and $H_{c1\parallel}$ can be evaluated as 0.9~1.1 Oe and 0.2~0.3 Oe in $(SN)_x$. $H_{c1\perp}$ and $H_{c1\parallel}$ values in $(SNBr_{0.4})_x$ are thus evaluated as 3~4 Oe and 0.9~1.3 Oe, respectively, which roughly agree with theoretical consideration. The Meissner effect was also confirmed by Oda *et al.* [79] with the measurement of *ac* magnetic susceptibility.

Superconductivity in $(SN)_x$ is also confirmed from the measurement of thermoelectric power S by Azevedo *et al.* [67]. S drops to 0 sharply at about 0.3 K with decreasing temperature, corresponding to the vanishing entropy in the pair condensation. The dependence of S on magnetic field and its anisotropy were consistent with those of conductivity on H.

Dependence of T_c on applied pressure was studied in detail [61, 74, 80–82]. T_c is found to increase linearly with increasing pressure up to 8~9 Kbar and the transition width ΔT_c increases with pressure. At around 9 Kbar a significant drop of T_c is observed as indicated in Figure 20.

Gill *et al.* [80] discussed the pressure dependence in terms of McMillan theory for T_c resulting from the change of electronic band structure under pressure.

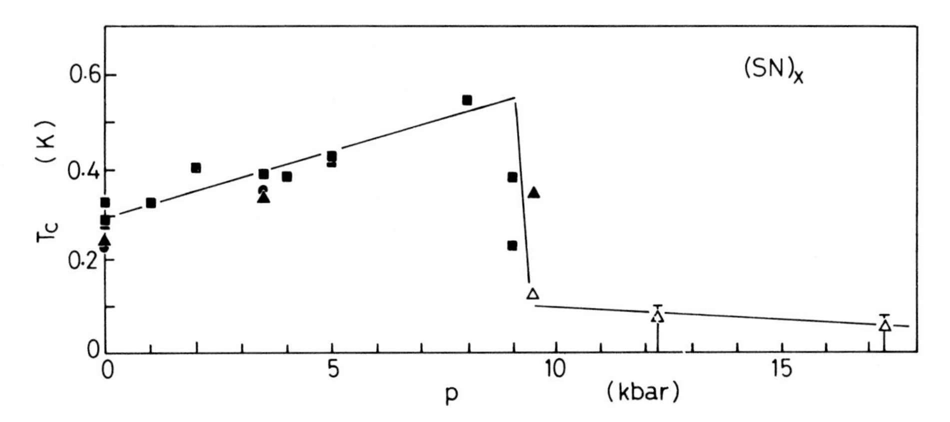

Fig. 20. Pressure dependence of superconducting temperature T_c in $(SN)_x$ [74, 81].

Muller *et al.* [81] pointed out that not only the intrinsic properties of $(SN)_x$ such as the increase of the density of states at Fermi level E_F, $N(E_F)$, and of the electron–phonon scattering parameter by pressure, but also the fibrous nature of $(SN)_x$ crystal should have significant influence. The increase of T_c with pressure can be also caused by the same mechanism.

Bickford *et al.* [74] observed just the same pressure dependence of T_c with that of Muller *et al.* [81] and explained in terms of the fibrous nature of $(SN)_x$ crystal. Namely, the fibers with dimension of several 100 Å are considered to be decoupled, and only weak Josephson coupling exists. When pressure increases, the coupling between fibers increases as derived by Deutcher *et al.* [82] from following relation

$$1/T_c \simeq 1/T_{c0} + 0.1N(E_F)/cT_{c0}^2 \tag{5-1}$$

(T_{c0}: transition temperature in the strong coupling limit) just as the case of dichalcogenide superconductor such as $NbSe_2$. The estimation of coherence length $\xi_\perp$ and $\xi_\parallel$ indicates that $\xi_\perp$ increases remarkably with pressure, in contrast to the small change of $\xi_\parallel$. From these results, it is considered that at lower pressure fibers are only weakly coupled to each other and $H_{c2\parallel}(T)$ is determined by the paramagnetic limit of isolated fibers. Above 9.5 Kbar fibers become no longer completely decoupled and the paramagnetic limit is not the determining factor for $H_{c2\parallel}(T)$. Thus, the superconducting properties of $(SN)_x$ can be explained in terms of the collection of weakly coupled filaments.

The basal anisotropy of Figure 16 may be also influenced by the anisotropy of coupling between fibers, if it is in range of the decoupling temperature. Generally, however, such anisotropy is of the order of 0.1%, being much smaller than the observed anisotropy of 50%. Gill *et al.* [80] suggested that, even in weak coupling between fibers, the basal anisotropy may originate in the intrinsic anisotropy of the band system in each fiber. The similarity of the ratio of the plasma tensor in the principal axis with the ratio of the critical fields is considered to be one evidence.

Upward curvature of $H_{c2\perp}(T)$ already mentioned in Figure 17 becomes less pronounced at higher pressure, which can be explained satisfactorily in terms of fibrous model as following. At atmospheric pressure, as the temperature is lowered, a progressive decoupling of fibers occurs, resulting in the upward curvature. On the other hand, above 9.5 Kbar, at all temperature ranges, the fibers are coupled, resulting in the linear change with pressure.

From experiments under extremely high pressure (up to 400 Kbar), Dunn *et al.* [61] found that T_c increased from 2 to 3 K at pressures from 130 to 400 Kbar, which is remarkably higher than that under atmospheric pressure. The strong increase of T_c cannot be explained by the enhancement of coupling between fibers at such a high pressure, but by the appearance of new phases at such pressure, and the change in the electronic band structure, resulting in the increase of electron–phonon coupling.

As has already been mentioned, the transition from a normal state into a superconducting state is not so sharp. Such behaviour sometimes can be explained by thermodynamic fluctuations in the superconducting state. The excess conductivity above T_c was also studied by Civiak *et al.* [83, 84] and discussed in terms of fluctuation conductivity by the theory of Aslamazov and Larkin (AL) [85], namely by the superconducting pairs. Aslamazov and Larkin indicated that the fluctuation conductivity $\Delta\sigma_{\mathrm{AL}}$ can be related to temperature by the equation:

$$\Delta\sigma_{\mathrm{AL}}/\sigma_n = [(T - T_c)/\epsilon T_{c0}]^{(4-D)/2} \tag{5-2}$$

where σ_n, D and ϵ are the normal state conductivity, number of dimensions in which the size of superconductor is larger than the temperature-dependent coherence length ξ, and a constant which is dependent upon dimensionality. T_{c0} is the transition temperature in the absence of pair breaking effect.

The high quality sample showed $(T - T_c)^{-3/2}$ dependence, as shown in Figure 21, which can be explained by 1-D LA theory. 1-D behaviour indicates that the sample is behaving as bundles of weakly-connected superconducting filaments, the size of which is larger than the coherence length in only one direction. On the other hand, the excess conductivity of low quality sample with small $\sigma_{4.2}/\sigma_{\mathrm{R}}$ ratio can be best fitted with $(T - T_c)^{-2}$ dependence, indicating a 0-D particle model. By fitting with 1-D model, T_{c0} = 1.4 K and T_c = 0.43 K are evaluated, which corresponds to average diameter of 1-D units of 240 Å.

As is evident from the above considerations, the relation between coherence length and fiber diameter seems to be very important, which is consistent with Azevedo's *et al.* analysis. $(SN)_x$ seems to consist of bundles of segmental fibers of many polymer chains with diameter smaller or comparable to the coherence length in the direction perpendicular to the fiber. In low quality samples, the typical length of polymer chain should be shorter than the longitudinal coherence length $\xi_{\parallel}$, resulting in zero dimensional characteristic of fluctuation conductivity.

Br_2 doping also influences strongly the superconductivity of $(SN)_x$. Kwak *et al.* [33] studied detailed superconducting characteristics in $(SNBr_{0.4})_x$. The phase transition temperature increased remarkably and the spread of transition temperature also decreased remarkably by Br_2 doping as indicated in Figure 15. The upward curvature of temperature dependence of $H_{c2\perp}$ of pristine $(SN)_x$ disappeared and linear dependence is observed in $(SNBr_{0.4})_x$. These facts seem to suggest the increase of coupling between fibers in a doped state, resulting in the increased dimensionality in $(SNBr_{0.4})_x$.

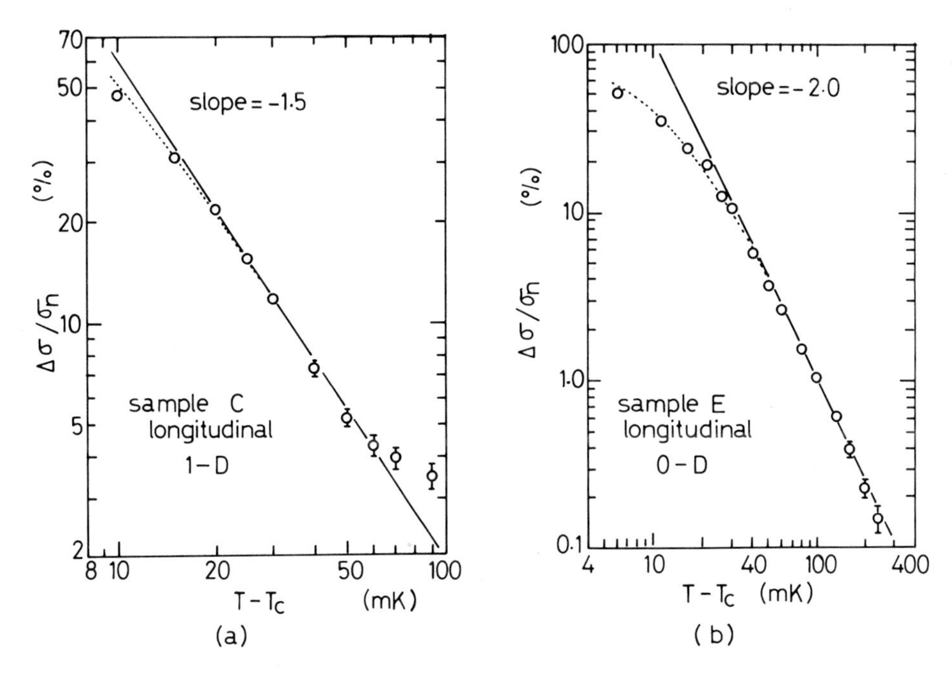

Fig. 21. Normalized fluctuation conductivity of $(SN)_x$, (a) high quality sample [84], (b) low quality sample.

Contrary to the case of $(SN)_x$ which has characteristics of low dimensionality, Br_2-doped $(SN)_x$ seems to be treated by anisotropic three-dimensional Ginzburg–Landau theory [86]. In $(SN)_x$ Josephson coupling between fibers should be necessary.

Magnetic measurements of Dee *et al.* [78] also indicated stronger interfiber coupling in brominated $(SN)_x$. Namely, nearly complete diamagnetization was confirmed by the observation of the Meissner effect (Figure 18).

Br_2-doped $(SN)_x$ is now considered to be a three-dimensionally coupled superconductor. This result also indicates that some of the bromine must lie between fibers and it interacts strongly with the $(SN)_x$ chain. Br_3^- ions residing around fibers may be one possibility.

A tunnelling investigation of superconducting $(SN)_x$ has been studied by Binning *et al.* [70] with $(SN)_x$/In and $(SN)_x/(SN)_x$ junctions. The Josephson effect is observed in $(SN)_x$/In junctions. From the differential tunnelling conductance of an $(SN)_x/(SN)_x$ tunnelling junction, the superconducting gap Δ was estimated to be 80 ± 10 μeV. Namely, $2\Delta/k_BT_c = 6.5 \pm 1$. This is considered to be the energy gap $\Delta_\parallel$ for tunnelling parallel to the fiber axis. On the other hand, a transverse gap $\Delta_\perp$ is unobserved. This Δ/k_BT_c ratio is quite large for a superconductor. Generally, a large Δ/k_BT_c is observed in strong coupling superconductors, but this is ruled out in $(SN)_x$ because of the small ratio of $T_c/\Theta_D \sim 0.002$. This contradiction is explained by the reduction of T_c by keeping $\Delta(0)$ unchanged. However, the details are not clear at this stage.

In conclusion, the superconductivity of $(SN)_x$ is not related to the excitonic theory of superconductivity proposed by Little [88] but can be explained by BCS theory [89]. Namely, $(SN)_x$ is a highly anisotropic type II superconductor being influenced strongly by the weakly coupled fibrous structure of crystalline $(SN)_x$.

6. Galvanomagnetic Effects of $(SN)_x$ and $(SNBr_y)_x$

6.1. Normal Positive Magnetoresistance in $(SN)_x$

Measurements of galvanomagnetic effects such as magnetoresistance (MR) and Hall effect are useful methods to clarify the mechanism of carrier transport. However, the fibrous structure of crystalline $(SN)_x$ causes difficulties in the analysis of the experimental data. Several reports [29, 38, 57, 63–65, 90, 91] on MR and Hall effect have appeared with a qualitative discussion.

Figure 22(a) and (b) shows the temperature dependences of the transverse MR at 1.7~290 K and below 1 K, respectively, in which the configuration of the magnetic field B, current J and the b axis was $B \perp J \parallel b$. MR is defined by $\Delta\rho/\rho_0 = (\rho(B) - \rho_0)/\rho_0$, where $\rho(B)$ and ρ_0 are resistivity under the magnetic field B and at zero field, respectively. A quadratic dependence of $\Delta\rho/\rho_0$ on B was observed at temperatures between 77 and 290 K throughout the applied magnetic field. Below 4.2 K, the negative MR at lower magnetic field and the positive MR with quadratic dependence of MR at higher magnetic field were observed as shown in Figure 22(a).

Carrier mobility can be estimated from these quadratic dependences by the relation of $\Delta\rho/\rho_0 \propto \mu_{eff}^2 B^2$, where μ_{eff} is the effective MR mobility. In the case of anisotropic material, μ_{eff} may be expressed by $(\mu_\parallel \mu_\perp)^{1/2}$, where $\mu_\parallel$ and $\mu_\perp$ are mobilities parallel and perpendicular to the high-conducting axis, respectively. Assuming that conductivities $\sigma_\parallel$ and $\sigma_\perp$, obtained from the dc measurement, can be given by the simple relation $\sigma_\parallel = ne\mu_\parallel$ and $\sigma_\perp = ne\mu_\perp$ (n; carrier density, in this case the carrier should be the electron, judging from the negative thermopower.), $\mu_\parallel$, $\mu_\perp$ and n are estimated and summarized in Table I. The carrier density n is found to be the order of 10^{19} cm^{-3} which is considerably smaller, when compared with the theoretical prediction [38, 51]. This may be due to the apparent small value of observed $\sigma_\perp$ originating from the fibrous structure of $(SN)_x$.

TABLE I

Electrical conductivities of $\sigma_\parallel$ and $\sigma_\perp$, and carrier mobilities μ_{eff}, $\mu_\parallel$ and $\mu_\perp$ in $(SN)_x$ crystal obtained from dc and MR measurements.

Temp. (K)	290	77	4.2	1.7
$\sigma_\parallel$ (S/cm)	1.5×10^3	3×10^4	1×10^5	1×10^5
$\sigma_\perp$	5	6	3	3
μ_{eff}	40	60	120	300
$\mu_\parallel$ (cm^2/Vsec)	6.9×10^2	4.2×10^3	2.2×10^4	5.5×10^4
$\mu_\perp$	2.3	0.85	0.66	1.6

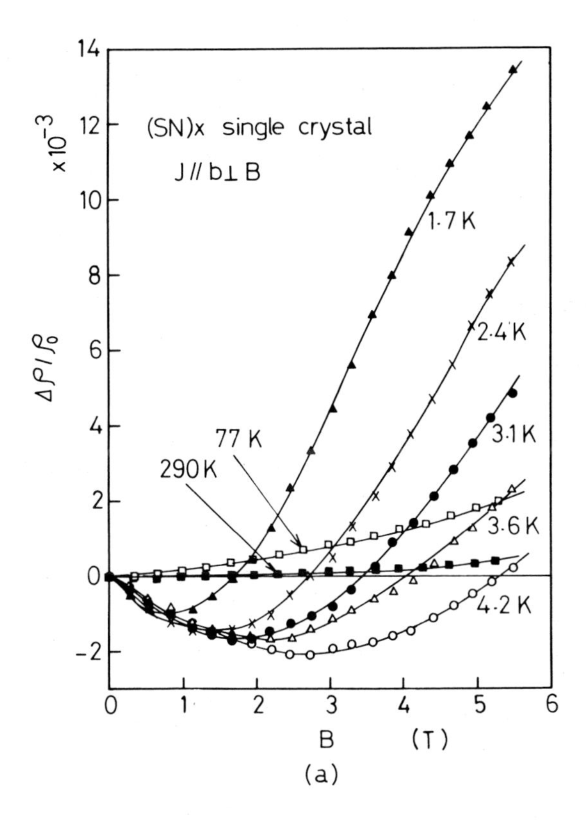

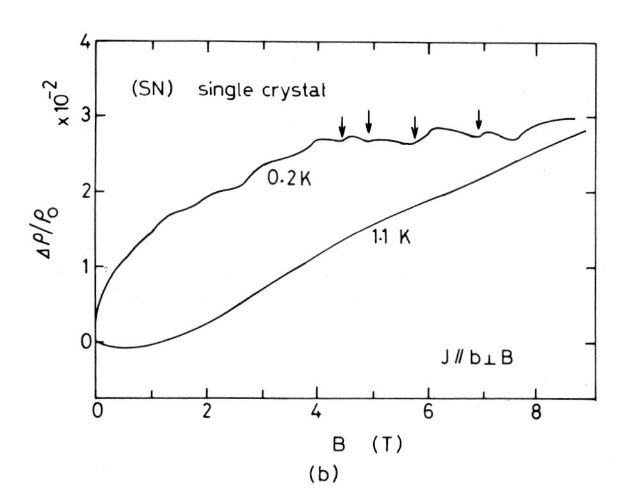

Fig. 22. Magnetic field B dependence of $\Delta\rho/\rho_0$ at various temperatures, (a) $T \geqslant 1.7$ K, (b) $T \leqslant 1.1$ K in $(SN)_x$ crystal for $J \parallel b \perp B$ configuration.

Below 2 K, at higher magnetic fields, MR deviated from the quadratic dependence, indicating clear saturation with several shoulders at 0.2 K, as is shown in Figure 22(b). Whether this saturation of MR has its origin either in the boundary effect of the thin fiber bundle or in the intrinsic nature of the $(SN)_x$ crystal is not known. However, saturation with several oscillation signals may suggest that the Fermi surface is closed [92], together with Shubinkov de Haas oscillations.

Although this oscillation seems to be complicated, tentative analysis gives periodic peaks against B^{-1} with the fraction of reciprocal magnetic field $\Delta(1/B)$ of *ca.* 3×10^{-2} tesla^{-1}. Since $\Delta(1/B) = 2\pi e/\hbar S_F$, where $\hbar$ and S_F are the Planck constant and the extremal area of cross section of the Fermi surface, respectively. Hence, S_F is estimated to be 3×10^{13} cm^{-2}. This is considerably smaller compared with the theoretical value [6] of $S_F \simeq 3 \times 10^{14}$ cm^{-2} for the innermost electron or hole pocket derived from Figure 7(b).

Transverse MR depends on the magnetic field direction rotated within the *ac* plane at $J \parallel b$ configuration, as shown in Figure 23. The relation between the applied magnetic field direction and the crystal axes within *ac* plane is illustrated in the inset of Figure 23, where the crystal axes are determined by the X-ray diffraction technique. The

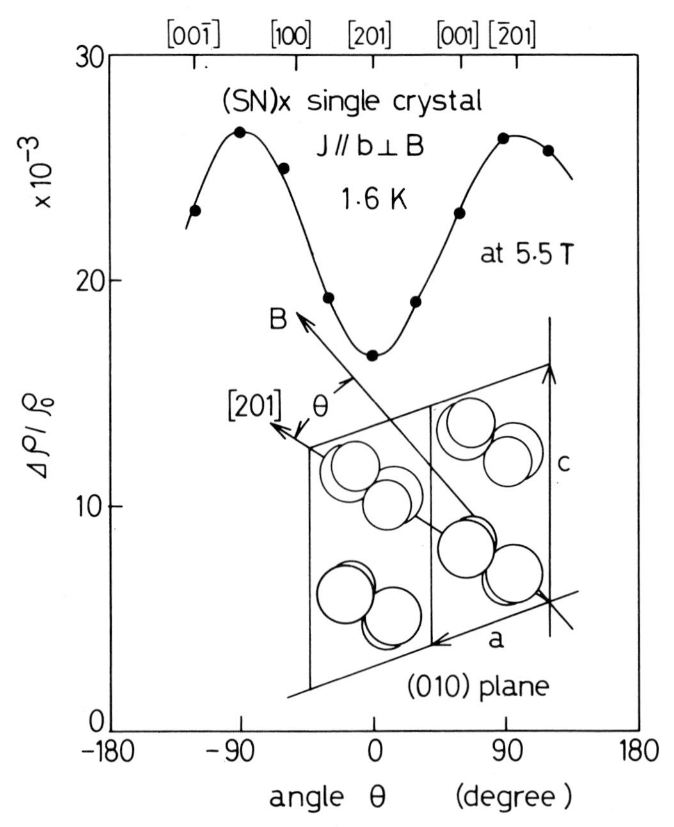

Fig. 23. Angular dependence of $\Delta\rho/\rho_0$, B rotated within the *ac* plane at 1.6 K and 5.5 T for $J \parallel b$ in $(SN)_x$ crystal.

maximum and minimum MRs were observed periodically at the magnetic field direction normal and parallel to $(\bar{1}02)$ plane, respectively, indicating the anisotropy of effective mobility μ_{eff} within the *ac* plane. Namely, the carrier mobility along the [201] direction is slightly larger than that of $[\bar{2}01]$ direction by 25~30% [65]. This result suggests that the interchain coupling is larger in the $(\bar{1}02)$ plane with shorter interchain distance compared with (102) plane, being consistent with the theoretically proposed tubular Fermi surface of ellipsoidal cross-section [6, 38].

The anisotropy of MR was also measured by Beyer *et al.* [91] for the configuration of $J \perp b$ and the magnetic field rotated in the plane normal to the *a* axis. The anisotropy ratio of μ_{eff}^2 along the *b* axis to that of normal to the *ab* plane was found to be 3.1, being comparable with the calculated ratio of 5.7 [91].

6.2. NEGATIVE MAGNETORESISTANCE IN $(SN)_x$

The negative magnetoresistance of $(SN)_x$ crystal observed below 4.2 K and lower magnetic field (<3T) is probably associated with defects, since the mechanical deformation [57] enhances the negative MR. Negative MR is also closely related to the resistivity mimimum against temperature observed at around 30 K in low quality samples ($\sigma_{4.2}/\sigma_R < 50$).

As shown in Figure 22, the initial slope of $\Delta\rho/\rho_0$ becomes steeper with decreasing temperature, and negative MR shows saturation at relatively low magnetic fields, as in doped semiconductors [93, 94]. According to Toyozawa's theory [93], the negative MR in doped semiconductor below 4.2 K can be expressed by the equation,

$$\Delta\rho/\rho_0 = -f\left[\left(\frac{CB}{T+\Theta}\right)^2\right] + \mu_{eff}^2 B^2, \tag{6-1}$$

where f is a function which saturates at relatively low magnetic field, C and Θ are constants depending on the concentration of magnetic spins. The first and second terms of Equation (6-1) correspond to the negative and the normal positive components of MR, respectively. For the low field limit, the coefficient S of the negative MR is given by the equation [94],

$$S = \lim_{B \to 0} \frac{1}{B^2} f\left[\left(\frac{CB}{T+\Theta}\right)^2\right] \simeq \left(\frac{C}{T+\Theta}\right)^2 \tag{6-2}$$

and S is related to the anti-ferromagnetic susceptibility χ [94], as given by the equation:

$$S^{-1/2} \propto \frac{T+\Theta}{C} \propto \chi^{-1}. \tag{6-3}$$

Plots of negative MR component or $(\Delta\rho/\rho_0 - \mu_{eff}^2 B^2)$ against B^2 at low magnetic field (below 2.5 T) are shown in Figure 24. The coefficient S can be obtained from the initial slope of those curves. Figure 25 shows $S^{-1/2}$ against temperature for different samples. The straight line corresponds to the Curie–Weiss law of antiferromagnetism (Equation 6-3) with Θ = 2 K, which is obtained from the intercept of lines on the horizontal axis in Figure 25. The saturation of the negative MR corresponds to the saturation of magnetization of localized magnetic spins [94].

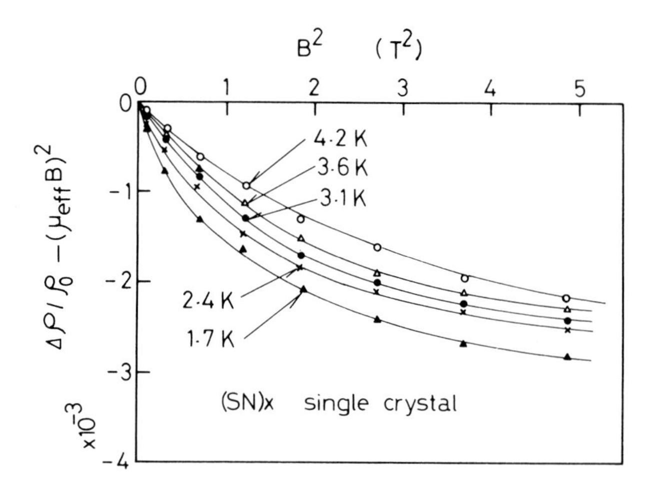

Fig. 24. Negative component of MR *vs.* B^2 for various temperatures in $(SN)_x$ crystal at $J \parallel b \perp B$.

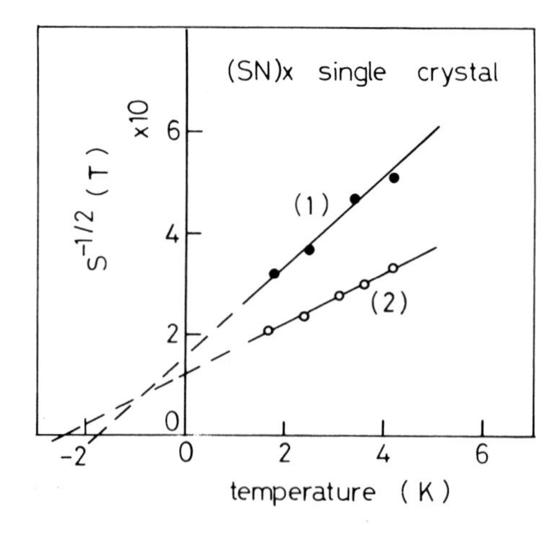

Fig. 25. Temperature dependence of $S^{-1/2}$ for different samples (1) lower and (2) higher concentration of magnetic spins.

The existence of localized magnetic spins at defects in the $(SN)_x$ crystal is confirmed by direct observation of the Curie–Weiss behaviour of the magnetic susceptibility in the γ-ray irradiated sample [58] and by EPR [95] measurements. For the γ-ray irradiated sample ($\sim 10^7$ Röntgen from a Co^{60} source), the magnetic susceptibility followed the Curie–Weiss law and gave a magnetic spin density of 2.5×10^{20} cm^{-3}.

EPR measurements were carried out by Cohen *et al.* [14] and by Love *et al.* [95] to investigate the polymerization process of $(SN)_x$. The EPR signal first increased and then decreased during the polymerization process [14], indicating the formation of free radical S_2N_2 at an intermediate stage of the polymerization. Love *et al.* [95] claimed that, under certain conditions, the completion of polymerization took more than several months, and that a large number of localized spins were still observed in the crystalline $(SN)_x$ and in the sublimed thin film.

An alternative explanation for the negative MR may be qualitatively possible as follows. The magnetic field assists the carrier hopping to neighbouring chains at chain breaks or defects by the transverse Lorentz force, where the thermal energy of the carrier is not sufficient, at lower temperatures, for hopping to the next chain, resulting in the negative MR at relatively lower magnetic field, followed by the normal positive MR.

6.3. Hall Effect in $(SN)_x$

Experimental results on the Hall effect do not contradict those of MR measurements to any great extent, although the fibrous structure of $(SN)_x$ crystal complicates their interpretation. Figure 26 shows the typical result of induced Hall voltage against the magnetic

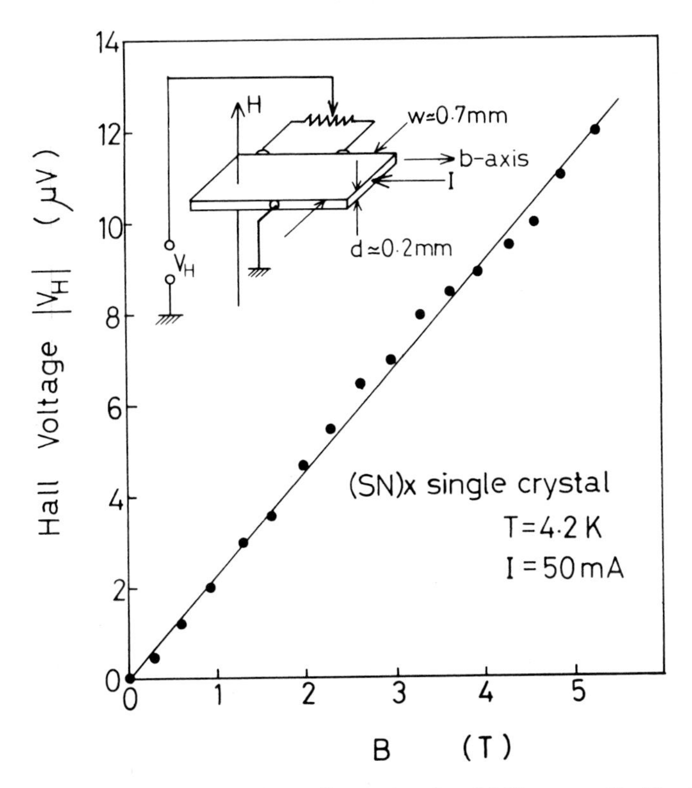

Fig. 26. Induced Hall voltage $|V_H|$ against B at 4.2 K, current I is 50 mA.

field at 4.2 K. In this experiment, the Hall coefficient R_H was -9×10^{-2} cm^3/coul, corresponding to an electron density of *ca.* 7×10^{20} cm^{-3}. From the simple relation of $\mu_H = R_H\sigma_{\parallel}$, the Hall mobility μ_H at 4.2 K was estimated to be 180 cm^2/V sec, being comparable to the value of μ_{eff} obtained by the MR measurement (Table I). However,

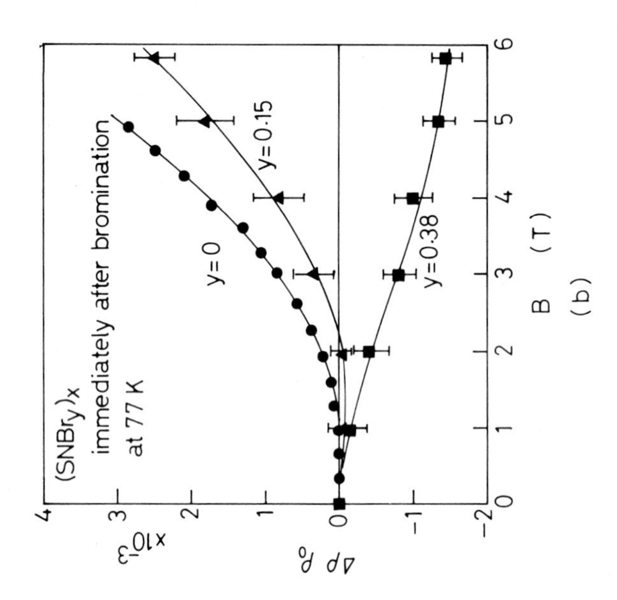

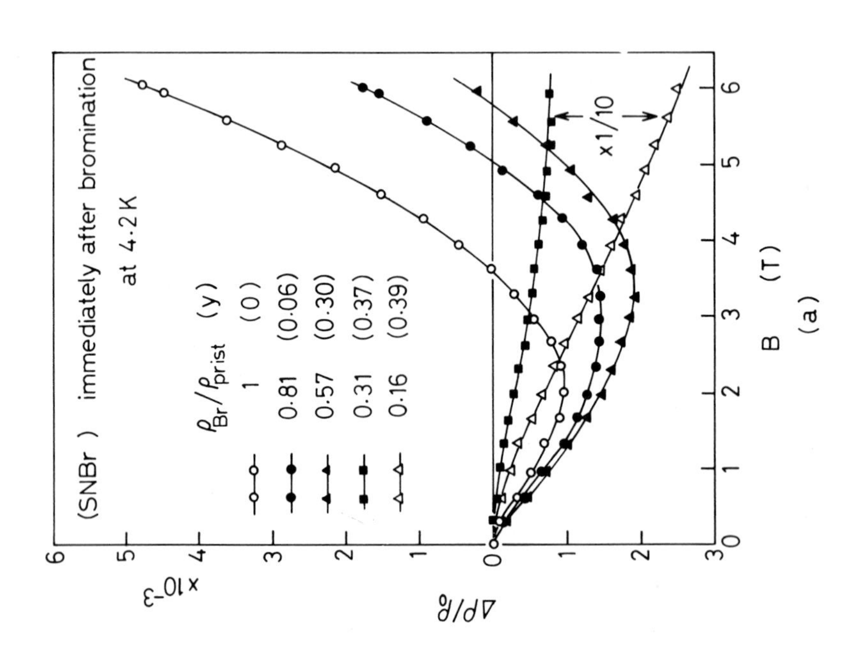

Fig. 27. $\Delta\rho/\rho_0$ *vs.* magnetic field B in $(SNBr_y)_x$ immediately after bromination for various y, (a) at 4.2 K (b) at 77 K.

R_H was remarkably sample-dependent and varied over two orders of magnitude [65].

Both Hall effect and MR in a thin film of $(SN)_x$ were also measured by Beyer *et al.* [29]. They obtained a positive Hall coefficient, which suggests that the dominant carriers are holes, with a concentration of *ca.* 10^{21} cm^{-3} for thin films prepared on suitable substrates at various temperatures. The remarkable difference of carrier sign between the single crystal and thin film may indicate the semimetal character of $(SN)_x$.

6.4. Magnetoresistance in $(SNBr_y)_x$

The magnetoresistance of $(SNBr_y)_x$ crystals does not simply show the enhancement of the normal positive component, which is to be expected from the increased conductivity, as discussed in Section 4.3, but indicates the enhanced negative MR [7].

Curves in Figure 27(a) show the transverse MR of $\Delta\rho/\rho_0$ in $(SNBr_y)_x$ against B at the configuration of $B \parallel b$ at 4.2 K for various bromine contents y immediately after bromination. The negative MR increased remarkably with increasing y; for example, the negative $\Delta\rho/\rho_0$ value in $(SNBr_{0.39})_x$ was larger by 10~20 times than that of the pristine $(SN)_x$ at *ca.* 2 T as shown in Figure 27(a). The MR of $(SNBr_y)_x$ at 77 K also shows apparently negative characteristic, as shown in Figure 27(b) for a heavily brominated sample, although the negative component was not observed at all for the pristine $(SN)_x$, as was stated in Section 6.1.

The brominated $(SN)_x$, which was kept for several days under reduced pressure, showed slight recovery in the positive MR, due to the decrease of bromine content, indicating no permanent decomposition of $(SN)_x$ upon bromination [7]. The longitudinal MR of $(SNBr_y)_x$ measured for the $J \parallel b \parallel B$ configuration at 4.2 K showed a comparable magnitude of the negative MR to the transverse MR [7].

Although there have been many discussions [96, 97] on the origin of negative MR, for heavily doped semiconductors [93, 94], glassy carbon [98, 99] and silicon inversion layers [100], at the present time it is not clear which mechanism plays the most important role for the appearance of the negative MR in $(SNBr_y)_x$. The large Curie–Weiss behaviour at low temperatures was observed by the static susceptibility measurement [101] of $(SNBr_y)_x$, which may account for the enhanced negative MR. The enhanced interchain interaction or increased 3-D behaviour upon bromination may be related to the negative MR too.

7. Optical Properties of $(SN)_x$ and $(SNBr_y)_x$

7.1. Reflectance spectra in $(SN)_x$

Both single crystal and thin film $(SN)_x$ display a shiny golden or brassy reflection, indicating the metallic nature, or the plasma reflection by free carriers. This metallic behaviour is remarkable for light polarized parallel to the b axis ($E \parallel b$) compared with the perpendicular direction ($E \perp b$). This obviously suggests the intrinsic anisotropy of electronic properties in single crystals, in directions parallel and perpendicular to the b axis. Although dc $\sigma_\perp$ is restricted by fiber bundles, high frequency measurement in the visible light region will be less influenced by the interbundle boundary.

The measurements of optical reflectance for the crystalline $(SN)_x$ were reported by Pintschovius *et al.* [102], Grant *et al.* [60] and Kaneto *et al.* [103]. Reflectance spectra for the thin film were presented by Kamimura *et al.* [51] and for the orientated thin film by Bright *et al.* [23, 104] and by Möller *et al.* [28].

In the observed reflectance spectra at $E \parallel b$ as shown in Figure 28, reflectance increases steeply below the photon energy at *ca.* 2.7 eV, being consistent with other workers' reports. However, the accurate measurement of absolute reflectance in the crystal was fairly difficult, due to light scattering by the fiber boundary. By selecting a smooth surface, polishing the $(SN)_x$ crystal [102], or using the orientated thin film, the reflectance obtained experimentally at around 2 eV ranged around 50~70% for $E \parallel b$.

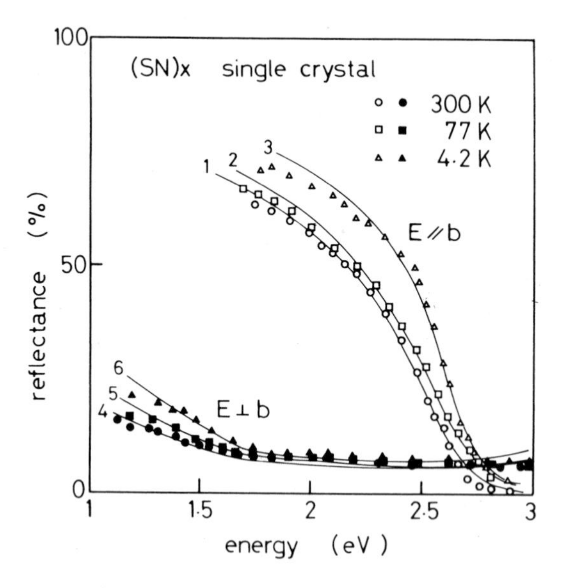

Fig. 28. Reflectance spectra of $(SN)_x$ for $E \parallel b$ and $E \perp b$ at various temperatures, numbered solid curves represent the calculated curves derived from Table II.

The reflectance spectra for the $E \perp b$ were not indicative of the strong plasma edge, as in the case of $E \parallel b$. However, some increase of reflectance below *ca.* 1.5 eV is observed [28, 60, 103] for the $E \perp b$, as shown in Figure 28 and seems to be due to the weak plasma reflection.

The Drude approximation of the dielectric constant $\epsilon(\omega)$ in metals is currently adopted to analyze the reflectance spectra at around the plasma edge for both $E \parallel b$ and $E \perp b$. The dielectric function,

$$\epsilon(\omega) = \epsilon_{\text{core}} - \frac{\omega_p^2}{\omega^2 + i\omega/\tau} = \epsilon_1 + i\epsilon_2 \tag{7-1}$$

is used to calculate the reflectance of

$$R = \frac{1 + |\epsilon| - [2(|\epsilon| + \epsilon_1)]^{1/2}}{1 + |\epsilon| + [2(|\epsilon| + \epsilon_1)]^{1/2}}, \tag{7-2}$$

where ϵ_{core} is the dielectric constant of the core polarizability, τ is the relaxation time of the free carrier and ω_p is the plasma frequency,

$$\omega_p = 4\pi ne^2/m^*m_e, \qquad |\epsilon| = (\epsilon_1^2 + \epsilon_2^2)^{1/2},$$

and m^* is the effective mass.

Although a detailed analysis of reflectance spectra was carried out [60] using the two carrier model, an essentially similar result was obtained.

The Drude parameters of ϵ_{core}, ω_p and τ were determined by fitting the calculated curve of Equation (7-2) to the experimental curves in Figure 28 at various temperatures, and are summarized in Table II. In Figure 28, numbered solid curves correspond to calculated curves using the parameters of Table II.

The obtained values of ω_p in crystalline $(SN)_x$ and orientated thin film $(SN)_x$ agree fairly well, ranging around $8.3 \sim 10.5 \times 10^{15}$ sec^{-1}. The relaxation time of $\tau = 1 \sim 4 \times 10^{-15}$ sec and the conductivity given by $\omega_p^2\tau/4\pi$ at optical frequencies of the order of 10^4 S/cm can be deduced from Table II, being about one order of magnitude larger than that of dc $\sigma_{\parallel}$.

For the $E \perp b$ orientation, $\omega_p \simeq 4.3 \times 10^{15}$ sec^{-1}, $\hbar\omega_p/\epsilon_{core}^{1/2} \simeq 1.5$ eV and $\tau \simeq 3.6 \times 10^{-15}$ sec at room temperature are average values. Judging from parameters in Table II, the mass anisotropy of $m^*_{\perp}/m^*_{\parallel}$ at optical frequencies is *ca.* 6, which is discussed by Greene *et al.* [38] comparing it with their calculated values of 8 based on an OPW model.

The reflectance spectra both for $E \parallel b$ and $E \perp b$ show a blue shift with decreasing temperature, as shown in Figure 28, which should be responsible for the decrease of ϵ_{core} and/or increase of ω_p. The parameters fixing ϵ_{core} are listed in Table II and also those of fixed ω_p in parentheses [103]. The τ obtained also shows a temperature dependence; however, the increase of τ by a factor of 1.5 from 300 K to 4.2 K is much smaller than the expected value of $10^2 \sim 10^3$ from the temperature dependence of dc $\sigma_{\parallel}$. Detailed mechanisms of these temperature dependence are not known.

Electron energy loss spectroscopy (ELS) was performed by Chen *et al.* [105] to clarify the plasmon dispersion and anisotropy. Angular dependence of the plasmon energy from the wave vector parallel to or perpendicular with respect to the b axis showed a variation of plasmon energies from 2.5 to 1.5 eV. These energies are in good agreement with the optical reflectance results. The relatively large plasmon energy observed in the perpendicular direction to the b axis was interpreted as the evidence that the $(SN)_x$ is a 2- or 3-D metal rather than a 1-D metal.

This experimental result was analyzed by Ruvalds *et al.* [106] in terms of the mass anisotropy in the electron gas. When the mass anisotropy is assumed to be 1.9, the anisotropy of plasmon energy is well explained. However, the mass anisotropy obtained by this method is considerably smaller than that estimated from the optical reflectance data [60]. This discrepancy is attributed to the core polarizability which is neglected for the calculation using electron gas model.

7.2. REFLECTANCE SPECTRA IN $(SNBr_y)_x$

The reflectance spectra of $E \parallel b$ and $E \perp b$ were remarkably changed in the plasma edge, with a red shift caused by the bromination [8, 31, 32] as shown in Figure 29 in

TABLE II

Summary of Drude parameters from reflectance spectra for $E \parallel b$ and $E \perp b$ in $(SN)_x$ by various workers.

	No. or workers	Temp (K)	ϵ_{core}	ω_p ($\times 10^{15}$ sec^{-1})	$\hbar\omega_p/\epsilon_{core}^{1/2}$ (eV)	τ ($\times 10^{-15}$ sec)	$n/(m^*/m_e)$ ($\times 10^{22}$ cm^{-3})	σ_{opt} ($\times 10^3$ S/cm)
$E \parallel b$	1	300	4.4	8.3	2.16	1.3	2.2	7.8
	2	77	4.4	8.5	2.67	1.4	2.3	8.9
	3	4.2	4.4	8.6	2.70	1.9	2.3	16
			(4.2)	(8.3)	(2.67)	(2.2)	(2.2)	(13)
	[a] Bright *et al.* [104]	room	1.46	4.2	2.3	1.9	0.58	3
	(reexamined by Grant *et al.* [60])	room	4.1	7.7	2.5	1.2	1.9	5.7
	[a] Kamimura *et al.* [51]	room	6.8	10.3	2.6	1.5	3.3	15
	Pintschovius *et al.* [102]	room	6.5	10.5	2.7	3.5	3.4	34
	Grant *et al.* [60]	room	6.5	10.5	2.7	2.6	3.4	25
	[b] Möller *et al.* [28]	room	6.5	10.4	2.7	2.6	3.4	25
$E \perp b$	4	300	4.2	4.5	1.5	0.48	0.68	0.85
	5	77	4.2	4.9	1.8	0.50	0.75	1.1
	6	4.2	4.2	5.2	1.9	0.52	0.85	1.2
	Grant *et al.* [60]	room	3.6	3.6	1.3	0.33	0.42	0.38
	[b] Möller *et al.* [28]	room	4.0	4.8	1.6	0.26	0.74	0.54

[a] Thin film
[b] Orientated thin film.

TABLE III

Drude parameters in $(SNBr_y)_x$ for $E \parallel b$ and $E \perp b$.

	workers	y	ϵ_{core}	ω_p ($\times 10^{15}$ sec^{-1})	$\hbar\omega_p/\epsilon_{core}^{1/2}$ (eV)	τ ($\times 10^{-15}$ sec)	$n/(m^*/m_e)$ ($\times 10^{22}$ cm^{-3})	σ_{opt} ($\times 10^3$ S/cm)
	Tanimura [32]	0.25	5.0	7.1	2.1	0.92	1.6	3
$E \parallel b$	Gill *et al.* [8]	0.4	6.48	7.6	2.0	1.2	1.8	5.5
	Chiang *et al.* [31]	0.4	8.7	7.3	1.6	1.6	1.7	6.8
$E \perp b$	Tanimura [32]	0.25	4.5	2.3	0.7	1.1	0.16	0.46

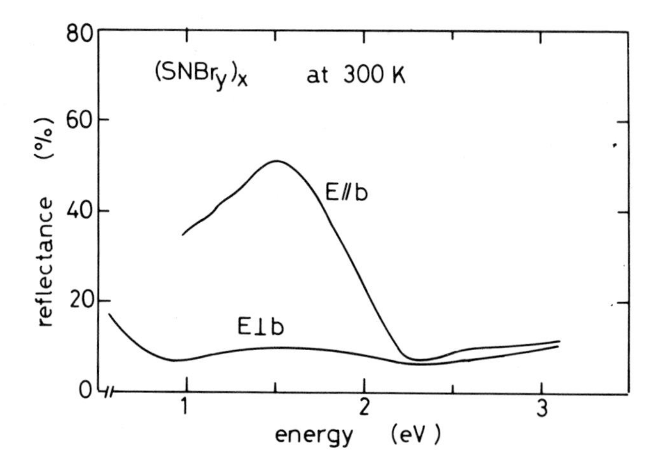

Fig. 29. Reflectance spectra of $(SNBr_y)_x$ for $E \parallel b$ and $E \perp b$ at room temperature.

comparison with Figure 28. The Drude parameters of brominated $(SN)_x$ are summarized in Table III with reported values by Gill *et al.* [8] and by Chiang *et al.* [31].

The red shift of the plasma edge upon bromination should be due to the increased ϵ_{core} [8, 31, 32]. By the bromination, in spite of remarkable increase of dc $\sigma_{\parallel}$ by more than one order of magnitude, the evaluated optical $\sigma_{\parallel}$ shows a decrease of several times. This fact, and unchanged τ upon bromination, should definitely suggest that the scattering mechanisms of carrier and/or kind of mobile carriers (electron or hole) are not identical between the optical frequency and dc regime.

References

1. V. V. Walatka Jr., M. M. Laves and J. H. Perlstein, *Phys. Rev. Lett.* **31**, 1139 (1973). C. H. Hsu and M. M. Labes; *J. Chem. Phys.* **61**, 4640 (1974).
2. C. M. Mikulski, P. J. Russo, M. S. Saran, A. G. MacDiarmid, A. F. Garito and A. J. Heeger, *J. Am. Cem. Soc.* **97**, 6358 (1975).
3. R. L. Greene and G. B. Street and L. J. Suter, *Phys. Rev. Lett.* **34**, 577 (1975).
4. H. Kamimura, A. M. Glazer, A. J. Grant, Y. Natsume, M. Scheiber and A. D. Yoffe, *J. Phys. C, Solid State Phys.* **9**, 291 (1976).
5. W. E. Ruge and P. M. Grant, *Phys. Rev. Lett.* **35**, 1799 (1975).
6. A. Oshiyama, H. Kamimura, *J. Phys. C, Solid State Phys.* **14**, 5091 (1981).
7. K. Kaneto, S. Sasa, K. Yoshino and Y. Inuishi, *J. Phys. Soc. Japan* **49**, 1902 (1982).
8. P. C. Bernard, A. Herold, M. Lelaurian and G. Robert, *C. R. Acad. Sci. Paris* C283, 625 (1976); W. D. Gill, W. Bludan, R. H. Geiss, P. M. Grant, R. L. Greene, J. J. Mayerls and G. B. Street, *Phys. Rev. Lett.* **38**, 1035 (1977).
9. F. P. Burt, *J. Chem Soc.*, 1171 (1910).
10. M. Boudeulle, A. Dovillard, P. Michel, and G. Vallet, *C. R. Acad., Sci. Ser. C.* **272**, 2137 (1971).
11. G. B. Street, H. Arnal, W. D. Gill, P. M. Grant and R. L. Greene, *Mat. Res. Bull.* **10**, 877 (1975).
12. M. M. Labes, P. Love and L. F. Nicholos, *Am. Chem. Soc. Rev.* **79**, 1 (1979).
13. K. Kaneto, K. Tanimura, K. Yoshino and Y. Inuishi, *Oyo Butsuri* **46**, 299 (1977) [in Japanese]. *Technol. Repts. Osaka Univ.* **26**, 455 (1976). K. Yoshino, K. Kaneto and Y. Inuishi; *Proceeding of 8th Mol. Cryst. Symp.*, p. 424, (1977).

14. A. G. MacDiarmid, C. M. Mikulski, P. J. Russo, M. S. Saran, A. F. Garito and A. J. Heeger, *J. Chem. Soc. Chem. Comm.*, 476 (1975). M. J. Cohen, A. F. Garito, A. J. Heeger, A. G. MacDiarmid, C. M. Mikulski, M. S. Saran and J. Kleppinger, *J. A. Chem. Soc.* **98**, 3844 (1976).
15. R. H. Baughman, R. R. Chance and M. J. Cohen, *J. Chem. Phys.* **64**, 1869 (1976).
16. M. H. M. Arnold, J. A. C. Hugill and J. M. Hutson, *J. Chem. Soc.*, 1645 (1936).
17. M. V. Blanco and W. L. Jolly, *Inorg. Syn.* **9**, 98 (1967).
18. J. Bragin and M. V. Evans, *J. Chem. Phys.* **51**, 268 (1969).
19. H. Karlart and B. Kundu, *Mat. Res. Bull.* **11**, 967 (1976).
20. G. B. Street and R. L. Greene, *IBM J. Res. and Develop.* **21**, 98 (1977).
21. C. K. Chiang, M. J. Cohen, A. F. Garito, A. J. Heeger, C. M. Mikulski and A. G. MacDiarmid, *Solid State Comm.* **18**, 1451 (1976).
22. M. Boudeulle, *Cryst. Struct. Comm.* **4**, 9 (1975). Ph.D Thesis, Claude Bernard Univ., Lyon (1974).
23. A. A. Bright, M. J. Cohen, A. F. Garito and A. J. Heeger, *Appl. Phys. Lett.* **26**, 612 (1975).
24. K. Yoshino, Y. Yamamoto, K. Tanimura, K. Kaneto and Y. Inuishi, *Japan. J. Appl. Phys.* **18**, 841 (1979).
25. N. Ueno, K. Sugita, O. Koga, S. Suzuki, K. Yoshino, K. Kaneto and Y. Inuishi, *Japan. J. Appl. Phys.* **18**, 1597 (1979).
26. R. D. Smith, J. R. Wyatt, J. J. Decorpo, F. E. Saarfelt, M. J. Moran and A. G. MacDiarmid, *Chem. Phys. Lett.* **41**, 362 (1976).
27. E. J. Louis, A. G. MacDiarmid, A. F. Garito and A. J. Heeger, *J. Chem. Soc. Chem. Comm.*, 426 (1976).
28. W. Möller, H. P. Geserich and L. Pintschovius, *Solid State Comm.* **18**, 791 (1976).
29. W. Beyer, W. D. Gill and G. B. Street, *Solid State Comm.* **27**, 343 (1978). W. Beyer, H. Mell and W. D. Gill, *Solid State Comm.* **27**, 185 (1978).
30. K. Yoshino, S. Ura, S. Sasa, K. Kaneto and Y. Inuishi, *Japan. J. Appl. Phys.* **21**, L507 (1982).
31. C. K. Chiang, M. J. Cohen, D. L. Peebles, A. J. Heeger, M. Akhtar, J. Kleppinger, A. G. MacDiarmid, J. Milliken and M. J. Moran, *Solid State Comm.* **23**, 607 (1977).
32. K. Tanimura; Thesis, Master's Degree, Osaka Univ. (1978).
33. J. F. Kwak, R. L. Greene and W. W. Fuller, *Phys. Rev. B* **20**, 2658 (1979).
34. Z. Iqbal, R. H. Baughman, J. Kleppinger and A. G. MacDiarmid, *Solid State Comm.* **25**, 409 (1978).
35. H. Temkin and G. B. Street, *Solid State Comm.* **25**, 455 (1978).
36. G. B. Street and W. D. Gill, *Molecular Metals*, W. E. Hatfield, ed. (Plenum Publishing Corp. 1979), p. 301.
37. K. Kaneto, S. Sasa, K. Yoshino. Y. Inuishi and T. Yamabe, *Solid State Comm.* **40**, 889 (1981).
38. R. L. Greene and G. B. Street, *Chemistry and Physics of One-Dimensional Metals*, H. J. Keller, ed. (Plenum Press, 1977).
39. D. E. Parry and J. M. Thomas, *J. Phys. C, Solid State Phys.* **8**, L45 (1975).
40. W. I. Friesen, A. J. Berlinsky, B. Bergersen, L. Weiler and T. M. Rice, *J. Phys. C, Solid State Phys.* **8**, 3549 (1975).
41. A. A. Bright and P. Soven, *Solid State Comm.* **18**, 317 (1976).
42. D. R. Salahub and R. P. Messmer, *Phys. Rev. B* **14**, 2592 (1976).
43. R. P. Messmer and D. R. Salahub, *Chem. Phys. Lett.* **41**, 73 (1976).
44. L. Mihich, *Sold State Comm.* **28**, 52 (1978).
45. V. T. Rajan and L. M. Falicov, *Phys. Rev. B* **12**, 1240 (1975).
46. M. Schlüter, J. R. Chelikowsky and M. L. Cohen, *Phys. Rev. Lett.* **35**, 869 (1975).
47. A. Oshiyama, Ph.D. Thesis, University of Tokyo (1980).
48. A. Oshiyama and H. Kamimura, *Solid State Phys.* **16**, 317 (1981) [in Japanese].
49. R. E. Peierls, *Quantum Theory of Solids*, (Clarendon Press, Oxford, 1955) Chap. 5.
50. A. J. Epstein and E. M. Conwell (eds.) *Molecular Crystals and Liquid Crystals*, p. 77 (1981) part A.
51. H. Kamimura, A. J. Grant, F. Lévy, A. D. Yoffe and G. D. Pitt, *Solid State Comm.* **17**, 49 (1975).

52. P. Mengel, P. M. Grant, W. E. Rugde, B. H. Schechtman and D. W. Rice, *Phys. Rev. Lett.* **35**, 1803 (1975).
53. L. Ley, *Phys. Rev. Lett.* **35**, 1796 (1975).
54. R. L. Greene, P. M. Grant and G. B. Street, *Phys. Rev. Lett.* **34**, 89 (1975).
55. K. Kaneto, K. Yoshino and Y. Inuishi, *Proceeding of Symp. on Design of Inorg. and Org. Materials of Technological Importance*, Kyoto (1979).
56. P. L. Kronick, H. Kaye, E. F. Chapman, S. B. Mainthia and M. M. Labes, *J. Chem. Phys.* **36**, 2235 (1962).
57. H. Kahlert and K. Seeger, *Proc. 13th Int. Conf. Phys. Semicond.* Rome (Tipographia Marves 1976) p. 353.
58. K. Kaneto, K. Tanimura, K. Yoshino and Y. Inuishi, *Solid State Comm.* **22**, 383 (1977).
59. C. H. Hsu and M. M. Labes, *J. Chem. Phys.* **61**, 4640 (1974).
60. P. M. Grant, R. L. Greene and G. B. Street, *Phys. Rev. Lett.* **35**, 1743 (1975).
61. K. J. Dunn, F. P. Bundy and L. V. Interrante, *Phys. Rev. B* **23**, 106 (1981).
62. J. Kondo, *Prog. Theor. Phys.* **32**, 37 (1964).
63. K. Kaneto, M. Yamamoto, K. Yoshino and Y. Inuishi, *Solid State Comm.* **26**, 311 (1978).
64. M. Yamamoto, K. Kaneto, K. Yoshino and Y. Inuishi, *Solid State Comm.* **29**, 541 (1979).
65. K. Kaneto, M. Yamamoto, K. Yoshino and Y. Inuishi, *J. Phys. Soc. Japan* **47**, 167 (1979).
66. J. M. E. Harper, R. L. Greene, P. M. Grant and G. B. Street, *Phys. Rev. B* **15**, 539 (1977).
67. L. I. Azevedo, P. M. Chaikin, W. G. Clark, W. W. Fuller and J. Hamman, *Phys. Rev. B* **20**, 4450 (1979).
68. K. Tanimura, K. Kaneto, K. Yoshino and Y. Inuishi, *Technol. Repts. Osaka Univ.* **29**, 435 (1979).
69. P. M. Chaikin, P. K. Hansma and R. L. Greene, *Phys. Rev. B* **17**, 179 (1978).
70. G. Binnig and H. E. Hoenig, *Z. Physik B* **32**, 23 (1978).
71. W. Jost, *Diffusion* (Academic Press, 1976) p. 46.
72. C. A. Kukkonen and P. F. Maldague, *Phys. Rev. Lett.* **37**, 782 (1976).
73. L. J. Azevedo, W. G. Clark, G. Deutcher, R. L. Greene, G. B. Street and L. J. Suter, *Solid State Comm.* **19**, 197 (1976).
74. L. R. Bickford, R. L. Greene and W. D. Gill, *Phys. Rev. B* **17**, 3525 (1978).
75. L. A. Turkevich and R. A. Klemm, *Phys. Rev. B* **19**, 2520 (1979).
76. R. H. Dee, A. J. Berlinsky, J. F. Carolan, E. Klein, N. J. Stone and B. G. Turrell, *Sold State Comm.* **22**, 303 (1977).
77. R. H. Dee, D. H. Dollard and B. G. Turrell and J. F. Carolan, *Solid State Comm.* **24**, 496 (1977).
78. R. H. Dee, J. F. Carolan and B. G. Turrell and R. L. Greene, *Phys. Rev. B* **22**, 174 (1980).
79. Y. Oda, H. Yakenaka, H. Nagano and I. Nakada, *Solid State Comm.* **32**, 659 (1979).
80. W. D. Gill, R. L. Greene and G. B. Street and W. A. Little, *Phys. Rev. Lett.* **35**, 1732 (1975).
81. W. H. G. Muller, F. Baumann, G. Dammer and L. Pintschovius, *Solid State Comm.* **25**, 119 (1978).
82. G. Deutcher, Y. Imry and L. Gunther, *Phys. Rev. B* **10**, 4598 (1974).
83. R. L. Civiak, C. Elbaum, W. Junker, C. Gough, L. F. Nichols, H. I. Kao and M. M. Labes, *Solid State Comm.* **18**, 1205 (1976).
84. R. L. Civiak, C. Elbaum, L. F. Nichols, H. I. Kao and M. M. Labes, *Phys. Rev. B* **14**, 5413 (1976).
85. L. G. Aslamazov and A. I. Larkin, *Sov. Phys. Solid State* **10**, 857 (1968).
86. V. L. Ginzgurg and L. D. Landau, *Zh. Eksperim. i. Teo. Fiz.* **20**, 1064 (1950).
87. S. Sasa, K. Kaneto, K. Yoshino and Y. Inuishi, *unpublished*.
88. W. A. Little, *Phys. Rev. A* **134**, 1416 (1964).
89. J. Bardeen, L. N. Cooper and J. R. Schrieffer, *Phys. Rev.* **108**, 1175 (1957).
90. W. Möller, R. Plante, R. Ranvaud, S. Roth and P. Rödhammer, *Solid State Comm.* **19**, 943 (1976).
91. W. Beyer, W. D. Gill and G. B. Street, *Solid State Comm.* **23**, 577 (1977).
92. C. Kittel, *Quantum Theory of Solids* (John Wiley, New York, 1963).
93. Y. Toyozawa, *J. Phys. Soc. Japan* **17**, 986 (1962).

94. W. Sasaki, *J. Phys. Soc. Japan* **21**, suppl. p. 543 (1966).
95. P. Love and M. M. Labes, *J. Chem. Phys.* **70**, 5147 (1979).
96. F. T. Hedgcock and T. W. Raudorf, *Solid State Comm.* **18**, 1819 (1970).
97. S. Fujita, *Phys. Lett.* **39** A, 429 (1972).
98. P. Delhaes, P. De Kepper and M. Uhlrich, *Phil. Mag.* **29**, 1301 (1974).
99. A. A. Bright, *Phys. Rev. B* **20**, 5142 (1979).
100. S. Hikami, A. I. Larkin and Y. Kawaguchi and S. Kawaji, *J. Phys. Soc. Japan* **48**, 699 (1980).
101. J. C. Scott, J. D. Kulick and G. B. Street, *Solid State Comm.* **28**, 723 (1978).
102. L. Pintschovius, H. P. Geserich and W. Möller, *Solid State Comm.* **17**, 477 (1975).
103. K. Kaneto, K. Yoshino and Y. Inuishi, *J. Phys. Soc. Japan* **43**, 1013 (1977).
104. A. A. Bright, M. J. Cohen, A. F. Garito, A. J. Heeger, C. M. Mikulshi, P. J. Russo and A. G. MacDiarmid, *Phys. Rev. Lett.* **34**, 206 (1975).
105. C. H. Chen, J. Silcox, A. F. Garito, A. J. Heeger and A. G. MacDiarmid, *Phys. Rev. Lett.* **36**, 525 (1976).
106. J. Ruvalds, F. Brosens, L. F. Lemmens and J. T. Devresse, *Solid State Comm.* **23**, 243 (1977).

OPTICAL INVESTIGATIONS OF ONE-DIMENSIONAL INORGANIC METALS

H. P. GESERICH

Institut für angewandte Physik, Universität Karlsruhe, Kaiserstr. 12, D–7500 Karlsruhe, Federal Republic of Germany.

1. Optical Properties of Metals

The optical properties of metals are determined in an energy range between zero and a few eV essentially by the absorption of the radiation by free charge carriers. This energy range, beginning at very small frequencies, comprises the range of radio frequencies, the entire range of infrared radiation, and extends over the visible spectrum up into the ultraviolet part of the spectrum. The strong influence of the free carriers in this large energy range manifests itself by a high reflectivity, which leads to the well-known metallic lustre for visible light. Other absorption processes, which may take place somewhere in this energy range, are absorption by infrared-active lattice vibrations with a few hundredth of an eV, and by electronic transitions to the conduction band in the range of several eV. These processes, if not masked altogether, lead, however, merely to certain modifications of the properties as determined by the free carriers.

1.1. THE DIELECTRIC FUNCTION

The optical properties of matter are described by means of the complex dielectric function

$$\epsilon(\omega) = \epsilon_1(\omega) + i\epsilon_2(\omega). \tag{1}$$

The imaginary part $\epsilon_2(\omega)$ represents the dielectric losses. It is connected with the frequency-dependent conductivity $\sigma(\omega)$ by the relation

$$\sigma(\omega) = \epsilon_0 \omega \epsilon_2(\omega) \tag{2}$$

In the limit $\omega \to 0$ the quantity $\sigma(\omega)$ coincides with the dc-conductivity, and $\epsilon_1(\omega)$ becomes the static dielectric constant.

The contributions of the different absorption processes enter additively into the dielectric function. This allows the dielectric function ϵ_M of a metal to be represented by a sum. It consists of a contribution from the vacuum polarization $\epsilon_V = 1$, a contribution from the optically active lattice vibrations $\epsilon_L(\omega)$, a contribution from band-to-band transitions $\epsilon_B(\omega)$, and finally of a dominant contribution from the free charge carriers $\epsilon_{FC}(\omega)$. It results in the sum

$$\epsilon_M(\omega) = 1 + \epsilon_L(\omega) + \epsilon_B(\omega) + \epsilon_{FC}(\omega), \tag{3}$$

in which each of the different contributions themselves are also complex.

P. Monceau (ed.), Electronic Properties of Inorganic Quasi-One-Dimensional Materials, II, 111–138.

1.2. THE LORENTZ–DRUDE MODEL

A comprehensive, and in many cases also sufficiently precise, description of the optical properties of metals, is given by the Lorentz–Drude model. Here the contributions from the lattice vibrations and from the band-to-band transitions are represented by a sum of Lorentz oscillators, and the contribution from the free carriers is given by a Drude term. Optical transitions, the energy of which is sufficiently far above the investigated energy range, furnish at lower energies merely a constant contribution to the real part of the dielectric function. This suggests that one should lump this contribution together with the vacuum polarization into a constant ϵ_∞. It follows from this procedure for the dielectric function $\epsilon_M(\omega)$ of a metal that

$$\epsilon_M(\omega) = \epsilon_\infty + \sum_1^n \frac{\omega_{pn}^2}{\omega_{on}^2 - \omega^2 - i\omega\Gamma_n} + \frac{\omega_p^2 \epsilon_\infty}{\omega^2 + i\omega/\tau} \tag{4}$$

The ω_{pn} denote the strength, the ω_{on} the eigenfrequencies, and the Γ_n the damping coefficients of the Lorentz oscillators [1].

The Drude term contains two parameters for the characterization of the free carrier gas. One of them is the plasma frequency ω_p, defined by the relation

$$\omega_p^2 = \frac{Ne^2}{\epsilon_\infty \epsilon_0 m^*} . \tag{5}$$

It depends on the concentration N and the effective mass m^* of the free charge carriers. The other parameter is the collision time τ, which depends on the electron–phonon interaction.

The plasma frequency is a quantity that characterizes the spectral distribution of the reflectivity of a metal. Above the plasma frequency the reflectivity decays strongly. Particularly in metallic conductors with a large value of the collision time the decrease of reflectivity may reach several orders of magnitude and therefore the reflectivity decay is usually termed the plasma edge. The Figures 1a to 1c exhibit schematically the spectral distribution of the real part $\epsilon_1(\omega)$, of the frequency dependent conductivity $\sigma(\omega)$ and of the reflectivity $R(\omega)$ of a metal. It is assumed that oscillator-like absorption processes may be neglected in the chosen energy range.

The figures show once again that the free charge carriers form a broad absorption band with the eigenfrequency zero and the half width $1/\tau$. The plasma frequency ω_p marks the transition through zero of the real part of the dielectric function, that means the spectral position of the reflection edge.

1.3. MEASURING AND EVALUATION PROCEDURES

The most appropriate, and in most cases the only applicable, measuring method for optical investigations of metals appears to be the determination of the spectral distribution of the reflectivity. In general this is done at near-normal incidence in order to simplify the corresponding Fresnel equation. In order to determine the parameters ω_p

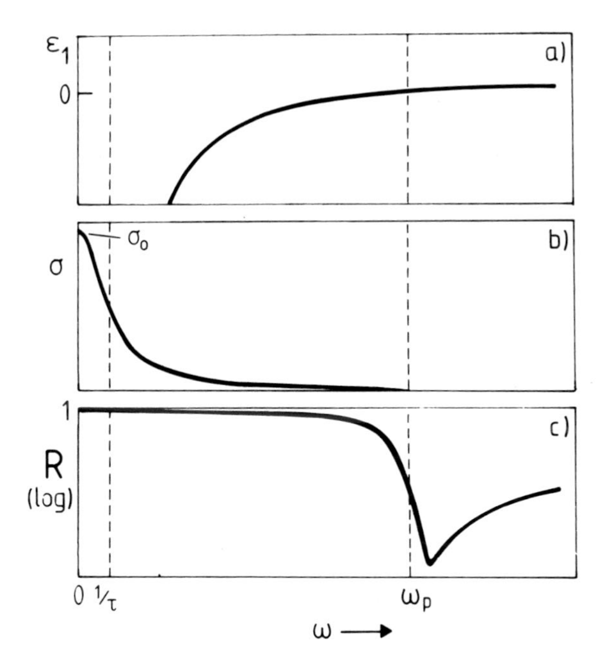

Fig. 1. Spectral distribution of the real part of the dielectric function $\epsilon_1(\omega)$, of the conductivity function $\sigma(\omega)$, and of the reflectivity $R(\omega)$ for a metallic conductor.

and τ of the free charge carrier gas, it is essential to know the reflectivity in the vicinity of the plasma edge. To be certain that the absorption is really by free carriers, it is advisable, however, to extend the reflection measurements as far as possible to low frequencies. This ensures the exclusion of the possibility that the high reflectance values diminish again towards low energies, as is the case for a reflection band caused by an oscillator-like absorption process (compare Section 4.1.2). Towards high energies, the re-rise of the reflectivity should be followed up until ϵ_∞ and with it the influence of optical transitions at higher energies on the position of the plasma edge may be determined.

1.3.1. *Method of Fitting*

If the Lorentz–Drude model is chosen for the evaluation of the experimental data the parameters of interest may be determined by fitting. Their values are varied until the reflectivity distribution, as calculated from the model for perpendicular incidence, coincides with the measured reflectivity as a function of frequency. From the parameters ω_p and τ one obtains by, means of the relation

$$\sigma_0 = \omega_p^2 \epsilon_0 \epsilon_\infty \tau = \frac{Ne^2 \tau}{m^*}, \tag{6}$$

the dc-conductivity as well.

With respect to a simple dc-measurement of the conductivity, the optical method for the determination of the conductivity offers a number of advantages. First, there do not occur any problems with contacts. Second, the problem of the geometry and the shape of the specimen does not exist. Finally, in the optical determination of the conductivity the product σ_0 is split into two factors. The first factor, ω_p, is proportional to the concentration of the free charge carriers. The second factor, τ, represents the interaction of the free charge carriers with the lattice.

1.3.2. *Kramers–Kronig Relations*

The description of the free electron gas by a Drude term is limited by the approximation of the collision time of the free carriers as being considered as constant. This approximation is good as long as the energy of the incident photons is small with respect to the Fermi energy. For photon energies comparable with the Fermi energy or even larger, the electron–lattice interaction will in general depend on the energy of the incident photons, this dependence being determined by the predominant scattering process.

One may refine the evaluation of the optical measurements in order to obtain from the measured spectral distribution of the reflection information on the predominant scattering process. In this case one must refrain from using the Drude–Lorentz model and turn to a procedure of evaluation of the dielectric function that is independent of any model. In this situation one is confronted with the problem that the usual measurement of reflection at normal incidence yields only the square of the absolute amount $R = |r|^2$ of the complex reflectivity $r^{i\phi}$. In order to obtain the phase ϕ, knowledge of which is needed for the complex dielectric function, one may make use of the fact that, in all complex optical functions, the real and the imaginary part, or absolute amount and phase, respectively, are connected by Kramers–Kronig dispersion relations.

The phase of the reflectivity at a certain frequency ω_0 may be computed from the dispersion relation [1]

$$\phi(\omega_0) = \frac{\omega_0}{\pi} \int_0^\infty \frac{\ln R(\omega) - \ln R(\omega_0)}{\omega_0^2 - \omega^2} \, d\omega \tag{7}$$

To account for the requirement to know the entire reflection distribution between $\omega = 0$ and $\omega \to \infty$, one extends the measurement of the reflection over as far a spectral range as possible. For the remaining spectral regions at high and low frequencies one looks out for appropriate extrapolation expressions.

After, for instance, the spectral distribution of the imaginary part of the dielectric function $\epsilon_2(\omega)$ or the conductivity function $\sigma(\omega)$, respectively, have been determined in this way, one may compare this experimental distribution with the results of quantum mechanical calculations. This comparison yields information about the predominant scattering processes (compare Section 4.1.1).

2. Application to One-Dimensional Metals

So far we have discussed only isotropic media. If these methods of investigation are to be extended to one-dimensional metals – and that means to optically extremely anisotropic materials – the experimental procedure and its evaluation has to be modified in several respects.

2.1. Consequences of Crystal Optics

The dielectric function becomes, because of the dependence of the transport properties in one direction, a tensor quantity. It hence becomes necessary to use linearly polarized light for the determination of its components. For the selection of the direction of incidence and of the orientation of the polarization plane of the incident radiation one takes advantage of the symmetry properties of the crystals under investigation. For crystals belonging to the hexagonal, trigonal, or tetragonal system, the situation is quite clear. The dielectric tensor has the form of a rotational ellipsoid whose rotational axis coincides with the direction of the metallic conductivity and with the optical axis of the crystal. One finds the principal axes of this ellipsoid by carrying out reflectivity measurements at normal incidence with the electric field vector once orientated parallel and once perpendicular to the optical axis [2]. The result of these measurements will in general yield, in the case of the polarization direction parallel to the metallic axis, a plasma edge. From this follows the spectral distribution of the conductivity function $\sigma_{\parallel}(\omega)$. On the other hand for the polarization direction perpendicular to the optical axis, in most cases one will not observe an influence of the free carriers on the optical properties. In those cases an upper limit may be given for the transverse conductivity.

If the one-dimensional metals to be investigated belong to crystal systems with lower symmetry, namely to the orthorhombic, monoclinic or triclinic system, then the dielectric tensor is represented by a triaxial ellipsoid. Crystals with this symmetry are optically biaxial, with the optical axes in general not coinciding with either of the principal axes of the tensor ellipsoid. For the complete determination of such a triaxial tensor one obviously needs more information than for crystals with only one optical axis. One has to measure the reflectivity at normal incidence of differently orientated crystal faces in two directions of polarization. Hence it could be necessary to produce the various faces by cutting the crystals.

In orthorhombic and monoclinic crystals one may – by profiting from the symmetry properties – choose the optical set-up in such a way that incident linearly polarized light propagates into the crystal and is reflected in this very same state of polarization. In triclinic crystals this is no longer possible because of the lack of symmetry elements. Incident linearly polarized light propagates in any event as an elliptically polarized wave into the crystal, and is also reflected as such. In this case the ellipticity of the reflected light has to be measured, too [3].

There may occur a further complication in crystals belonging to either the orthorhombic, monoclinic, or triclinic system. The directions of the principal axes of the dielectric tensor may exhibit a spectral dependence. This happens if for instance the dielectric tensor for the electronic transitions and for optically active lattice vibrations is orientated differently from the tensor accounting for the absorption by free charge carriers. This has been observed so far, however, only at a few organic one-dimensional metals, in which the normal of the molecular planes is tilted with respect to the metallic axis.

In practice, the exact optical investigation of one-dimensional metals with low crystal symmetry meets with considerable difficulty, because these compounds are often available only in the form of thin platelets. Hence the reflection measurements are limited only to one crystal face. This principal draw back is, however, compensated by the fact that due to the extreme anisotropy of the electrical properties in most crystals, it suffices also in one-dimensional metals of low symmetry to investigate the directions parallel and perpendicular to the metallic axis and to desist from further differentiations of the optical anisotropy.

2.2. Investigation of the Electrical Anisotropy

As far as the information content of optical investigations is concerned that is additional to that of a simple dc-current measurement, the advantages of the optical method stand out even more clearly in one-dimensional metals than in the three-dimensional case discussed earlier.

First, information about the electrical anisotropy of the crystals is obtained in a simple and reliable way by rotation of the plane of polarization. Because of the needle-like shape of the crystals this is not easily done for many one-dimensional metals by dc-measurements and reliable results may not readily be obtained.

2.3. Determination of Effective Mass, Bandwidth, Fermi Energy and of Mean Free Path

Second, in one-dimensional metals, the splitting of the electrical conductivity into a factor ω_p, that depends on the ratio N/m^*, and a factor τ, that accounts for the electron–lattice interaction, may be pursued even further. In one-dimensional metals the free carrier concentration N may in general be computed from the amount of charge ρ transferred from the donor-type to the acceptor-type atoms or molecules, respectively, and the charge transfer itself follows obviously in most cases from the chemical composition. Then Equation (5) yields the effective mass directly. On the basis of a tight-binding model it follows from the relation

$$4t = \frac{2\hbar^2 k_F}{m^* a \sin(k_F a)} \tag{8}$$

where the width, $4t$, is of the conduction band. The Fermi vector k_F is given by the relation

$$k_F = (\rho/2)(\pi/a) \tag{9}$$

with a denoting the distance between neighboring metal atoms, or the corresponding molecules or constituent elements, respectively.

It follows further from this band model, that one can find a value for the Fermi energy E_F as:

$$E_F = 2t(1 - \cos k_F a), \tag{10}$$

and for the Fermi velocity v_F:

$$v_F = \frac{2t\, a \sin(k_F a)}{\hbar}. \tag{11}$$

Finally by this procedure also the mean free path

$$\Lambda = v_F\tau \tag{12}$$

is determined.

2.4. Distinction between Intrinsic and Extrinsic Properties

A number of examples may be mentioned in favor of the optical determination of the conductivity where, by means of optical investigations, intrinsic transport properties could be separated from extrinsic effects. Extrinsic effects in one-dimensional metals may be caused by structure defects within the metallic chains, but also by defects in-between metallically conducting fiber bundles. An example will be given in Section 4.2.1.

2.5. Optical Investigation of Metal–Semiconductor Transitions and of Charge Density Waves

A most remarkable phenomenon that occurs on the transition from three-dimensional to one-dimensional metals, is the enhanced instability of the metallic state with respect to lattice distortions. This instability leads in most one-dimensional metals to the formation of a superstructure upon cooling. It is connected with an energy gap originating at the Fermi level. Such metal–semiconductor transitions of the Peierls type have been the subject of thorough physical investigations for years. They are also treated in other contributions in this volume. It is obvious that a Peierls transition also has an effect on the optical properties, although in a less spectacular way than on the dc-conductivity. The reason for this is that the dc-conductivity of a metal is determined essentially by the electronic density of states within a range of a few kT at both sides of the Fermi level, while optical transitions take place between states up to a distance of $\hbar\omega$ at both sides of the Fermi level. The formation of a gap at the Fermi level thus to a large extent suppresses the single-particle excitation of electrons by an electrical dc-voltage, thereby completely destroying the metallic dc-conductivity. In contrast to the dc-properties the metal–semiconductor transition affects the optical properties only to the extent that merely a part of those transitions which are observed in the metallic state are suppressed, namely the transitions with an excitation energy less than the gap energy ΔE. Since the Peierls gap has in general a low ΔE value of the order of a tenth of one eV, its influence becomes noticeable only at long wave lengths, in the range of the middle and far infrared and beyond, where the photon energy is smaller than the Peierls gap. This leads, despite the destruction of the metallic state, to the conservation of the 'typically metallic' plasma edge, usually located in an energy range between 1 and 2 eV.

Phase transitions in one-dimensional metals not only suppress single-particle excitations of the conduction electrons but, at the same time, collective excitations in the form of charge density waves or spin density waves are rendered possible. If such charge density waves may move freely within the crystal, they contribute to the dc-conductivity. In the far more frequent case that they are pinned at defects in the lattice, they may be excited to optically active oscillations. Their eigenfrequency ω_{CDW} is very low because of the large number of electrons that are rigidly coupled in this case and because of the small restoring forces.

Summarizing, one may state the following: above the metal–semiconductor transition the optical behaviour of a one-dimensional metal for a direction of polarization parallel to the metallic axis is similar to that of a three-dimensional metal, as is depicted in Figure 1a to 1c. For the direction of polarization perpendicular to the optical axis, the optical behavior of an insulator or semiconductor may be expected. Passing through the phase transition, the optical properties at higher energies, particularly in the range of the plasma edge, change only little. At very low energies the influence of the electronic gap and of pinned charge density waves takes effect. The Figure 2a to 2c show, again schematically, the spectral distribution of the real part $\epsilon_{1\,\parallel}(\omega)$, of the frequency-dependent conducting $\sigma_{\parallel}(\omega)$, and of the reflectivity $R_{\parallel}(\omega)$ of a one-dimensional metal below the phase transition. Because the characteristic frequencies ω_{CDW}, $\Delta E/\hbar$, and ω_p may differ by orders of magnitude, and because the frequency $\omega = 0$ in this case has lost its particular importance, a logarithmic frequency scale was chosen in the figures.

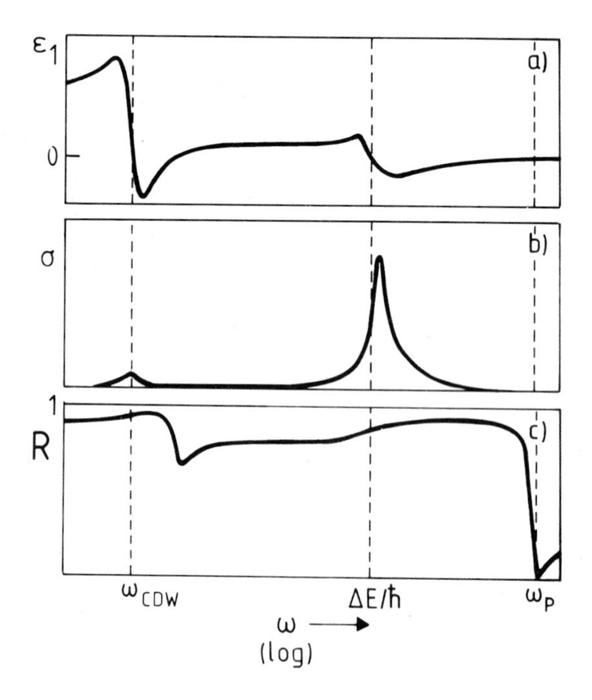

Fig. 2. Spectral distribution of $\epsilon_{1\parallel}(\omega)$, of $\sigma_{\parallel}(\omega)$, and of $R_{\parallel}(\omega)$ for a 1-D metal below the Peierls transition according to [33] and [62]. (∥ refers to the direction of polarization with respect to the metallic axis).

3. Survey of Synthesized One-Dimensional Inorganic Metals

3.1. Planar Transition Metal Complex Compounds

The physical realization of an one-dimensional metal was discovered for the first time by the structural investigations by Krogmann and coworkers in the 'sixties on various partially oxidized planar platinum complex compounds like the bisoxalatoplatinate $Mg_{0.82}Pt(C_2O_4)_2 \cdot 5.3H_2O$ [4] and the tetracyanoplatinate $K_2Pt(CN)_4Br_{0.3} \cdot 3H_2O$ [5, 6]. Since then, the research in this field common to both solid-state chemistry and physics has received an enormous impetus. In the beginning it had seemed as if one-dimensional metallic conduction was closely connected to the structural pecularities of planar transition metal complex compounds, thereby exhibiting a rare phenomenon among solids. Very soon, however, exciting results obtained on organic charge transfer salts, especially on tetrathiafulvalene tetracyanoquinodimethane (TTF–TCNQ) showed that there exist quite different possibilities for the realization of one-dimensional metals [7]. These compounds, in contrast to the transition metal complex compounds, did not even contain chains of metal atoms, and this was their novelty. The intense efforts of synthesis and investigation of a great number of new one-dimensional metals having an organic basis has culminated so far in the well-known discovery of organic superconductors [8, 9].

In the field of one-dimensional *inorganic* metals, which we are dealing with here, research has remained concentrated for a good while on the planar transition metal complex compounds. These compounds could be modified in a variety of ways. Besides the cations, the ligands as well as the central atoms of the complexes were exchanged. Most of the compounds which were obtained by this method are, however, merely very anisotropic semiconductors. It is true that they fullfill one important condition for the formation of one-dimensional metallic conduction, namely a configuration of metal atoms in the form of linear chains. In these compounds, the one-dimensional metal–metal bonds may, however, be saturated. One then is confronted with completely filled and completely empty bands, and electrical conduction in such a case is only effected by thermal activation of charge carriers. Metallic conduction, on the other hand, occurs only in such planar complex compounds, which in addition exhibit a composition having a non-integral stoichiometry. Those compounds exhibit partially filled bands. This condition may be achieved in two ways. The first is the incorporation of acceptors, as in the case of the bromine-doped $K_2Pt(CN)_4Br_{0.3} \cdot 3H_20$. The second consists in a deficit of cations, as in the case of $K_{1.75}Pt(CN)_4 \cdot 1.5H_2O$. Both instances lead to a partial depletion of the highest occupied band.

Most of the metallically conducting transition metal complex compounds dissolve in water and also contain water of crystallization in most cases. That means that they are held together by hydrogen bonds. This has an unfavorable effect on the stability of the lattice and also on the reproducibility of the electrical data. For this reason these compounds hardly come into consideration for technical applications.

3.2. The Chain Compounds $(SN)_x$ and $Hg_{3-\delta}XF_6$ (X = As or Sb)

A breakthrough in the direction of a totally different structure was brought about in the beginning of the 'seventies by detailed investigations on the polymeric sulfur nitride $(SN)_x$. Here one was concerned for the first time with an *inorganic*, and in addition a chemically very simple compound, which exhibited metallic conduction in the same way as the one-dimensional organic metals, namely without containing metal atoms.

This compound showed, furthermore, for the first time in the field of quasi-one-dimensional metals, the peculiarity that its metallic conduction was not destroyed by a phase transition [10]. As the most spectacular success, this material was observed to be superconducting [11]. Somewhat later it was shown, mainly by optical investigations, that $(SN)_x$ is not a one-dimensional, but merely a highly anisotropic metal (compare Section 4.2.1). The initially very great interest in this compound decreased strongly, however, when it became clear that $(SN)_x$ remained a rather unique compound that did not admit any essential chemical variations: it did not turn out to be the first example of a new class of substances.

A similarly unique phenomenon to $(SN)_x$ is represented by the compounds $Hg_{3-\delta}XF_6$ (X = As or Sb), which were for the first time prepared in the early 'seventies [12]. In these compounds mercury atoms are embedded in a matrix of AsF_6 or SbF_6 complexes, respectively, in such a way that they form linear chains with metal–metal distances which are even smaller than in pure mercury. This leads to an electrical conductivity of approximately the same value as in mercury metal. $Hg_{3-\delta}AsF_6$ and $Hg_{3-\delta}SbF_6$ do not exhibit phase transitions upon cooling to low temperatures, either, but retain their metallic properties and become superconducting as well. The metallic chains in these compounds occur, however, in two crystallographic directions, which are perpendicular to each other. Therefore they are expected to behave, at least optically, as two-dimensional metals in spite of the one-dimensional coordination of the mercury atoms.

3.3. Transition Metal Chalcogenides

From the middle of the 'seventies a new wide field for the investigation of the one-dimensional metallic state has been opened. At that time the synthesis of a new class of one-dimensional metals started with the trichalcogenides of certain transition metals, above all $NbSe_3$ [13, 14]. Their transport properties, characterized in a most striking way by charge density waves, are extensively discussed in this book. In these trichalcogenides the metallic conduction is effected by transition metal atoms in a chain-like configuration. These metal atoms are surrounded by triangular configurations of chalcogen atoms, thus screening the metallic chains from each other.

Recently the class of one-dimensional conducting compounds based on transition metal and chalcogen atoms has been enlarged by a new type of compound. It is that of the tetrachalcogenides $(MX_4)_nY$ with M = Nb or Ta, X = S or Se, and Y = Cl, Br, or I. This type of chemical structure offers the interesting possibility to synthesize crystals with different charge carrier concentrations, and thereby different positions of the Fermi level, by varying the halogen content [15]. In addition, the incorporation of halogen atoms enhances the distance between the metallic chains compared to that in the trichalcogenides, thereby changing the electrical anisotropy.

A further interesting variant in the field of one-dimensional structures containing transition metal and chalcogen atoms are the molybdenum compounds of the type $M_2Mo_6X_6$ with M = In, Tl, Ba, Na, K, Rb or Cs, and with X = S or Se. In these compounds, derived from the Chevrel-phases, the metallic filaments are not formed by linear metallic chains but by columnar structures consisting of molybdenum clusters. As in the other one-dimensional conducting chalcogen compounds, these metallic filaments are screened from each other by configurations of chalcogen atoms. These compounds are superconductors. It is not yet quite clear whether they constitute real one-dimensional or merely highly anisotropic metals.

3.4. OTHER ONE-DIMENSIONAL INORGANIC METALS

The survey given here contains the essential groups of the hitherto known one-dimensional inorganic metals. The search for new compounds is by no means yet brought to an end. One rather has the impression that in most recent times the efforts are even still growing.

One does not, of course readily dispose of variation possibilities as vast and rich as for the one-dimensional organic metals. Neither may one expect the abundance of materials already obtained with organic substances. Nonetheless even in the inorganic realm there have always been new starting points either for synthesis of new materials or for the discovery of one-dimensional conduction in known substances. As an example, very recently, metallic conduction has been found in certain molybdenum bronzes [17].

4. Discussion of Experimental Results

In Section 2 it was shown that optical investigations may furnish important data on one-dimensional metals. Furthermore one may have plenty of materials at one's disposal for those investigations, because the different classes of one-dimensional inorganic materials, outlined in Section 3, comprise altogether dozens of different substances. In view of this situation the amount of existing optical data may not be called very extensive, particularly in comparison with the results stemming from structural, electrical and other physical investigations. There is, in fact only one one-dimensional inorganic metal that has been examined systematically and rather thoroughly, namely $K_2Pt(CN)_4Br_{0.3} \cdot 3H_2O$. In this material, on one hand, the region of the plasma edge had been carefully investigated; besides the effective mass, also the predominant scattering processes were determined from these measurements. On the other hand, reflection measurements were carried out also in the middle- and far-infrared in order to determine the Peierls gap and to investigate the formation of charge density waves. Beyond this material measurements in the region of the plasma edge exist for a number of other compounds. For most one-dimensional inorganic metals there are, however, no optical data available at all.

This lack of results is surprising at first glance. It is based on the fact that one-dimensional metals exhibit in most cases a rather 'one-dimensional' crystal growth. For this reason most specimens are too tiny for the usual optical measuring equipment, in particular for commercially available spectrometers. This drawback may be rather well compensated by means of special optical equipment for measurements in the spectral region of the plasma edge, hence for photon energies in the order of one eV. But the problems resulting from the small dimensions of the specimen, in particular that of a sufficient radiation intensity, are much more difficult to solve in the middle and far infrared region, where the information about the Peierls gap and the charge density waves should be obtained. The precise reflection measurements which have been carried out on $K_2Pt(CN)_4Br_{0.3} \cdot 3H_2O$ in this spectral region were only possible because crystals of good quality with a volume of about half a cubic centimeter could be grown from this compound in a relatively simple way.

In what follows, the available results of optical investigations on one-dimensional inorganic metals will be described and discussed in detail.

4.1. PLANAR TRANSITION METAL COMPLEX COMPOUNDS

4.1.1. $K_2Pt(CN)_4Br_{0.3} \cdot 3H_2O$

This compound was the first one-dimensional metal, for which the anisotropy of optical properties, typical for this class of solids could be experimentally demonstrated [18, 19]. Figure 3 shows the polarized reflectance spectra of a $K_2Pt(CN)_4Br_{0.3} \cdot 3H_2O$ single

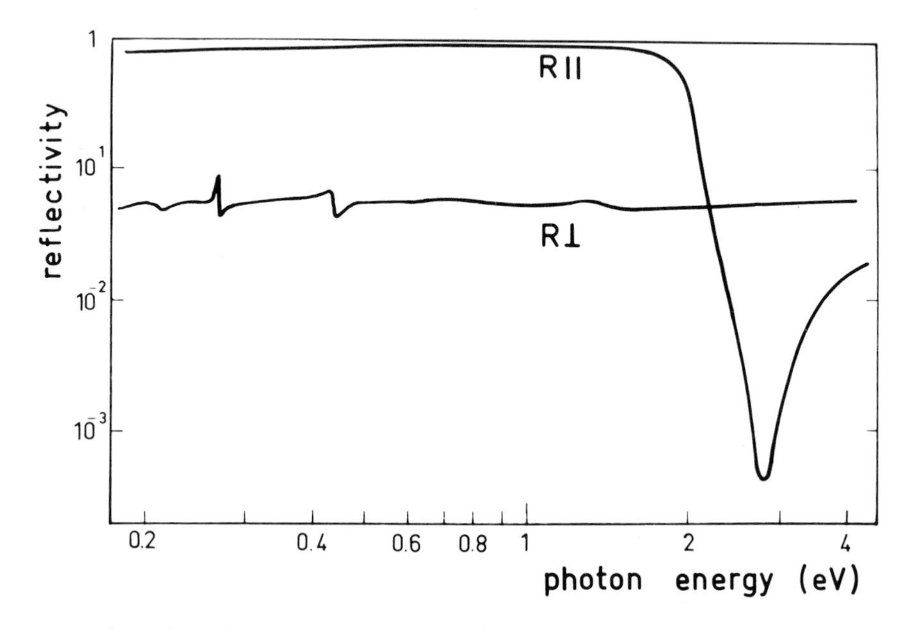

Fig. 3. Polarized reflectance spectra of $K_2Pt(CN)_4Br_{0.3} \cdot 3H_2O$ at 300 K [20].

crystal at room temperature. For the polarization direction perpendicular to the metallic axis the optical behavior of an insulator is observed with low and nearly constant values of the reflectivity, modulated by vibration modes in the infrared region. For the polarization direction parallel to the metallic axis the reflectance spectrum exhibits an exceedingly steep plasma edge at 2 eV with reflectance values extending over several decades. This plasma edge is responsible for the metallic lustre of this compound, which resembles that of copper. The evaluation of these measurements yields, with $\sigma_0 = 8 \times 10^3$ $(\Omega\text{cm})^{-1}$, an optical value for the dc-conductivity which is higher by about a factor of 30 than the electrically determined value [21]. The results of the two measuring procedures differe much more than, e.g., those for one-dimensional organic metals, for which the optical value is, at a maximum, twice as high as the electrical one. This suggests the conclusion that the electrically determined value of the conductivity in $K_2Pt(CN)_4Br_{0.3}$ $3H_2O$ is either drastically reduced by the influence of interruptions of the metallic chains, or that the electrical conduction even in the range around room temperature is affected by a developing Peierls gap. This discrepancy between the optically and electrically determined values of the dc-conductivity is even larger for the other optically investigated planar complex compounds.

The computation of the effective mass from the value of the plasma frequency by means of Equation (5) has up to now been controversial in the case of the planar transition metal complex compounds with respect to the model that should be applied. It appears suggestive, in the case of $K_2Pt(CN)_4Br_{0.3} \cdot 3H_2O$ with a charge deficit of $0.3e^-$ per platinum ion with respect to the divalent status, to speak of a band that is depleted by 15 percent. This implies that one start with the assumption of a p-type metal. Nonetheless, many authors use the model of an n-type conductor with correspondingly higher values of the charge carrier concentration. The main argument advanced for the second model is that, in this case, one obtains a value for the effective mass roughly that of a free electron [22, 23, 24]. The p-type conduction model, on the other hand, yields much smaller values for the effective mass, for $K_2Pt(CN)_4Br_{0.3} \cdot 3H_2O$ that of $m_p{}^* = 0.18\ m_0$ [20]. Whereas measurements of the thermopower yield results which are contradictory in this respect [25, 26, 65], the essential features of the p-type model are confirmed by band structure calculations [27]. Finally, a quantitative analysis of the imaginary part of the dielectric function supports the conduction by holes. In [20] the function $\epsilon_{2\|}(\omega)$ was computed from the reflectivity shown in Figure 3 by means of the Kramers–Kronig relations. The experimental curve could be described quantitatively, without fitting the values of any parameters, by scattering of holes on bromine ions. The n-type model, on the other hand, yields theoretically determined values of $\epsilon_{2\|}(\omega)$ that are by a factor 20 to 30 lower than the values obtained from experiment (Figure 4). We therefore conclude that the description of $K_2Pt(CN)_4Br_{0.3} \cdot 3H_2O$ and of the other partially-oxidized transition metal complex compounds by a hole model is more adequate than that by an electron model.

As Figure 4 shows, and as is confirmed by other measurements [28, 30], the monotonic decay of $\epsilon_{2\|}(\omega)$ is superimposed by an absorptive structure directly above the plasma energy of 2eV. This structure has been related to the excitation of plasma oscillations [29, 30]. In fact, this structure can quantitatively be explained by inelastic scattering of the charge carriers under emission of plasmons [20].

This indirect excitation of plasmons must be clearly distinguished from direct excitation of plasma oscillations by p-polarized light at non-normal incidence. In that case the electric field vector lies in the plane of incidence and has a component normal to the surface. It was shown by experimental and theoretical investigations on $K_2Pt(CN)_4Br_{0.3} \cdot 3H_2O$ that this effect, well-known from thin metal films, may also be observed in the case of one-dimensional metals on bulk material [31, 32].

Of considerable importance among the optical investigations on one-dimensional metals are the reflection measurements of Brüesch and coworkers concerning the Peierls transition in $K_2Pt(CN)_4Br_{0.3} \cdot 3H_2O$ [33, 56]. In Figure 5 is shown the spectral distribution of the reflectivity up to the far-infrared, both at room temperature and at 40 K. These measurements demonstrate the great advantage which the planar transition metal

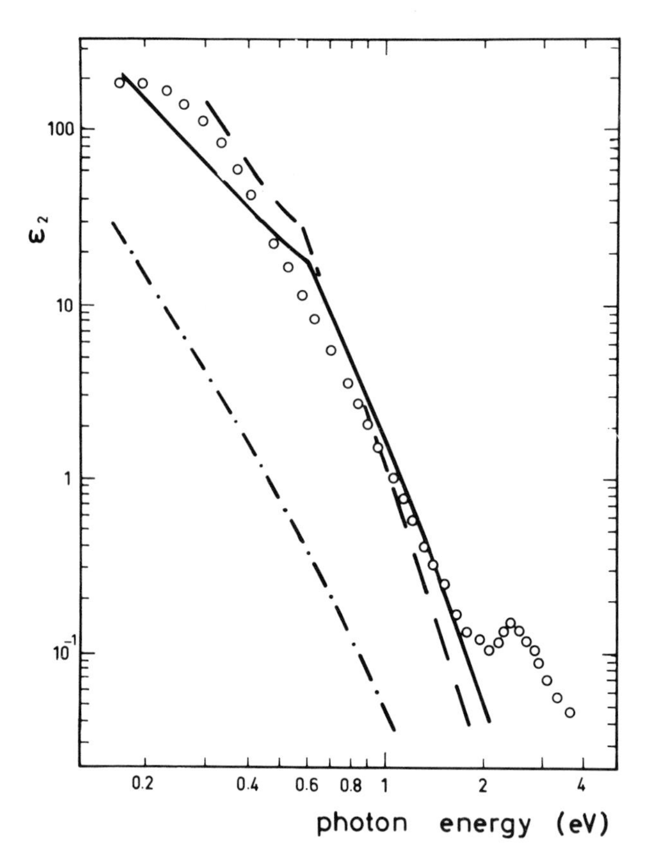

Fig. 4. Spectral distribution of $\epsilon_{2\parallel}(\omega)$ of $K_2Pt(CN)_4Br_{0.3} \cdot 3H_2O$ at 300 K. Experimental data (open circles), theoretical results for scattering by Br ions: hole model without (dashed line) and including shielding of the Coulomb potential (full line); electron model (dashed and dotted line) [20].

complex compounds possess, based on the high degree of symmetry of their crystal structure, and based on their high optical conductivity. The optical properties for the direction of polarization parallel to the metallic axis are in practice not modified at all by the influence of molecular vibrations. This causes the electronic transport properties to stand out much more clearly. It is obvious from Figure 5 that, even at room temperature, the reflectance spectrum $R_{\parallel}(\hbar\omega)$ exhibits marked deviation from the normal metallic behavior, because below the plasma edge it does not show the expected flat monotonous rise toward low energies. Rather, in the energy range between 10 meV and 0.3 eV, a broad and flat minimum is observed. Upon cooling down to 40 K, deviations become more pronounced, and in the range around 4 meV an oscillator-like structure appears.

The physical meaning of these experimental results becomes clear from the presentation of the conductivity function $\sigma_{\parallel}(\hbar\omega)$ in Figures 6 and 7, which has been obtained by a Kramers–Kronig analysis and by an oscillator fit from $R_{\parallel}(\hbar\omega)$. The conductivity function exhibits a resonance-like dependence at 0.2 eV even at room temperature, instead of the Drude behavior otherwise typical for metals. This is interpreted by a

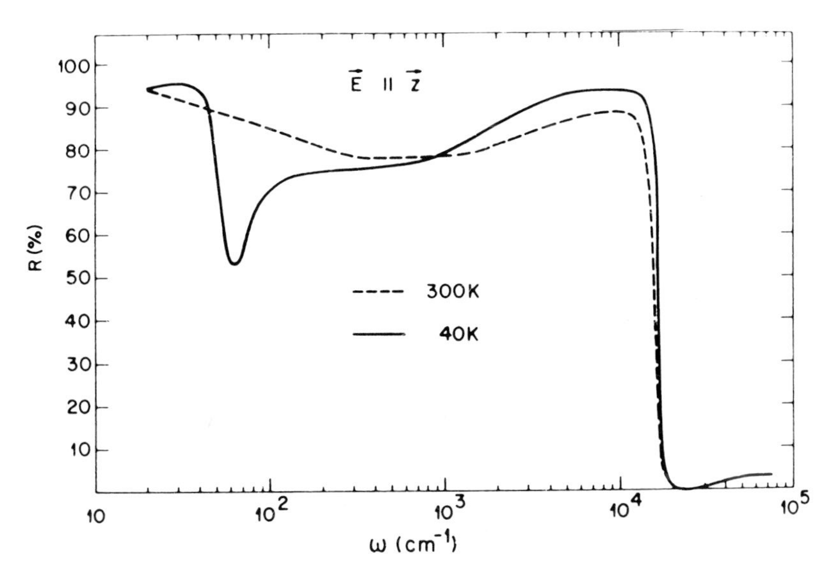

Fig. 5. Far-infrared to uv reflectance spectrum of $K_2Pt(CN)_4Br_{0.3} \cdot 3H_2O$ for light polarized parallel to the metallic axis at 300 K (dashed line) and at 40 K (solid line) [33].

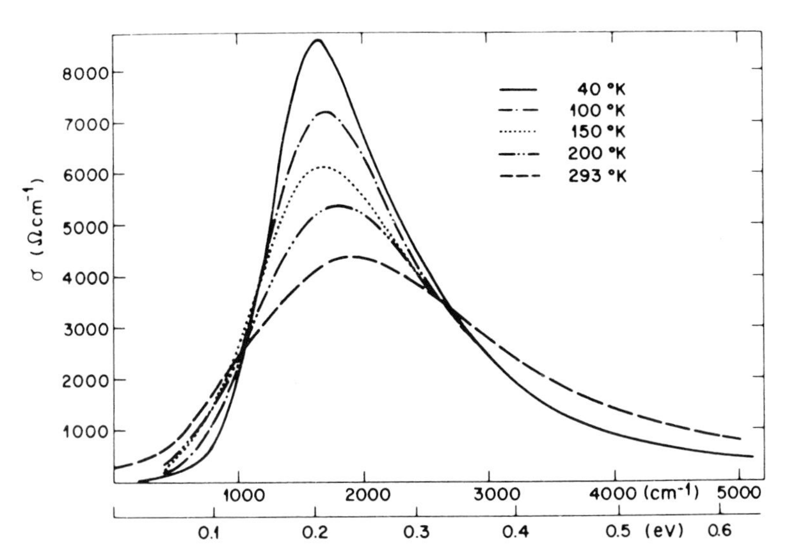

Fig. 6. Temperature dependence of $\sigma_{\parallel}(\omega)$ of $K_2Pt(CN)_4Br_{0.3} \cdot 3H_2O$ in the infrared. The resonance at high temperatures originates from a strong mobility pseudogap which transforms into the real Peierls energy gap as $T \rightarrow 0$ without significantly changing the gap energy [33].

dynamical Peierls distortion leading, even at this temperature, to a mobility gap [33]. Upon cooling the crystal below the three-dimensional ordering temperature, this pseudogap is eventually transformed into a true energy gap. The pronounced structure of the pertinent absorption peak at low temperatures may mainly be ascribed to the singularity

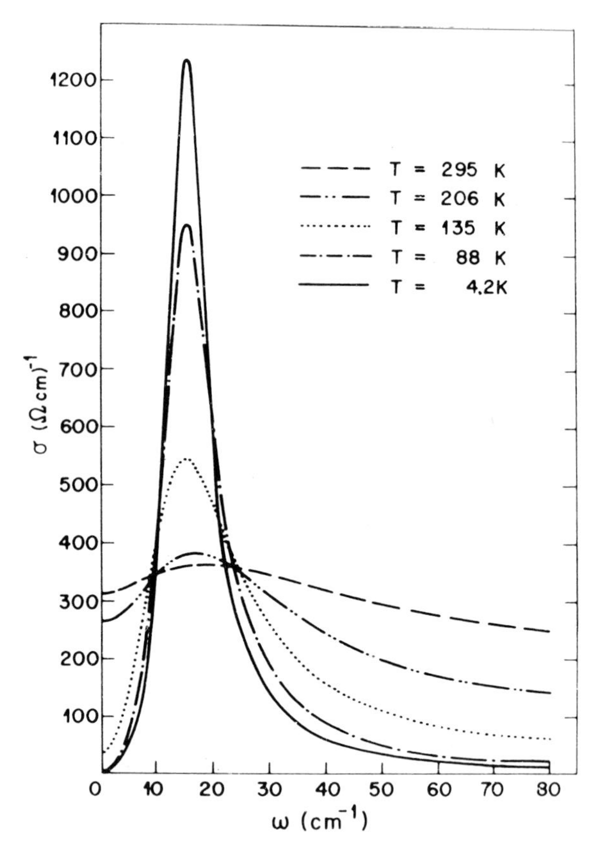

Fig. 7. Temperature dependence of $\sigma_{\parallel}(\omega)$ of $K_2Pt(CN)_4Br_{0.3} \cdot 3H_2O$ in the far-infrared. The structure at $\omega = 15\ cm^{-1}$ is assigned to an oscillation of the pinned charge density wave [33].

of the one-dimensional density of states at the band gap edges. The existence of a mobility gap at room temperature would at the same time explain why the electrically measured conductivity parallel to the chains reaches only values of 300 $(\Omega cm)^{-1}$, although the electronic transport parameters determined from the position and the shape of the plasma edge yield a conductivity value almost 30 times as high.

The energetically lower lying resonance structure at 15 cm^{-1}, or 2 meV, respectively, which forms only at low temperature (Figure 7), is based on the pinned charge density wave described in Section 2.5, due to the Peierls distortion. The oscillator mass $m^*_{CDW} = 1000\ m_0$, determined by model fitting, shows that one is dealing in fact with a pinned charge density wave and not with a low-frequency lattice vibration. This pinned charge density wave was also found by other authors in the infrared [34] as well as in the Raman spectrum [35, 36].

4.1.2. *Other Planar Transition Metal Complex Compounds*

As has already been mentioned, besides the extensive optical investigations on $K_2Pt(CN)_4Br_{0.3} \cdot 3H_2O$, there also exist reflection measurements in the region of the plasma edge on other partially oxidized transition metal compounds. These compounds are the tetracyanoplatinates of the type $M_{1.75}Pt(CN)_4 \cdot 3H_2O$ (M = K, Rb, or Cs) [37, 38] and the dioxalatoplatinate $Zn_{0.81}Pt(C_2O_4)_2 \cdot 6H_2O$ [39]. For these substances, the partial oxidation of which is caused by a cation deficit, reflectance spectra very similar to those for $K_2Pt(CN)_4Br_{0.3} \cdot 3H_2O$ are observed, with the plasma frequency shifted markedly towards lower energy. The more recent investigations on $K_{1.75}Pt(CN)_4 \cdot 1.5H_2O$ [38], extending to longer wavelengths, have yielded two interesting results besides. First, in this compound even at room temperature a static Peierls gap is formed which, in addition, is larger than that in $K_2Pt(CN)_4Br_{0.3} \cdot 3H_2O$. The stronger stability of the Peierls distortion is quite plausible, because the Fermi vector in $K_{1.75}Pt(CN)_4 \cdot 1.5H_2O$ is commensurate with the periodicity of the lattice. Furthermore the distortion of the platinum chains in form of a zig-zag leads, via a reduction of the Brillouin zone, to a half-filled conduction band. For half-filled bands the metallic state is particularly unstable. The second interesting result lies in the observation of a pronounced structure in the spectrum $R_{\|}(\hbar\omega)$ at a photon energy of 0.18 eV. This structure may be accounted for by a coupling between the charge carriers in the platinum chains and the C≡N stretching vibration of the tetracyanoplatinate complex. It would be the first example of the observation of an interaction between conduction electrons and intramolecular vibrations in the infrared spectrum of a one-dimensional inorganic metal. So far this coupling had only been detected for inorganic linear conductors by Raman scattering [35, 36, 40], while for organic conductors it has been more thoroughly investigated by means of the infrared spectrum.

Finally, we wish to cite here an example of the fact that a 'metallic' lustre in the visible part of the spectrum does not present a sufficient criterion for the occurrence of metallic conduction. To this end, in Figure 8, are contrasted the polarized reflectance

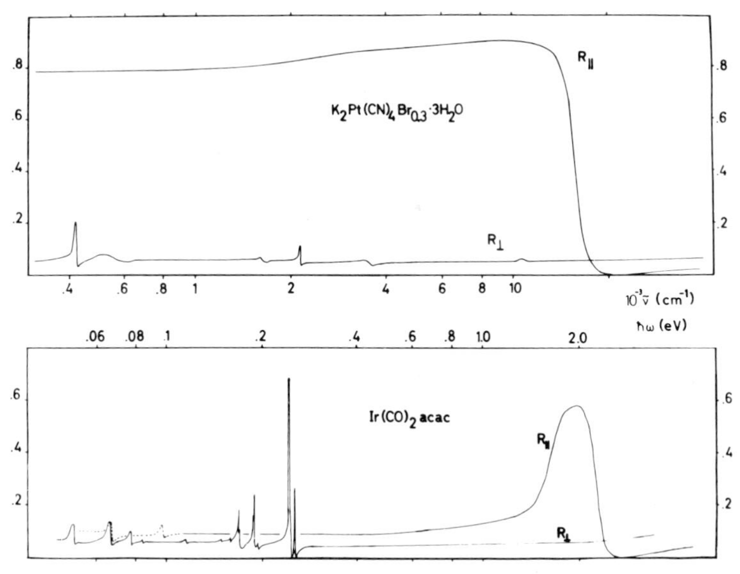

Fig. 8. Polarized reflectance spectra of $K_2Pt(CN)_4Br_{0.3} \cdot 3H_2O$ (1-D metal) and $Ir(CO)_2acac$ (1-D semiconductor) [41].

spectra of two planar transition metal complex compounds which, upon visual inspection, both appear to be metallic. It is the $K_2Pt(CN)_4Br_{0.3} \cdot 3H_2O$ with its copper-like lustre and the $Ir(CO)_2$acac with a gold-like lustre. The dependence of the reflectance $R_{\parallel}(\hbar\omega)$ at low energies shows that the iridium compound, with its integral stoichometric composition, is not a one-dimensional metal but, contrary to the visual impression, only a one-dimensional semiconductor.

4.2. The Chain Compounds $(SN)_x$ and $Hg_{3-\delta}XF_6$ (X = As or Sb)

4.2.1. *$(SN)_x$*

There is now general agreement that the conducting polymer $(SN)_x$ is not a one-dimensional, but merely a highly anisotropic metal. This has, however, not been clear since the beginning. On the contrary, there were at first strong indications that $(SN)_x$ is a particularly good example of a one-dimensional metal. These indications were the strong anisotropy of both the dc-conductivity and the optical properties in the visible spectral range. When $(SN)_x$ single crystals are inspected under polarized light, they exhibit a gold-yellow lustre if the electrical vector of the incident light is orientated parallel to the polymeric axis. They appear almost black when the plane of polarization is rotated by 90°. For this reason the first reflectance spectrum, taken on polycrystalline films, was interpreted as implying a one-dimensional metal [42]. Even the first reflectance spectra taken on single crystals seemed to corroborate this concept [43, 44]. Only when the reflection measurements on orientated $(SN)_x$ films and on single crystals were extended further into the infrared spectral region [45–49], did it become apparent that a plasma edge occurs, and hence that the substance is metallic, even when the plane of polarization is orientated perpendicular to the polymeric axis (Figure 9). It is true, however, that in this case the plasma edge is shifted towards lower energies. Further, it is damped more strongly and depends on the direction of the incidence of the radiation within the plane perpendicular to the polymeric axis.

This implies that the reflection measurements yield a smaller degree of electrical anisotropy than the value of 1000 : 1 following from the dc-current measurements for the ratio of the longitudinal to the transversal conductivity [10]. Besides, there was the minor slur on the optical results that, indeed, the dependence of $R_{\parallel}(\hbar\omega)$ was excellently reproduced by a Drude model, but that the experimental dependence of $R_{\perp}(\hbar\omega)$ below 0.7 eV deviated drastically from the Drude behavior. The discrepancies between the optically and electrically determined transport parameters, as well as the deviations from the Drude behavior may be ascribed to the real structure of $(SN)_x$. As microscopical investigations show, particularly those using the scanning electron microscope [10], even quite perfect $(SN)_x$ single crystals consist of parallel orientated fibers with an average diameter of some 100 Å. The axis of these fibers coincides with the polymeric axis of the $(SN)_x$ chains and, therefore, also with the direction of the maximum electrical conductivity. This means that the transversal electrical conduction in $(SN)_x$ single crystals is determined not only by the intrinsic transport properties within the $(SN)_x$ fibers but also by the charge transport between the fibers. It hence becomes necessary to treat the transversal conduction in $(SN)_x$ as that of a system consisting of metallic particles embedded in a dielectric matrix. Examples of systems composed

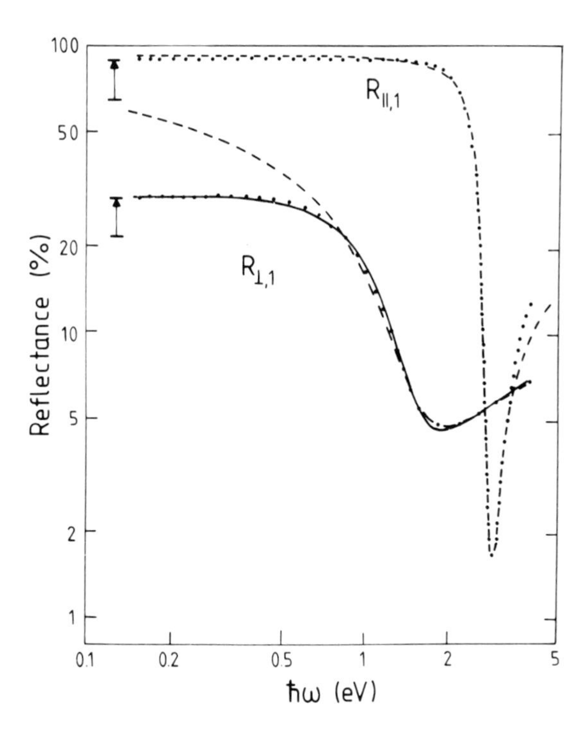

Fig. 9. Polarized reflectance spectra of $(SN)_x$ for the (100)-surface at 300 K. Experimental data, corrected by a factor of 1.37 to account for light scattering at surface imperfections (dotted line), Drude fits (dashed lines), Maxwell–Garnett fit with $q = 0.83$ (full line) [47].

in this way are amply known from the field of regular three-dimensional metals. To these systems belong, for instance, very thin metal films with island structures or the so-called Cermet films, which consist of spherical metallic particles embedded in a ceramic matrix. With the theory of Maxwell–Garnett [50] proved on these systems, one may also describe the transversal optical properties of $(SN)_x$, if the modified geometry of the conducting particles is accounted for by adapting the polarization factor [47]. The following relation holds between the transversal dielectric function of a single fiber $\epsilon_{\perp f}(\omega)$ and that of the aggregated system $\epsilon_{\perp}(\omega)$:

$$\frac{\epsilon_{\perp}(\omega) - 1}{\epsilon_{\perp}(\omega) + 1} = q \, \frac{\epsilon_{\perp f}(\omega) - 1}{\epsilon_{\perp f}(\omega) + 1} \tag{13}$$

The factor q indicates the volume fraction of the conducting material. The dielectric constant of the matrix was assumed to be unity. For the dielectric function of a single fiber, which is responsible for the intrinsic transport properties of $(SN)_x$, a Drude model was used. Figure 9 shows the excellent fit of the Maxwell–Garnett model to the experimental values of $R_{\perp}(\hbar\omega)$. In Figure 10 the conductivity functions resulting from this fit are plotted. While the longitudinal conductivity function $\sigma_{\parallel}(\hbar\omega)$ of a $(SN)_x$ crystal may be described by a Drude model, the transversal conductivity function $\sigma_{\perp}(\hbar\omega)$ of

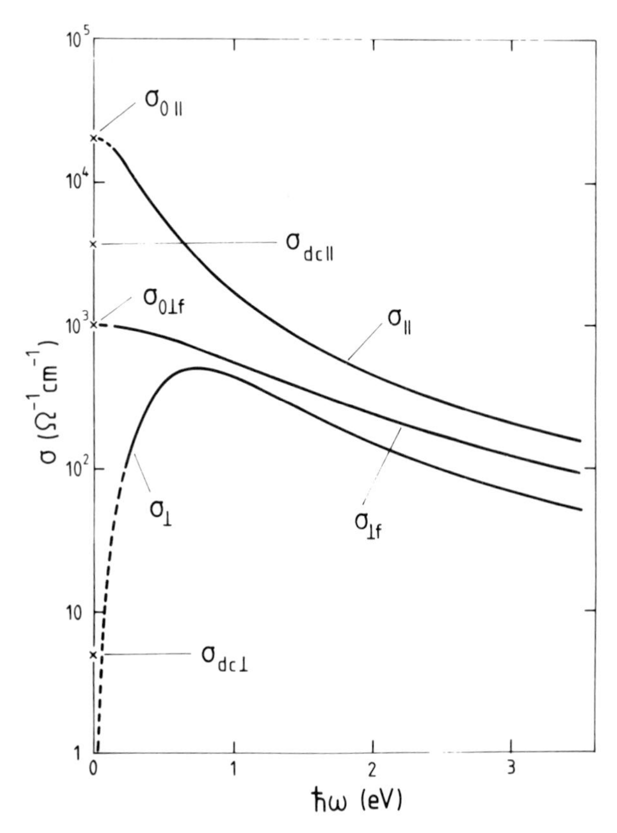

Fig. 10. Spectral distribution of the longitudinal $\sigma_{\parallel}(h\omega)$ and of the transverse conductivity function $\sigma_{\perp}(h\omega)$ of a $(SN)_x$ single crystal and of the transverse conductivity function $\sigma_{\perp f}$ of a single fiber. For a comparison the dc values are given too [47].

the entire crystal deviates from this behavior. These deviations increase drastically with decreasing photon energy. They lead to a resonance-like form of the conductivity function. Despite of the high value of 10^3 $(\Omega cm)^{-1}$, obtained for the transverse intrinsic dc-conductivity of $(SN)_x$, the transversal conductivity function of the entire crystal has decreased already in the far infrared to the electrically measured value of 5 $(\Omega cm)^{-1}$. Approaching the dc-situation, the Maxwell–Garnett model fails, however, completely. It accounts excellently for the conductivity in the range of optical frequencies because it implies the screening of the metallic fibers by the dielectric matrix but not the charge transport between the fibers.

Table I shows the electronic transport parameters of $(SN)_x$ as obtained from optical measurements. They exhibit quantitatively the difference between the intrinsic electrical anisotropy of $(SN)_x$ and the anisotropy of the dc-conductivity due to the fiber structure of the crystals.

The intrinsic metallic conduction observed perpendicularly to the polymeric chains of $(SN)_x$ exists probably only in the plane of the chains [49]. In any event, it has the

TABLE I

Intrinsic transport properties of $(SN)_x$ [47].

	σ_0 (Ω^{-1} cm^{-1})	τ (sec^{-1})	m^*
$\parallel$:	2×10^4	2.2×10^{-15}	$0.9\ m_0$
$\perp$:	1×10^3	5.8×10^{-16}	$5\ m_0$
$\parallel/\perp$:	20 : 1	3.8 : 1	1 : 5.5

consequence that $(SN)_x$ does not possess plane Fermi surfaces like a one-dimensional metal. From this stems the stability of the metallic state of this compound at low temperature with respect to a Peierls distortion.

Attempts to chemically modify the $(SN)_x$ were successful in incorporating halogen atoms and thereby changing the electrical properties. Thus, incorporation of bromine leads to an increase of the conductivity by one order of magnitude without affecting the value of the plasma frequency [51]. The increase in the conductivity must, therefore, be ascribed to a drastic increase of the mean free path.

4.2.2. $Hg_{3-\delta}XF_6$ (X = As or Sb)

Optical investigations were carried out also on the linear chain compounds $Hg_{3-\delta}AsF_6$ and $Hg_{3-\delta}SbF_6$ [52, 53]. The silvery-golden lustre of these compounds lead us to expect metallic behavior of the reflectance. The suboxidation state, however, in which the mercury atoms exist, makes the crystals highly unstable in the presence of polarizable molecules such as water. Thus the optical measurements were done on crystals placed *in vacuo* or in an inert atmosphere. Figure 11 shows the reflectance obtained for incidence in the direction of the tetragonal axis, hence for investigation of a (001)-face. The electrical vector of the incident light in this case lies in a plane parallel to the mercury chains. Because of the symmetry of the crystals these chains lie in the direction of both the **a**-axis and the **b**-axis. These directions are orthogonal and the mercury chains intersect without penetrating into each other. As expected, for this geometry of illumination, one observes, independent of the direction of polarization, the same pronounced plasma edge, since the electrical vector stays within the plane of metallic conduction. If, on the other hand, the reflectance of surfaces is measured, which are inclined with respect to the metallically conducting (001)-face, the electrical anisotropy of these compounds becomes readily noticeable. While for a direction of polarization perpendicular to the **c**-axis the same plasma edge results as for the (001)-plane, for the direction of polarization within the plane of incidence the reflectance spectrum depends on the angle between the illuminated face, and with it between the electrical vector of the incident light and the **c**-axis (Figure 12). In the limiting case of this vector coinciding with the **c**-axis one finds the optical behavior of an insulator. The reflectance curves, measured for planes oblique to the **c**-axis, may quantitatively be explained by taking into account the fact that the optical excitation does not take place along the main axes of the tensor ellipsoid. The evaluation of the reflectance spectra yields for the optically determined value of the

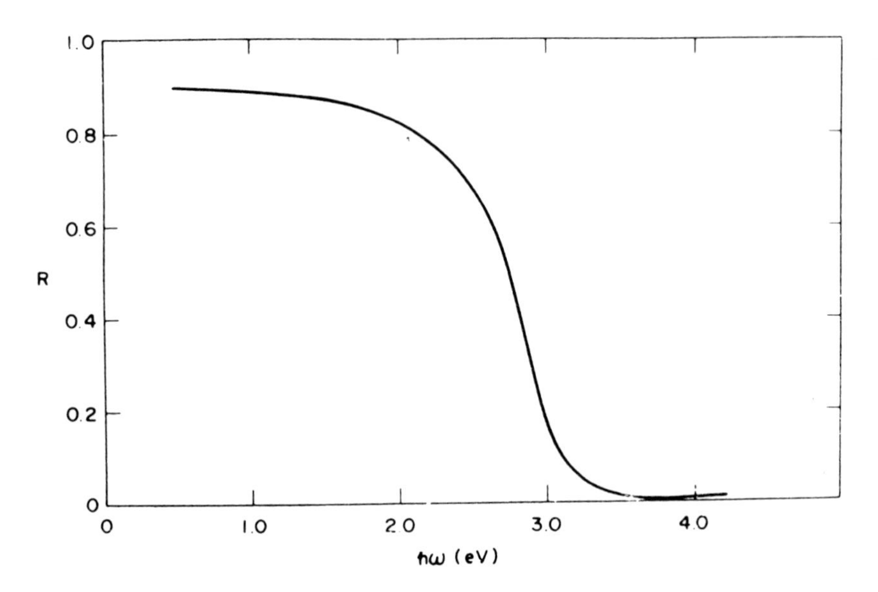

Fig. 11. Reflectance spectrum of the isotropic (001)-face of $Hg_{2.86}AsF_6$ at 300 K [53].

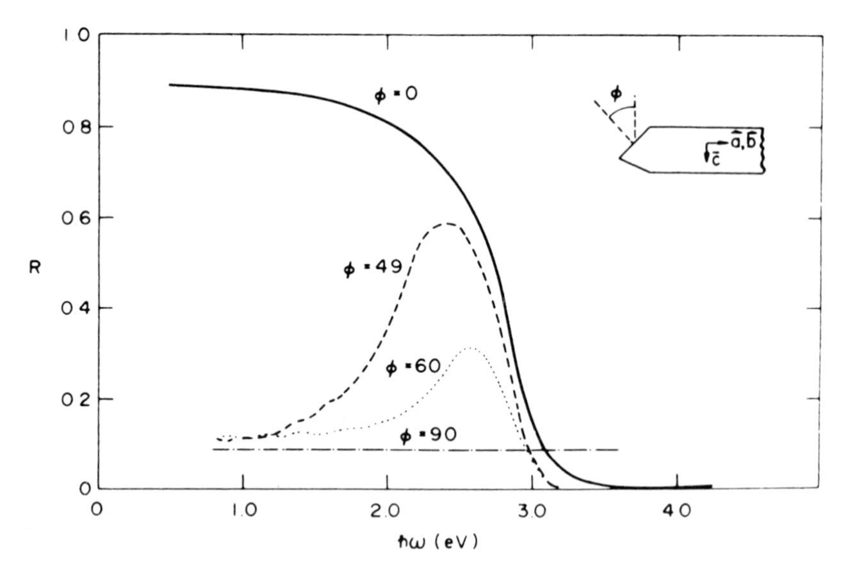

Fig. 12. Polarized reflectance spectra of the anisotropic faces of $Hg_{2.86}AsF_6$ for the plane of polarization including the c-axis [53].

dc-conductivity 1.2×10^4 $(\Omega cm)^{-1}$ in very good agreement with the value of 1×10^4 $(\Omega cm)^{-1}$ obtained from measurements under application of a dc-voltage. On the basis of an *n*-type conduction model one finds for the effective mass the value of roughly the free electron mass. As in the case of the planar transition metal complex compounds it must be doubted that the description by an n-type conduction model of a band filled to a degree of 5/6 is plausible. In addition, the positive sign of the thermopower is at variance with this description [12].

4.3. Transition Metal Chalcogenides

There exist only a few optical studies on compounds of this group, because of the minuteness of the needle- or ribbon-shaped crystals. Figure 13 exhibits the polarized

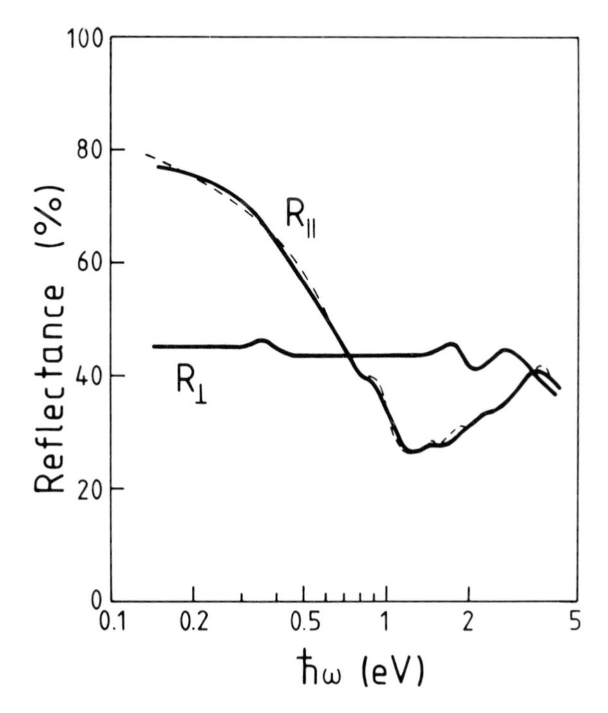

Fig. 13. Polarized reflectance spectra of $NbSe_3$. Experimental data (full line) and Lorentz–Drude (dashed line) [54].

reflectance spectra of $NbSe_3$ at room temperature. These spectra are distinguished from the reflectance spectra shown so far, primarily by the high reflectance values of 45 percent for the direction of polarization perpendicular to the metallic chains. The $NbSe_3$ owes its 'metallic' lustre to this reflectance dependence, which is typical for a semiconductor. For the direction of polarization parallel to the metallic chains, the reflectance in the visible range of the spectrum is distinctly lower. The rise to high reflectance values in the region of the strongly damped plasma edge becomes efficient only in the infrared spectral range.

For this compound with a conduction band filled by 1/4 the application of an n-type conduction model is not controversial. It yields, for the value of the effective mass, 0.4 m_0. This value is twice as high as for the planar transition metal complex compounds, but only almost half as high as for many one-dimensional organic metals. The dc-conductivity of 2.6×10^3 $(\Omega cm)^{-1}$, as determined from the optical measurements, is somewhat lower than the highest values determined electrically [55].

For comparison with the measurements on $NbSe_3$, Figure 14 shows the reflectance of

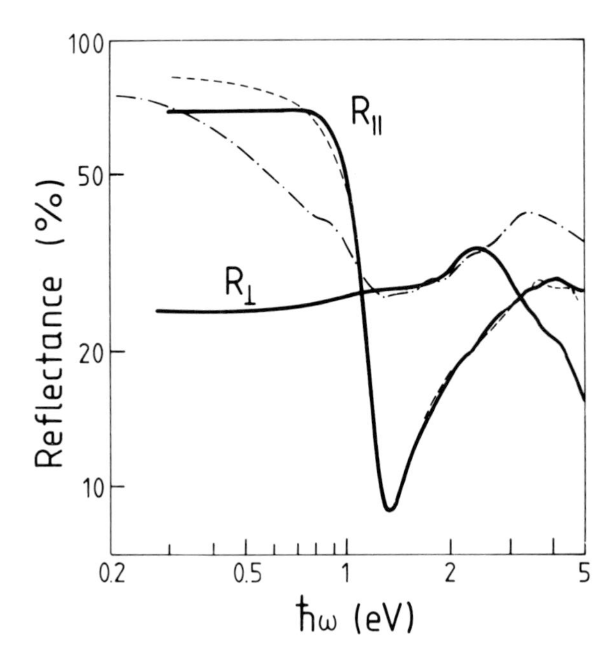

Fig. 14. Polarized reflectance spectra of $(TaSe_4)_2I$. Experimental data (full line) the Lorentz–Drude fit (dashed line). For a comparison the reflectance spectrum $R_{\parallel}(\hbar\omega)$ for $NbSe_3$ is also given (dashed and dotted line) [57].

the chemically and structurally related tetrachalcogenide $(TaSe_4)_2I$. This compound exhibits along the metallic axis a shorter metal–metal distance than $NbSe_3$. This leads to a much steeper plasma edge, hence to a higher value of the collision time and a higher optical value for the dc-conductivity. The optically determined conductivity of 4.7×10^3 $(\Omega cm)^{-1}$ of $(TaSe_4)_2I$ is almost twice as high as that of $NbSe_3$. This result is surprising in so far as electrical measurements showed the opposite trend, because the directly measured dc-conductivity of $(TaSe_4)_2I$ is smaller by about a factor of 1/5 than that of $NbSe_3$ [58]. The situation in $(TaSe_4)_2I$, with an optically determined value for the dc-conductivity exceeding the electrically determined value by almost an order of magnitude (see Table II), reminds one of that in the planar transition metal complex compounds. There this discrepancy is even more pronounced. In the case of $K_2Pt(CN)_4Br_{0.3} \cdot 3H_2O$ the low dc-conductivity was explained by the stability of the metallic state with respect to lattice distortions being so weak even at room temperature that a mobility gap is formed as a consequence of fluctuations. Because of the larger distance between the metallic chains in comparison with $NbSe_3$, $(TaSe_4)_2I$ exhibits a more pronounced structural anisotropy. It therefore approaches the extreme anisotropy of the planar

TABLE II

Transport properties of $NbSe_3$ and $(TaSe_4)_2I$ [54, 57].

	σ_{dc} (Ω^{-1} cm^{-1})	σ_0 (Ω^{-1} cm^{-1})	τ (sec^{-1})	m^*	a (Å)
$NbSe_3$:	4×10^3	2.6×10^3	1×10^{-15}	$0.4\ m_0$	3.48
$(TaSe_4)_2I$:	7×10^2	4.7×10^3	2.1×10^{-15}	$0.4\ m_0$	3.21

transition metal complex compounds. This suggests application of this interpretation to $(TaSe_4)_2I$, too. Also in favor of the influence of fluctuations on the transport properties of this substance in the range of room temperature is the fact that $(TaSe_4)_2I$ undergoes a metal–semiconductor transition just below this temperature, namely at 265 K [59].

4.4. Further Compounds

Recently it has been proved, by reflection measurements, that the blue molybdenum bronze $K_{0.3}MoO_3$ is a one-dimensional metallic conductor at room temperature [17]. Also, the metal–semiconductor transition at 180 K, which has been known for some time, was identified as a Peierls transition [60]. The optical measurements gave a Peierls gap of 0.15 eV [17]. Finally, in this material also, non-linear transport properties due to the formation of charge density waves were discovered [61].

The metallic conduction in the bronzes is not due to direct contact between the transition metal atoms. Rather, these atoms are surrounded by oxygen octahedra. Therefore, the oxygen atoms also participate in the metallic conduction. This means that a new structural principle has entered into the realm of research on one-dimensional inorganic metals. It possibly will open a new large field of activities.

5. Concluding Remarks

As this review shows, electronic transport properties of one-dimensional inorganic metals have, up to now, relatively seldom been investigated by optical methods. The examples given here may demonstrate that, nonetheless, optical measurements are capable of yielding a great variety of information and are, therefore, of use in answering quite different questions. It is the intention of this review to encourage workers in this field to apply optical methods more intensively in the study of this interesting class of solids.

Because the present article is limited to metallic conductors, work as important as the spectroscopical investigations of e.g., Yersin and coworkers on the relation between Pt—Pt distance and the band gap in semiconducting tetracyanoplatinates has not been covered [63]. It is dealt with for instance in the detailed article by Tanner on the optical properties of organic and inorganic one-dimensional systems [64].

Appendix: List of Symbols

Latin

a	metal specing along the metallic axis
a, b, c	crystallographic axes
1-D	one-dimensional
e	electronic charge
eV	electron Volt
E_F	Fermi energy
ΔE	Peierls gap
$\hbar$	Planck's constant
k_F	Fermi wave vector
m^*	effective mass
m_0	free electron mass
N	free carrier concentration
$r^{i\phi}$	complex reflectivity
$R(\omega)$	reflectivity
$R_{\|}(\omega); R_{\perp}(\omega)$	polarized reflectivity for the polarization direction parallel or perpendicular to the metallic axis
$4t$	band width
v_F	Fermi velocity

Greek

Γ_n	damping constant
ϵ_0	permittivity of free space
ϵ_∞	dielectric constant at high frequencies
$\epsilon(\omega)$	dielectric function
$\epsilon_1(\omega); \epsilon_2(\omega)$	real and imaginary part of the dielectric function
$\epsilon_{1\|}(\omega); \epsilon_{2\|}(\omega)$	real and imaginary part of the dielectric function for the polarization direction parallel to the metallic axis
$\epsilon_{\perp}(\omega)$	dielectric function of an $(SN)_x$ crystal for the polarization direction perpendicular to the polymeric chain axis.
$\epsilon_{\perp f}(\omega)$	dielectric function of a single $(SN)_x$ fiber for the polarization direction perpendicular to the polymeric chain axis
Λ	mean free path
ρ	amount of charge transfer
σ_0	optical value of the dc-conductivity
σ_{dc}	electrical value of the dc-conductivity
$\sigma(\omega)$	conductivity function
$\sigma_{\|}(\omega)$	conductivity function parallel to the metallic axis
$\sigma_{\perp}(\omega)$	conductivity function of an $(SN)_x$ crystal perpendicular to the polymeric chain axis
$\sigma_{\perp f}(\omega)$	conductivity function of a single $(SN)_x$ fiber perpendicular to the polymeric chain axis.
τ	collision time of the free carriers
ω	frequency
ω_p	plasma frequency
ω_{on}	eigenfrequency
ω_{CDW}	eigenfrequency of a pinned charge density wave.

References

1. F. Wooten, *Optical Properties of Solids*, Academic Press, New York (1972).
2. M. Born and E. Wolf, *Principles of Optics*, Pergamon Press, London (1959).
3. G. Schaack, *Phys. kondens. Materie* 1, 232 (1963).

4. K. Krogmann, *Z. anorg. allg. Chem.* **358**, 97 (1968).
5. K. Krogmann and H. D. Hausen, *Z. anorg. allg. Chem.* **358**, 67 (1968).
6. K. Krogmann, *Angew. Chem. internat. Edit.* **8**, 35 (1969).
7. See for a review: A. J. Heeger and A. F. Garito in: H. J. Keller (ed.), *Low-Dimensional Cooperative Phenomena*, (Nato ASI-Series B, vol. 7) Plenum Press, New York (1975).
8. D. Jerome, A. Mazaud, M. Ribault, and K. Bechgaard, *J. de Physique Lett.* **41**, L 95 (1980).
9. K. Bechgaard, K. Caneiro, M. Olsen, F. B. Rasmussen and C. S. Jacobsen, *Phys. Rev. Letters* **46**, 852 (1981).
10. See for reviews (a) H. P. Geserich and L. Pintschovius in: J. Treusch (ed.) *Adv. Solid State Physics*, (vol. 16, p. 65) Vieweg Braunschweig (1976); (b) R. L. Greene and G. B. Street in: H. J. Keller (ed.) *Chemistry and Physics of One-Dimensional Metals*, (Nato ASI-Series B, vol. 25) Plenum Press, New York (1977).
11. R. L. Greene, G. B. Street, and L. J. Suter, *Phys. Rev. Letters* **34**, 577 (1975).
12. See for a review: I. D. Brown, W. R. Daters, and R. J. Gillespie in: J. S. Miller (ed.) *Extended Linear Chain Compounds*, (vol 3, p. 1) Plenum Press, New York (1983).
13. A. Meerschaut and J. Rouxel, *J. Less-Common Metals* **39**, 197 (1975).
14. P. Haen, P. Monceau, B. Tissier, G. Waysand, A. Meerschaut, P. Molinie, and J. Rouxel, *Proc. 14th Int. Conf. on Low Temperature Physics, Otaniemi, Finnland*, (vol. 5 p. 445) (1975).
15. J. Rouxel, *Mol. Cryst. Liq. Cryst.* **81**, 31 (1982).
16. M. Potel, R. Chevrel, M. Sergent, J. C. Armici, M. Decroux, and Ø. Fischer, *J. Solid State Chem.* **35**, 286 (1980).
17. G. Travaglini, P. Wachter, J. Marcus, and C. Schlenker, *Solid State Comm.* **37**, 599 (1981).
18. D. Kuse and H. R. Zeller, *Phys. Rev. Lett.* **27**, 1060 (1971).
19. H. P. Geserich, H. D. Hausen, K. Krogmann, and P. Stampfl, *Phys. Stat. Sol. (a)* **9**, 187 (1972).
20. H. Wagner, H. P. Geserich, R. van Baltz and K. Krogmann, *Solid State Comm.* **13**, 659 (1973).
21. H. R. Zeller in: H. J. Queisser (ed.) *Adv. Solid State Physics*, (vol. 13, p. 31) Vieweg Braunschweig (1973).
22. D. Kuse, *Solid State Comm.* **13**, 885 (1973).
23. H. R. Zeller and P. Brüesch, *Phys. Stat. Sol. (b)* **65**, 537 (1974).
24. A. E. Underhill, D. M. Watkins, and C. S. Jacobsen, *Solid State Comm.* **36**, 477 (1980).
25. M. J. Minot and J. H. Perlstein, *Phys. Rev. Lett.* **26**, 371 (1971).
26. J. W. McKenzie, C. Wu, and R. H. Bube, *Applied Physics Lett.* **21**, 1 (1972).
27. D. W. Bullett, *Solid State Comm.* **27**, 467 (1978).
28. J. Bernasconi, P. Brüesch, D. Kuse, and H. R. Zeller, *J. Phys. Chem. Sol.* **35**, 145 (1974).
29. P. F. Williams and A. N. Bloch, *Phys. Rev.* **B10**, 1097 (1974).
30. P. F. Williams, M. A. Butler, and D. L. Rousseau, *Phys. Rev.* **B10**, 1109 (1974).
31. P. Brüesch, *Solid State Comm.* **13**, 13 (1973).
32. L. S. Agroskin, R. M. Vlasova, A. I. Gutman, R. N. Lynbovskaja, G. V. Papayan, L. P. Rautian, and L. D. Rosenstein, *Sov. Phys. Solid State* **15**, 1189 (1973).
33. P. Brüesch, S. Strässler and H. R. Zeller, *Phys. Rev.* **B12**, 219 (1975).
34. G. Winterling and T. P. Martin in: H. G. Schuster (ed.) *One-Dimensional Conductors*, (Lecture Notes in Physics, vol. 34), Springer Berlin (1975).
35. E. F. Steigmeier, R. Loudon, G. Harbeke, H. Auderset and G. Scheiber, *Solid State Comm.* **17**, 1447 (1975).
36. E. F. Steigmeier, D. Baeriswyl, G. Harbeke, and H. Auderset, *Solid State Comm.* **20**, 661 (1976).
37. R. L. Musselman and J. M. Williams, *J. C. S. Chem. Comm.*, 186 (1977).
38. L. H. Greene, D. B. Tanner, A. J. Epstein, and J. S. Miller, *Phys. Rev.* **B25**, 1331 (1982).
39. A. E. Underhill, D. M. Watkins, and C. S. Jacobsen, *Solid State Comm.* **36**, 477 (1980).
40. E. F. Steigmeier, D. Baeriswyl, H. Auderset and J. M. Williams, in: S. Barišić, A. Bjeliš, J. R. Cooper, and B. Leontić (ed.) *Quasi-One-Dimensional Conductors II*, (Lecture Notes in Physics, vol. 96, p. 229) (1979).
41. K. Krogmann and H. P. Geserich, in L. V. Interrante (ed.) *Extended Interactions between Metal Ions in Transition Metal Complexes*, (ACS Symposium Series, vol. 5, p. 350) American Chemical Society, Washington, D.C. (1974).

42. A. A. Bright, M. J. Cohen, A. F. Garito, and A. J. Heeger, *Phys. Rev. Lett.* **34**, 206 (1975).
43. L. Pintschovius, H. P. Geserich, and W. Möller, *Solid State Comm.* **17**, 477 (1975).
44. P. M. Grant, R. L. Greene, and G. B. Street, *Phys. Rev. Lett.* **35**, 1743 (1975).
45. A. A. Bright, M. J. Cohen, A. F. Garito, and A. J. Heeger, *Applied Physics Lett.* **26**, 612 (1975).
46. W. Möller, H. P. Geserich, and L. Pintschovius, *Solid State Comm.* **18**, 791 (1976).
47. H. P. Geserich, W. Möller, G. Scheiber, and L. Pintschovius, *Phys. Stat. Sol. (b)* **80**, 119 (1977).
48. K. Kaneto, K. Yoshino, and J. Inuishi, *J. Phys. Soc. Japan* **43**, 1013 (1977).
49. A. J. Gutman, L. S. Agroskin, G. V. Papayan, L. P. Rautian, and O. S. Schachnina, *Solid State Comm.* **49**, 187 (1984).
50. J. C. Maxwell-Garnett, *Phil. Trans. Roy. Soc. London* **203**, 385 (1904); **205**, 237 (1906).
51. W. D. Gill, W. Bludeau, R. H. Geiss, P. M. Grant, R. L. Greene, J. J. Mayerle, and G. B. Street, *Phys. Rev. Lett.* **38**, 1305 (1977).
52. E. S. Koteles, W. R. Datars, B. D. Cutforth, and R. J. Gillespie, *Solid State Comm.* **20**, 1129 (1976).
53. D. L. Peebles, C. K. Chiang, M. J. Cohen, A. J. Heeger, N. D. Miro, and A. G. MacDiarmid, *Phys. Rev.* **B15**, 4607 (1977).
54. H. P. Geserich, G. Scheiber, F. Lévy, and P. Monceau, *Solid State Comm.* **49**, 335 (1984).
55. P. Monceau, J. Richard, and M. Renard, *Phys. Rev.* **B25**, 931 (1982).
56. P. Brüesch and H. R. Zeller, *Solid State Comm.* **14**, 1037 (1974).
57. H. P. Geserich, G. Scheiber, P. Gressier, and P. Monceau to be published in: *Mol. Cryst. Liq. Cryst.*
58. Z. Z. Wang, M. C. Saint-Lager, P. Monceau, M. Renard, P. Gressier, A. Meerschaut, L. Guemas, and J. Rouxel, *Solid State Comm.* **46**, 325 (1983).
59. P. Gressier, A. Meerschaut, L. Guemas, J. Rouxel, and P. Monceau, *J. de Physique* **44**, C3-1741 (1983).
60. J. P. Pouget, S. Kagoshima, C. Schlenker, *J. de Physique Lett.* **44**, L 113 (1983).
61. J. Dumas and C. Schlenker, *Solid State Comm.* **45**, 885 (1983).
62. W. Denner, B. Schönfeld, and R. von Baltz, *Physics Lett.* **48A**, 313 (1974).
63. H. Yersin and G. Gliemann, *Ann. N.Y. Acad. Sci.* **313**, 539 (1978).
64. D. B. Tanner in: J. S. Miller (ed.), *Extended Linear Chain Compounds*, (vol. 2, p. 205) Plenum Press, New York (1982).
65. D. Kuse and H. R. Zeller, *Solid State Comm.* **11**, 355 (1972).

CHARGE-DENSITY WAVE TRANSPORT IN TRANSITION METAL TRI- AND TETRACHALCOGENIDES

P. MONCEAU

Centre de Recherches sur les Très Basses Températures,
C.N.R.S., BP 166 X, 38042 Grenoble-Cédex, France.

1. Introduction

Since the beginning of the seventies, intense research work has been undertaken in the study of the physical properties of systems of restricted dimensionality. Many chain-like compounds have been synthesized in which the electrical conductivity along the chains is much larger than that perpendicular to the chains. In a great number of compounds the interaction between ions and electrons, the so-called electron–phonon interaction, leads to structural instabilities at low temperature. The aim of the theoretical research was mainly to study the ground state of a one-dimensional electronic conductor at zero temperature as a function of the relative strength of several electron–electron couplings. Different ground states were predicted, a superconducting one or a spatially modulated insulating one: if the modulation involves the electronic charge density, the system is in a charge-density wave state (CDW), if the spin orientation is concerned, a spin density wave (SDW) occurs [1, 2].

The CDW one-dimensional instability was predicted by Peierls [3] fifty years ago. He showed that for a linear chain formed of atoms separated by the distance a, the electronic energy was lowered by a CDW formation:

$$\rho(x) = \rho_0 [1 + \alpha \cos(\mathbf{Q}x + \phi)] \tag{1-1}$$

where ρ_0 is the uniform electronic density, $\alpha\rho_0$ the charge modulation amplitude, $\mathbf{Q} = 2k_F$ the modulation wave vector and k_F the Fermi wave vector of the metal. The phase, ϕ, specifies the place of the CDW with regard to the lattice ions. Morever, in the electric field associated with $\rho(x)$, each ion is displaced in a new equilibrium position, the position for the nth ion being

$$u_n = u_0 \sin(n\mathbf{Q}a + \phi) \tag{1-2}$$

The amplitude of the displacement, u_0, is naturally small when compared to the lattice distance, a. The modulation of the ion positions can be detected by X-rays, neutrons or electron diffraction measurements [4]. Superlattice spots appear near the main Bragg spots corresponding to the unmodulated structure. Measurements of the components of these superlattice spots along the reciprocal axes give the CDW wave length. This new periodicity leads to a re-calculation of the band structure. The energy of a linear chain of atoms (with the very simple configuration of one electron per site or of a half-filled band) is drawn in Figure 1a as a function of the wave number. An energy gap, Δ, is opened at the Fermi level at $\pm\pi/2a$ which lowers the kinetic energy of the states below this level. All the states below the gap are occupied and all the states above the gap

P. Monceau (ed.), Electronic Properties of Inorganic Quasi-One-Dimensional Materials, II, 139–268.

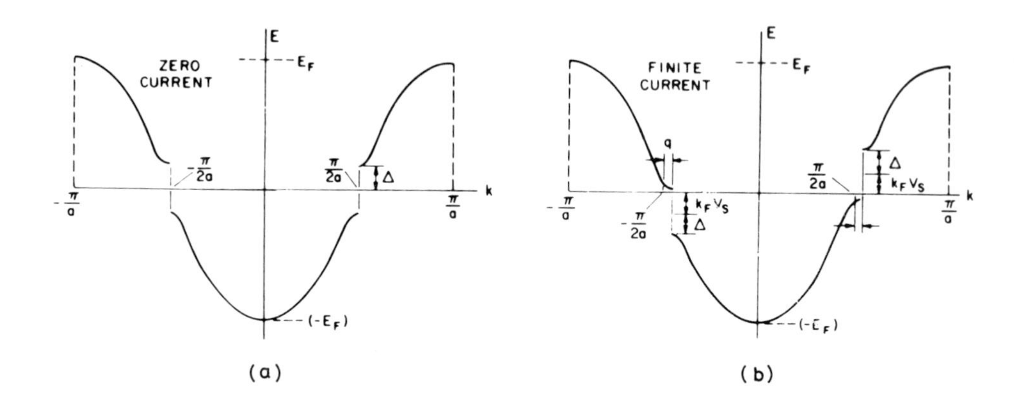

Fig. 1. (a) Tight binding band for a linear chain of atoms separated by a with one electron per atom. The Peierls instability gives rise to a distortion of period $2a$. An energy gap, Δ, is opened at the Fermi level $\pm\pi/2a$. (b) The same band as in (a) but displaced by q, which leads to a current $J = nev_s$ (after [7]).

are empty. The ground state of such a system is therefore insulating. This structural transition (or Peierls transition) occurs if the crystal distortion is energetically favourable when compared to the gain of electronic energy caused by the gap formation. Such a CDW formation has also been observed in two-dimensional compounds, namely transition metal dichalcogenides [5]. The Fermi surface of these compounds shows a particular shape in such a way that many states located on either side of the Fermi surface are connected by the same vector **Q** (the nesting condition). The low temperature ground state remains metallic. For a strictly one-dimensional conductor the Fermi surface consists of two parallel planes separated by **Q**. All the states are connected by the same **Q**. The energy gap involves the whole Fermi surface and the low temperature ground state is insulating.

Contrary to a semiconductor for which the energy gap at the Fermi level is due to the ionic potential, and therefore bound to the crystal frame, the Peierls energy gap is associated with a lattice deformation. The latter can slide as a propagating wave with a velocity, v_s, each ion being displaced as follows:

$$u = u_0 \cos \mathbf{Q}(x - v_s t) \tag{1-3}$$

and oscillating at the frequency $\omega = \mathbf{Q}v_s$. This model of a sliding charge density wave was proposed in 1954 by Fröhlich [6] in a jellium approximation as a mechanism which could lead to a superconducting state. The Fröhlich model has been studied again by Allender *et al.* in 1974 in a tight binding model [7]. Figure 1b shows the Fermi distribution of the same chain as in Figure 1a when the modulation is displaced with a uniform velocity, v_s. The two planes of the Fermi surface are at $(-\pi/2a) + q$ and $(+\pi/2a) + q$ with

$$m^* v_s = \hbar q \tag{1-4}$$

where m^* is the effective electronic mass. In the Galilean frame in motion with v_s, the system is again unstable with regard to a distortion which opens an energy gap at the new Fermi surface with the wave vector, $\mathbf{Q} = \pi/a$. All the electrons are below the gap and in this frame the electronic current is zero. If the velocity, v_s, is small, i.e., if the kinetic energy of the electrons in their translation, $\hbar k_F v_s$, is small compared to the Peierls gap, Δ, in the static reference frame, the current in the sample will be:

$$J = -nev_s \tag{1-5}$$

where n is the number of electrons per unit volume in the band affected by the CDW.

A comparison with the BCS superconductivity [8] can be established. In the latter case the energy gap in the excitations at the Fermi level does not at all prevent conductivity. This is so because the interaction involved does not require a specified reference frame and because Cooper pairs can be built either in states 'k and $-k$' or '$k + \kappa$ and $-k + \kappa$'. The latter state leads to a uniform velocity such as:

$$mv_s = \hbar\kappa$$

In the Fröhlich model, the energy gap reduces the elastic scattering of individual electrons because there is no state available for relaxing energy. The motion is therefore without dissipation and the compound becomes superconducting.

This Fröhlich mode is a direct consequence of the translation invariance, the CDW energy being independent of the phase, ϕ, with regard to its position along the chain axis. In fact, in real systems, as shown by Lee, Rice and Anderson [9] (LRA in the following) this translation invariance is broken because the phase, ϕ, is pinned to the lattice. The pinning can be provided by impurities, commensurability between the CDW wave-length and the lattice (when the wavelength of the CDW is not a simple multiple of the lattice distance, the CDW is said to be incommensurate), or by Coulomb interaction between adjacent chains. Oscillations of the CDW pinned mode is expected to lead to large low frequency ac conductivity and to a large dielectric constant. An applied dc electric field, however, can supply the CDW with an energy higher than the pinning one and above a threshold electric field, the CDW can slide and carry a current, but damping prevents superconductivity.

This chapter will be devoted to a review of physical systems exhibiting this sliding mode. In Section 2 we are going to describe some experiments performed on one-dimensional compounds which have been interpreted as a collective response to a CDW: experiments at infrared frequencies and in the temperature range above the Peierls transition where the CDW is in a fluctuating state. Nonlinear dc conductivity induced by an electric field is shown to occur at any temperature below Peierls transitions in three different families of inorganic compounds: the transition metal trichalcogenides, some bronzes and the halogenated transition metal tetrachalcogenides. Section 3 contains a description of the structure of these compounds and Section 4 a review of their physical properties in the limit of zero electric field. In Section 5 we describe the CDW current-carrying state and in Section 6 we analyze the different theoretical models. Analogy between CDW transport and other nonlinear systems is made in Section 7. The last section contains a critical analysis of these experiments compared to the above-mentioned theories.

2. Collective Response of a CDW in KCP and TTF–TCNQ

Over and above the Peierls [3] and Fröhlich [6] theories, most concepts used in subsequent research on one-dimensional conductors have been developed in the LRA paper [9]; especially the central concept of the pinned mode. LRA showed that the coupled electron–phonon mode which leads to the Kohn anomaly [10] at high temperature is split into two different modes below the incommensurate Peierls transition: an optical mode, Ω_+, and an acoustic mode, Ω_-, with the following frequency dependencies:

$$\Omega_+^2 = \lambda\omega_Q^2 + \frac{1}{3}\frac{m^*}{M^*} v_F^2 \,|q - \mathbf{Q}|^2 \tag{2-1}$$

$$\Omega_-^2 = \frac{m^*}{M^*} v_F^2 \,|\mathbf{Q} - q|^2 \tag{2-2}$$

with $\mathbf{Q} = 2k_F$, λ the electron–phonon coupling constant, ω_Q the bare phonon frequency at high temperature, m^* the band mass of the electrons and M^* the effective mass or Fröhlich mass with

$$\frac{M^*}{m^*} = 1 + 4\frac{\Delta^2}{\lambda\omega_Q^2} \tag{2-3}$$

Ω_+ mode is an amplitude mode and Ω_- mode is the phase mode, or phason as named by Overhauser [11]. From (2-2) the velocity, C_0, of the phase mode can be seen to be:

$$C_0 = (m^*/M^*)^{1/2} v_F \tag{2-4}$$

In fact, the splitting into these two modes is a general result for any incommensurate displacive transition [12]. The acoustic mode has recently been detected by neutron scattering investigation in phase III of biphenyl [13] and in thorium halides as $ThBr_4$ [14]. For ideal one-dimensional conductors the phason can exist at zero frequency which corresponds to the persistent dc current provided by the sliding motion of the CDW as proposed by Fröhlich. LRA pointed out that various mechanisms introduce a gap, ω_0, in the excitations of the phason mode which prevents the dc Fröhlich conductivity but leads to a large low frequency ac conductivity.

2.1. Far-Infrared Conductivity

Among the mixed-valency platinum compounds, or Krogmann salts, $K_2Pt(CN)_4Br_{0.3}(3H_2O)$, (in short KCP), was the first one-dimensional inorganic compound in which a Peierls transition was clearly identified [15–17]. The properties of these salts are reviewed by K. Carneiro in Chapter I of this book. Another family has aroused great interest, namely the organic transfer salts, the prototype of which is the tetrathiafulvanium–tetracyanoquinodimethanide, TTF–TCNQ (for a review see [18]). For these compounds the wavelength of the CDW is incommensurate with the lattice and the phase, ϕ, of the CDW is supposed to be pinned by impurities. The CDW response to a high frequency excitation has been calculated by LRA in terms of the dielectric function. They found that the dielectric constant comprises three parts:

(i) ϵ_∞ the high frequency dielectric constant;

(ii) the single particle oscillator in the presence of the Peierls gap, Δ. The oscillator strength is $\frac{2}{3}\Omega_p^2/\Delta^2$ with $\Omega_p^2 = 4\pi ne^2/m^*$ where ω_p is the conduction electron plasma frequency, n the total electron density and m^* the band electron mass;

(iii) the collective mode oscillator due to the oscillations of the pinned CDW mode. Rice *et al.* [19] have assumed that the equation of motion of the centre of mass X of the rigid CDW, displaced from its equilibrium position, is that of a damped harmonic oscillator. In general the equation of such as oscillator is:

$$M\ddot{X} + \Gamma\dot{X} + kX = F \tag{2-5}$$

where M is the inertia, Γ the damping constant, k the restoring force and F the driving force. Two characteristic times have to be considered:

$$\tau = \Gamma/k \tag{2-6}$$

which is the exponential damping time for the viscous pendulum when the term $\ddot{X}$ is neglected and

$$\omega_0^2 = k/M \tag{2-7}$$

which is the low amplitude natural oscillation frequency of the undamped pendulum.

Therefore the equation (2-5) can be written as [19]:

$$\ddot{X} + \Gamma\dot{X} + \omega_0^2 X = eE/M^* \tag{2-8}$$

Following (2-8) the dielectric constant arising from the oscillations of the CDW is:

$$\epsilon(\omega) = \epsilon_\infty \frac{\Omega_p'^2}{\omega_0^2 - \omega^2 - i\Gamma\omega}$$

where Ω_p' is the measure of the oscillator strength with $\Omega_p'^2 = 4\pi ne^2/M^*$. The conductivity $\sigma(\omega)$ can be deduced from $\epsilon(\omega)$ as follows:

$$\sigma(\omega) = \frac{ne^2}{i\omega M^*} \frac{\omega^2}{\omega_0^2 - \omega^2 - i\omega\Gamma} \tag{2-9a}$$

$$= \frac{ne^2}{\Gamma M^*}\left[\frac{\Gamma^2\omega^2}{(\omega_0^2 - \omega^2)^2 + \Gamma^2\omega^2} - \frac{i\Gamma(\omega_0^2 - \omega^2)}{(\omega_0^2 - \omega^2)^2 + \Gamma^2\omega^2}\right] \tag{2-9b}$$

As the CDW is pinned, $\sigma(\omega)$ contributes only at finite frequencies. The absence of pinning, $\omega_0 = 0$, results in a dc conductivity:

$$\sigma(0) = \frac{ne^2}{\Gamma M^*} \tag{2-10}$$

If, in addition, the damping is zero ($\Gamma = 0$) the conductivity, after (2-9a), is:

$$\sigma = -\frac{ne^2}{i\omega M^*}$$

which is exactly the result deduced from the London equation for a superconductor [20]. We once more find the Fröhlich supercurrent.

Infrared and far-infrared reflectivity measurements have been performed on KCP [21] and TTF–TCNQ [22–24]. (A review of the optical properties of one-dimensional conductors is given in Chapter 3 of this book by H. P. Geserich). The conductivity is deduced from a Kramers–Krönig analysis of the reflectivity. Structures in the far-infrared reflectivity have directly revealed the pinned mode of the CDW. The conductivity of KCP as a function of the frequency shows a sharp peak at low temperature around 15 cm^{-1} (see Figure 8 in Chapter 1 and Figure 7 in Chapter 3 in this book). The strength of the oscillator decreases when T is increased. From the experimental static dielectric constant, the oscillator CDW strength, Ω'_p, is deduced, which leads to a value of the Fröhlich M^*. For KCP, M^* was estimated to be $\sim 10^3$ the band electronic mass. Similarly, the conductivity of TTF–TCNQ has a strong peak at 40 cm^{-1} [23, 24]. This peak, again, has been associated with the oscillations of the phase mode of the CDW. The effective mass, M^*, deduced from the far-infrared measurements was found to be $M^* \sim 60\, m \sim 20\, m^*$ in [23] and $M^* \sim 24\, m \sim 6.3\, m^*$ in [24] which is much lower than the value commonly accepted, typically the one for KCP $\sim 10^3\, m$. Experiments [23] again show that the oscillator strength decreases when T is increased and disappears for temperatures higher than 60 K. At this temperature and above, a significant part of the dc conductivity takes its source from the collective mode, but in a temperature range in which the CDW is fluctuating. The conductivity associated with the CDW fluctuations is described in 2.2.

2.2. Fluctuating Conductivity in TTF–TCNQ

The measurements described in 2.1 clearly show the existence of a pinned mode at low temperatures well below the Peierls transition. However, above T_P, the CDW can fluctuate and does carry a current. This possibility has been raised by Bardeen [7, 25] to explain the huge electrical conductivity measured in highly controversial experiments in some TTF–TCNQ crystals in the vicinity of 60 K, just above the Peierls transition [18].

Another source of pinning, suggested by LRA, is the commensurability pinning. The pinning potential has the following form:

$$V = V_M(1 - \cos M\phi) \tag{2-11}$$

where M is the order of the commensurability. The amplitude V_M decreases strongly with increasing M as $1/M^2$. The pinned mode has a frequency calculated as follows:

$$\omega_p = \lambda^{1/2} M \left(\frac{e\Delta}{W}\right)^{\frac{1}{2}M - 1} \omega_Q \tag{2-12}$$

where W is the band width.

The study of TTF–TCNQ under pressure has shown the existence of a narrow pressure domain around 20 kbar in which the CDW wavelength becomes commensurate with the lattice with $M = 3$ [26]. Figure 2 shows the conductivity of TTF–TCNQ measured along the chain axis, **b**, as a function of the pressure at different temperatures [27, 28]. A drop in the conductivity is observed in the pressure range of 17–20 kbar. Such a drop

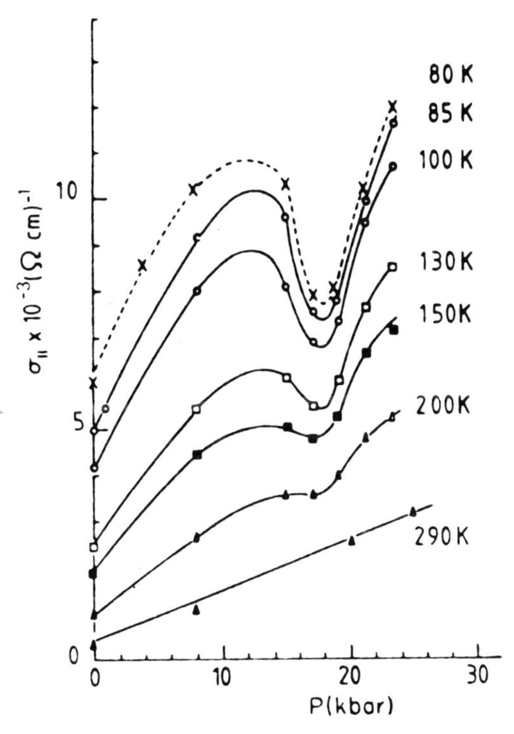

Fig. 2. Electrical conductivity of TTF–TCNQ as a function of pressure at different temperatures. The drop in the parallel conductivity in the vicinity of 20 kbar occurs in the pressure range in which the CDW becomes commensurate with the lattice. Such a drop is not detected in the transverse conductivity (after [27]).

does not exist when the conductivity is measured along the axes, **a** or **c**, perpendicular to the chain axis. The interpretation used to explain this drop was the commensurability locking of the CDW [29]. In the pressure range in which the CDW is incommensurate, sliding motion of the fluctuating CDW is possible, according to the Fröhlich model. When the CDW becomes commensurate, the phase of the fluctuating CDW is pinned and this restoring force prevents the translation of the fluctuating CDW. The conductivity has been expressed as follows:

$$\sigma = \sigma_{sp} + \sigma_F$$

with σ_{sp} the single particle conductivity and σ_F the collective conductivity due to the fluctuating CDW. σ_F is strongly suppressed by the commensurability. The same result has been measured in TSF–TCNQ although with a far smaller drop in conductivity [30].

2.3. Non-linear Excitations at Low Temperatures

Rice *et al.* [31] have predicted non-linear charged elementary excitations in the pinned

CDW condensate at low temperature. They have derivated the equation of motion for the condensate without damping, the phase being the generalized coordinate, as follows:

$$\frac{\partial^2 \phi}{\partial t^2} - C_0^2 \frac{\partial^2 \phi}{\partial x^2} + \omega_0^2 \frac{dV}{d\phi} = 0 \tag{2-13}$$

where ω_0 is the free oscillation frequency of the pinned CDW determined in far-infrared measurements as seen in Section 2.1, C_0 the phason velocity defined by (2-4). If the potential, V, has the cosinusoidal variation as given in (2-11), (2-13) is the Sine–Gordon equation. Its solutions are large amplitude solutions which correspond to solitons or 'kinks' of $\pm 2\pi/M$ phase, localized over a distance $d = C_0/\omega_0$ and propagating along the chains. K. Maki gives a review of the general properties of solitons in Chapter 4 of Part I of this work. In this section, therefore, we only intend to show some physical results involving solitons in transport properties.

(i) These solitons are local, mobile charges and they can carry a current although the whole CDW is pinned, preventing a Fröhlich-type conductivity. This contribution to the electrical conductivity follows an activated law in $\exp - (E_\phi/kT)$ where E_ϕ is the soliton creation energy [31]. If E_ϕ is lower than the Peierls gap, the solitons will dominate the transport properties. Experimentally, the conductivity in TTF–TCNQ [32] and in KCP(Br) [33] and KCP(Cl) [34] was found to deviate from the activation law in $\exp - (\Delta/kT)$ due to single particle excitation. The activation energy of TTF–TCNQ was found to be 14 K and 5 K in TSF–TCNQ at low temperature. The low temperature activation energy of KCP(Br) is 420 K which has to be compared to the Peierls gap, $2\Delta = 1840$ K. Besides the creation energy, E_ϕ, solitons are characterized by a temperature dependent diffusion time, t_D, which measures the time between soliton jumps. This diffusion time was also shown by NMR measurements to follow an activation law: $t_D = t_{D_0} \exp(E_m/kT)$ [35]. At low temperatures the conductivity has two channels: the soliton conductivity, σ^s, the temperature variation of which is $\sigma_0 [t_D \times \exp - (E_\phi/kT)]$ and the conductivity, σ^e, from single excitation through the CDW gap. Thomas [34] has measured the ratio between the soliton contribution, σ^s, and the total conductivity. At 40 K, the single electron and the soliton conductivity are roughly equal, but this ratio approaches the value of 1 when T is extrapolated towards zero temperature.

(ii) Cohen and Heeger have measured highly non-linear I(V) characteristics in TTF–TCNQ [32]. They have shown that the activation energy of the phase solitons solutions of (2 – 13), Δ_S^0, decreases linearly with the application of an electric field:

$$\Delta_S = \Delta_S^0(1 - E/E_0)$$

with the characteristic field $E_0 \sim 150 \pm 50$ V/cm. Figure 3 reproduces the variation of the current density, J, as a function of the electric field for TTF–TCNQ at low temperatures, in a logarithmic scale. The insert shows the linear plot of $J(E)$. Clearly, around 150 V/cm at the field which reduces the activation energy to zero, new excitations are created by the electric field contributing to the increase of the conductivity. As a function of T, all the curves, $J(E)$, coincide for fields higher than 600 to 1000 V/cm. This field is thought to unpin the whole CDW from the pinning potential, believed to be due to the interchain Coulomb pinning. Beyond this field which reduces the potential barrier to zero, the whole CDW is thought to slide freely.

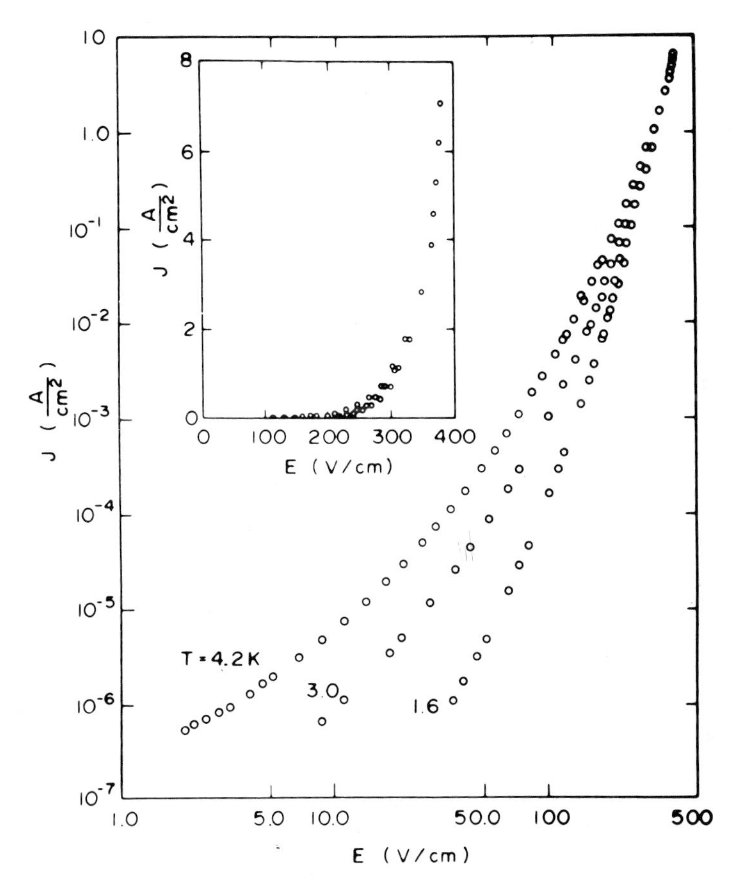

Fig. 3. Log–log variation of the current density measured along the chain axis of TTF–TCNQ as a function of the electric field at different low temperatures. The insert is a linear plot of J *vs* E (after [32]).

2.4. Summary

Measurements performed on one-dimensional compounds which belong to families such as Krogmann salts or organic charge transfer salts have revealed that frequencies for free oscillations of the CDW are in the far-infrared range. The CDW in the fluctuating state above the Peierls transition was also shown to carry a current. One explanation for the free motion of CDW fluctuations in TTF–TCNQ is to suppose that, because they are localized on a finite part of a single chain, they might be insensitive to pinning by defects [28]. Finally at low temperatures excitations or solitons are expected to dominate transport properties.

Completely different results will now be presented for some transition metal tri. and tetrachalcogenides. We will see that a dc extra conductivity is measured at any temperature *below* the Peierls transition when the electric field exceeds only typically 1 V/cm. For

$NbSe_3$ in a restricted temperature range it needs only a few mV/cm so that the extra-conductivity takes place.

3. Structural Description of Transition Metal Tri- and Tetrachalcogenides

We now wish to present only the general structural features of families of compounds exhibiting non-linear transport properties that we describe in Section 5. A detailed description concerning the chemical preparation and the structural characterization will be given by Meerschaut *et al.* in another book of this series [36].

3.1. MX_3 COMPOUNDS

MX_3 crystals are synthesized with the transition metal M atom which belongs to Group IV (Ti, Zr, Hf) or Group V (Nb, Ta) and with chalcogenide atoms X such as S, Se or Te. The basic constituent of the structure is a trigonal prism $[MX_6]$ with a cross-section close to an isosceles triangle. The transition metal atom is located roughly at the centre of the prism. These trigonal prisms are stacked on top of each other by sharing their triangular faces. Chains are staggered with respect to each other by half the height of the unit prism as shown in Figure 4b. Therefore besides the six chalcogen atoms of the $[MX_6]$ prism, each transition metal is bonded to two more X atoms from neighbouring chains, and its coordination number is eight. In the case of transition metal dichalcogenides, regular $[MX_6]$ trigonal prisms form infinite layers as shown in Figure 4a and the co-ordination number of the transition metal is really six [5].

The simplest unit cell is of the $ZrSe_3$ type [37] as shown in Figure 5a. It consists of a unique type of MX_3 chain with covalent Se_2^{2-} bonding between the nearest Se in the basis of the triangle of the prism as demonstrated by X-ray photoemission (XPS) by Jellinek *et al.* [38]. Among compounds formed with Group V transition metals, only NbS_3 (type I) retains this simple structure with one type of chains but with cation pairing along the chains leading to alternate long and short distances between the niobium atoms [39]. Therefore, the unit cell parameter along the chain axis is twice the height of the trigonal prism. Another polytype of NbS_3, called type II, has been found [40, 41]. Roucau has shown that its structure is monoclinic [42]. There is no more pairing along the chain axis but the lattice parameters perpendicular to the chains are twice those of NbS_3 type I. It seems that other polytypes of NbS_3 exist but they have not been characterized [43, 44].

Cross-sections of the unit cell perpendicular to the chain axis for $TaSe_3$ [45] and $NbSe_3$ [46, 47] are shown in Figures 5b and 5c. MX_3 chains can be distinguished according to the strength of the chalcogen—chalcogen bond in the basis of the triangle of the chain. $TaSe_3$ is made up of two groups of two chains with an intermediate bond (Se—Se distance of 2.576 Å) and a weak bond (Se—Se distance of 2.896 Å). The linkage between chains defines a set of weakly interacting layers parallel to the (110) face. The structure of $NbSe_3$ consists of units of four chains similar to those of $TaSe_3$ and of units of two chains similar to those of $ZrSe_3$. Chains with strong Se—Se pairing (Se—Se distance of 2.37 Å) are called chains III, with intermediate pairing (Se—Se distance of 2.49 Å) chains I, and with weak bond (Se—Se distance of 2.91 Å) chains II. $NbSe_3$ crystallises

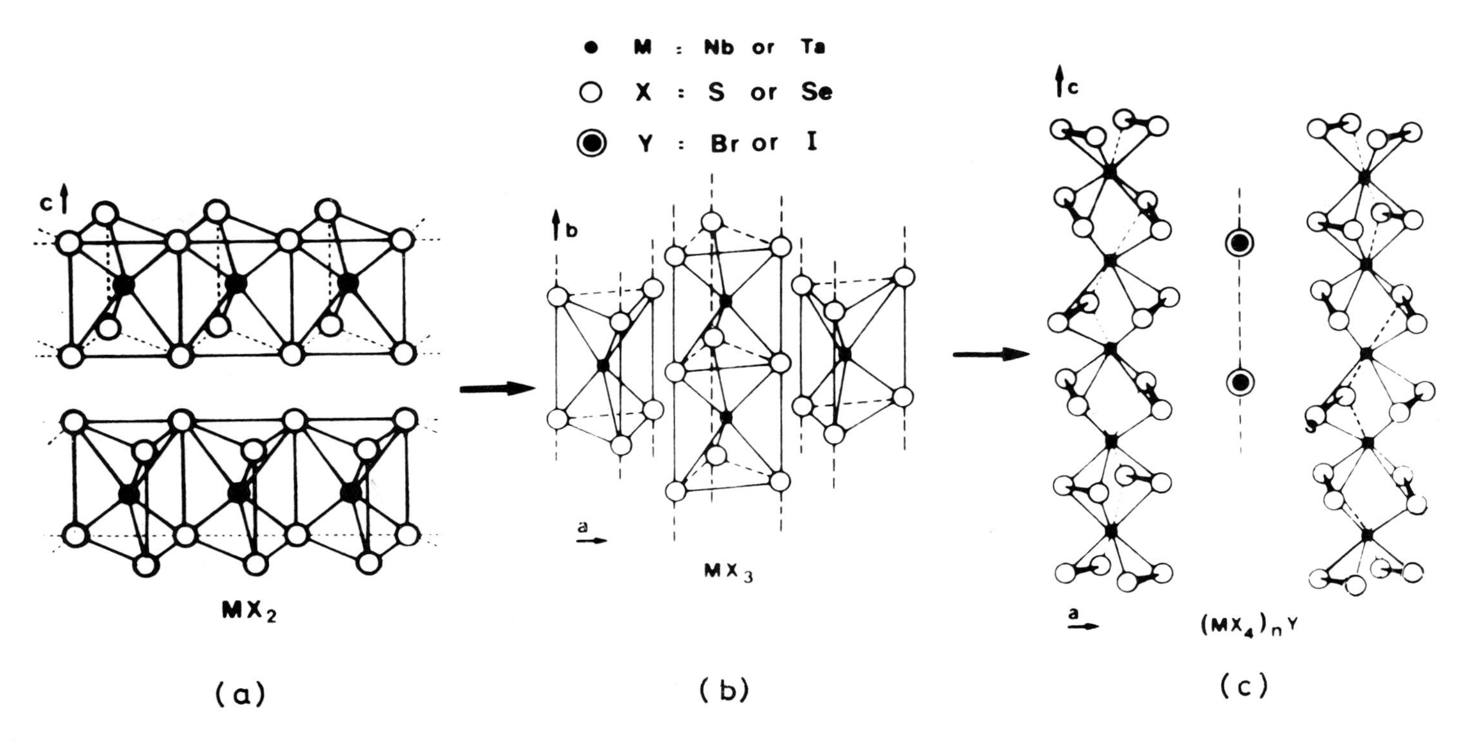

Fig. 4. Schematic representation of the transition metal atom environment. (a) In layered dichalcogenide MX_2 compounds; (b) in chain-like trichalcogenide MX_3 compounds; (c) in chain-like halogenated tetrachalcogenides $(MX_4)_nY$ compounds.

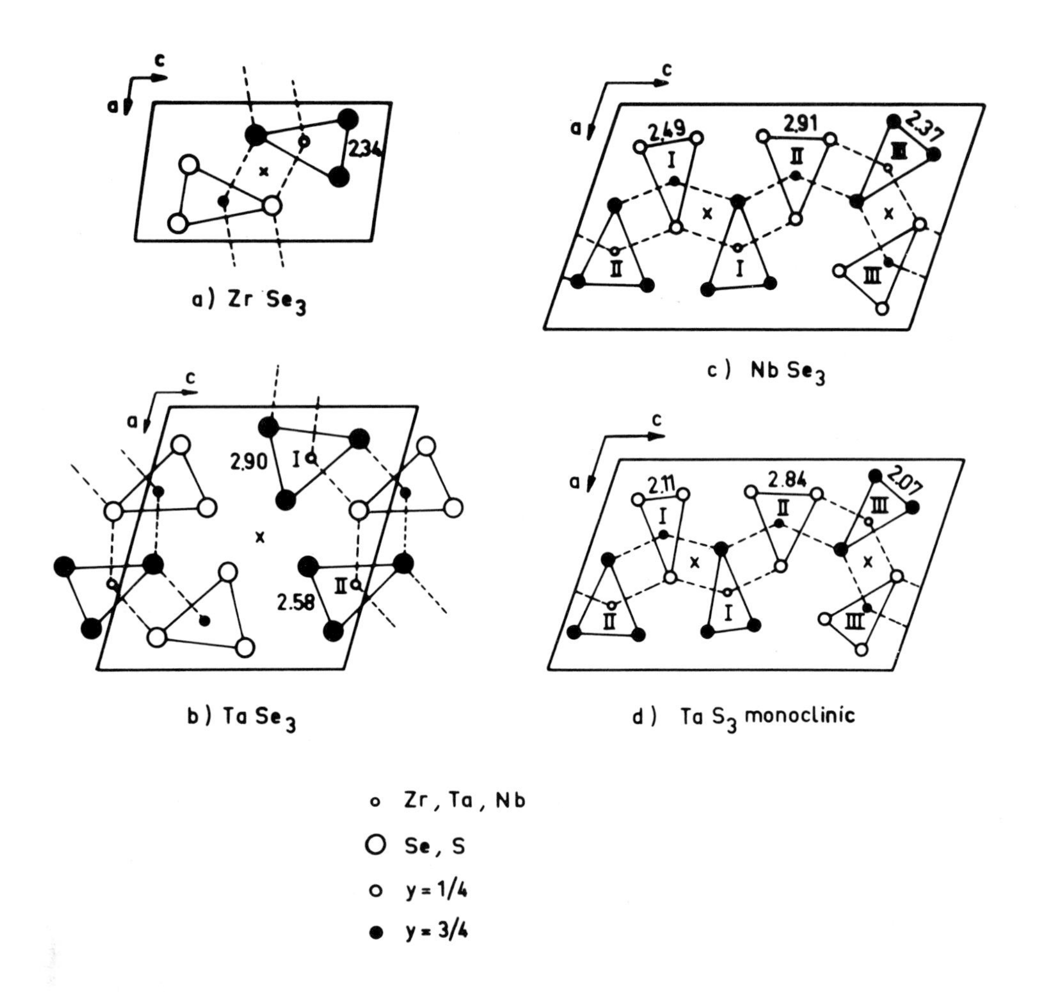

Fig. 5. Projection of the structure along the monoclinic **b** axis of (a) $ZrSe_3$, (b) $TaSe_3$, (c) $NbSe_3$ and (d) TaS_3 with the monoclinic unit cell.

in a thin ribbon-like shape. The plane of the ribbon was determined by X-rays to be the (**b**, **c**) plane.

Two polytypes of TaS_3 have been synthesized, one with a monoclinic [48], the other with an orthorhombic [49] unit cell. The monoclinic form of TaS_3 has a structure similar to the $NbSe_3$ one as shown in Figure 5d, the different short distances between the sulfur–sulfur pairs being 2.068 Å, 2.105 Å and 2.835 Å. The structure of the orthorhombic TaS_3 is as yet unknown. The unit cell is huge and comprises 24 chains.

Structurally the MX_3 compounds can be described as layers of chains two prisms thick weakly coupled through Van der Waals bonds which lead to a 'mixed' dimensionality: one-dimensional properties as far as the infinite chains are concerned and bidimensional behaviour depending on the strength of the lateral bonds between chains. This 'mixed' dimensionality has been clearly shown by lattice dynamics studies. Optical modes using

Raman techniques have been investigated in many of the Group IV MX_3 compounds [50, 51] and also in $TaSe_3$ and $NbSe_3$ [52]. (For a review, see A. Grisel and F. Lévy, Chapter V of this book). Anisotropy of the chemical bonding has also been demonstrated by the measurement of the anisotropy of compressibility in $TaSe_3$ [53] and in $NbSe_3$ [54]. For the latter it was found in kbar^{-1} $K_a = 13.7 \pm 1.5 \times 10^{-4}$, $K_b = 1.30 \pm 0.1 \times 10^{-4}$ and $K_c = 5.85 \pm 0.3 \times 10^{-4}$. Magnitudes of K_b and K_c are those of metals. K_a is large and comparable with the compressibility in layered compounds; the **a** axis is very compressive because it is nearly perpendicular to the Van der Waals gap.

3.2. $(Fe_{1+x}Nb_{1-x})Nb_2Se_{10}$ TYPE STRUCTURE

Attempts to dope $NbSe_3$ with iron has led to a new compound, firstly reported as $Fe_{0.25}Nb_{0.75}Se_{12}$ [55], but in fact $FeNb_3Se_{10}$ [56, 57]. Its structure is shown in Figure 6. It consists of two groups of chains running parallel to the monoclinic axis, **b**: the

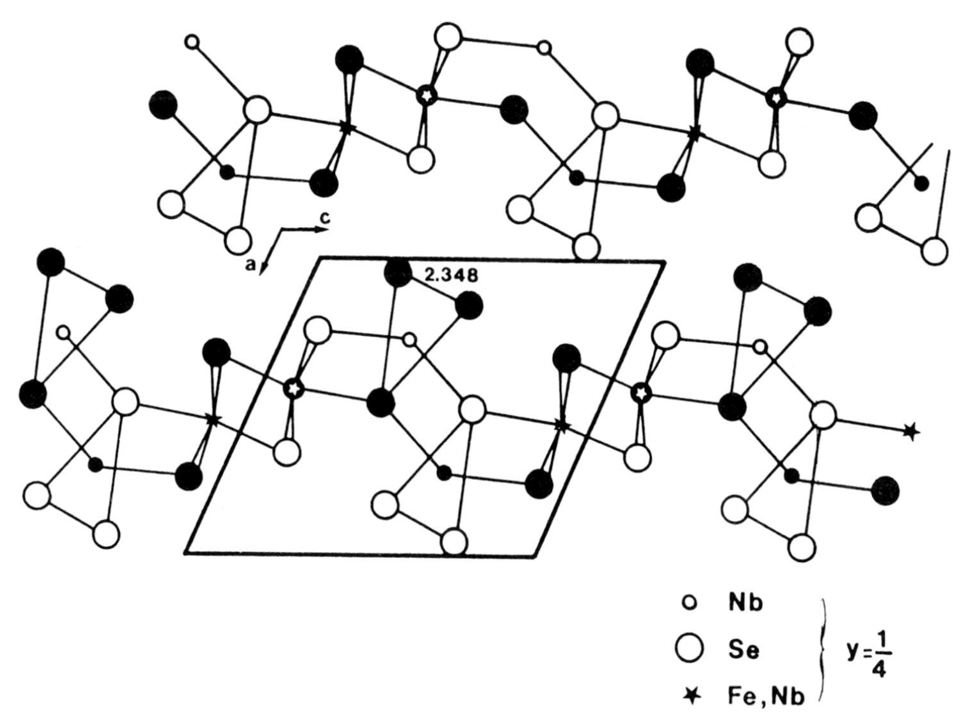

Fig. 6. Projection of the $(Fe_{1+x}Nb_{1-x})Nb_2Se_{10}$ structure along the monoclinic **b** axis. The unit cell consists of two trigonal prismatic chains and two octahedral chains.

first group is two $NbSe_6$ trigonal prismatic chains with the same interatomic distances (Se—Se distance of 2.34 Å) as the trigonal prismatic chains in $NbSe_3$ with the strongest chalcogen–chalcogen coupling (chains of type III in $NbSe_3$); the second group is a double chain of edge-shared octahedra of selenium around both iron and niobium, randomly distributed within the chain. As for MX_3 compounds, the chains are joined to form two-dimensional slabs separated by Van der Waals gaps perpendicular to the chain direction. The stoichiometry of these compounds remains an open question. Cava *et al.* [58] have found that the true stoichiometry was not $FeNb_3Se_{10}$ but $(Fe_{1+x}Nb_{1-x})Nb_2Se_{10}$ with $0.25 \leqslant x \leqslant 0.40$. A few other compounds have been synthesized with the same structural type: $(Fe, V)Nb_2Se_{10}$, $(Fe, Ta)Nb_2Se_{10}$ [58, 59] and $(Cr, Nb)Nb_2Se_{10}$ [59, 60].

3.3. $(MX_4)_nY$

A chain-like structure has also been found in a new family of transition metal tetrachalcogenides with the general formula $(MX_4)_nY$ where M = Ta, Nb; X = Se, S and Y = I [61] or Br [62]. Structural data have been obtained for $(NbSe_4)_3I$ [63], $(TaSe_4)_2I$ [64], and $(NbSe_4)_{10}I_3$ [65]. These compounds consist of parallel MX_4 chains with iodine atoms lying between them. In each chain the transition metal atom is located at the centre of a rectangular antiprism of eight Se atoms (Figure 4c). The dimensionals of each Se_4 rectangle of the MSe_4 chain are about 2.35–2.40 Å by 3.50–3.60 Å. The shortest Se—Se distance is similar to that found in chain type III of $NbSe_3$. Also each Se_4 rectangle can be seen as made up of two Se_2^{2-} dimers. According to the compounds, the unit cell **c** parameter along the chains comprises a different number of niobium atoms with a possible pairing between them. For instance, in $(NbSe_4)_3I$ six niobium atoms are located in the **c** parameter with the sequence a short Nb—Nb distance (3.06 Å) between two equivalent longer Nb—Nb distances (3.25 Å). In $(TaSe_4)_2I$ the distances between the four tantalum atoms of the unit cell along the chains are all identical to 3.206 Å. Finally there are 10 niobium atoms along **c** in $(NbSe_4)_{10}I_3$ with the following Nb—Nb sequence: —Nb–3.17–Nb–3.17–Nb–3.23–Nb–3.15–Nb–3.23–Nb–3.17— . . . .

3.4. OTHER CHAIN-LIKE COMPOUNDS

Other families of inorganic compounds also present a chain-like structure, namely such bronzes as $K_{0.30}MoO_3$, VS_4 or MX_4 compounds, pentachalcogenides as $HfTe_5$ and Nb_3X_4 compounds. The structure of $K_{0.30}MoO_3$, although different from that of $NbSe_3$, can be described as layers of infinite chains of MoO_6 octahedra, parallel to the high conductivity monoclinic **b** axis, sharing corners only, the K ions separating the layers [66].

VS_4 is isostructural with the $[MSe_4]$ chains of $(MSe_4)_nI$; it shows a V—V bond along the chain axis leading to alternating short and long distances between the vanadium atoms [67]. On the other hand, transition metal atoms in the chain of $NbTe_4$ [68, 69] and $TaTe_4$ [68] are located at the centre of an antiprism made up of two square Te_4 units (rectangular Se_4 units for $(MSe_4)_nI$). The distances between the transition metal atoms along the chain axis are equivalent but the shortest Te—Te distance does not represent the side of the square Te_4 unit but is found between two Te atoms from adjacent chains, which must lead to a more pronounced three-dimensional character. It was reported that the unit cell of these compounds is a $2\mathbf{a} \times 2\mathbf{b} \times 3\mathbf{c}$ super cell [68].

The structure of $ZrTe_5$ [70] consists of infinite chains of $ZrTe_6$ prisms parallel to the orthorhombic **a** axis, reminiscent of the MX_3 chains. These prismatic chains are linked along the **c** axis via zig-zag chains of tellurium atoms to form two dimensional slabs which are stacked along the **b** axis.

Finally Nb_3X_4 (with X = S, Se, Te) crystallizes with an hexagonal structure. The compound is formed with infinite zig-zag Nb chains parallel to the **c** direction [71].

3.5. Summary

In Table I we have collected structural data for the compounds the physical properties of which will be discussed in the following sections, namely the symmetry and the parameters of the unit cell, the space group and the number of chains in each unit cell, the underlined numbers indicating the number of inequivalent chains.

As far as transition metal tri- and tetrachalcogenides are concerned their general structural features can be summarized as follows:

(i) For MX_3 compounds, the essential character is the inequivalent strength of the chalcogen—chalcogen bond in different types of chains leading to a variable degree of *d*-band filling for transition atoms in these chains. Morever the intensity of the transverse coupling leads to a more or less one-dimensional behaviour. MX_3 compounds crystallize in form of long fibers with a length of a few mm up to a few cm. Techniques of growth of large and stoichiometric single crystals have been given in [72]. The maximum cross section is typically $50\,\mu \times 10\,\mu$ for $NbSe_3$ and much less for TaS_3.

(ii) For $(MX_4)_nY$ compounds, the interchain metal—metal distance (typically 6 ~ 7 Å) is much larger than the intrachain distance (~3.2 Å). The selenium atoms form a rectangular antiprism around the transition metal atom with a pairing between selenium. Between chains iodine atoms are isolated and therefore are in the oxidation state I^-. $(MX_4)_nY$ crystals grow in the shape of parallelepipeds a few mm long with a cross section of a few tenths of mm^2.

4. Static Properties of the CDW State

In this section, we show that most of the compounds listed in Table I undergo a Peierls transition, i.e., a coupled periodic lattice distortion (PLD) associated with a CDW formation. Band calculations reveal strong anisotropy in band dispersion between parallel and perpendicular directions to the chain axis which will favour a possible Peierls transition. A consequence of the lattice distortion is the occurrence a gap at the Fermi surface which affects all the physical properties of the compounds. The CDW state is shown to be sensitive to many external parameters such as pressure, doping, irradiation. Finally, for $NbSe_3$, the coexistence between the CDW state and the superconducting one is studied.

4.1. Peierls Transitions

4.1.1. *DC Electrical Resistivity*

The simplest method to detect a phase transition is to search for an anomaly in transport properties. Considering the size of the crystals, dc resistivity measurements along the

TABLE I

Structural data for transition metal tri-, tetra- and pentachalcogenides. The number of independent chains in each unit cell, when known, is indicated by a dotted figure.

	Symmetry	Space group	Unit cell parameters a(Å)	b(Å)	c(Å)	β(°)	Chains per unit cell	References
NbS_3 type I	triclinic	$P\bar{1}$	4.963	2 × 3.365	9.144	97.17; $\alpha = \gamma = 90$	(2 × $\underline{1}$)	[39]
NbS_3 type II	monoclinic		9.9	3.4	18.3	97.2	8	[42]
$TaSe_3$	monoclinic	$P2_1/m$	10.402	3.495	9.829	106.26	2 × $\underline{2}$	[45]
$NbSe_3$	monoclinic	$P2_1/m$	10.006	3.478	15.626	109.30	2 × $\underline{3}$	[46, 47]
TaS_3	monoclinic	$P2_1/m$	9.515	3.3412	14.912	109.99	2 × $\underline{3}$	[48]
	orthorhombic	$C222_1$	36.804	3.34	15.173		24	[49]
$(Fe_{1+x}Nb_{1-x})Nb_2Se_{10}$	monoclinic	$P2_1/m$	9.213	3.482	10.292	114.46	2 × $\underline{1}$ trigonal prismatic; 2 × 1 octahedral	[56, 57]
$(TaSe_4)_2I$	tetragonal	$I422$	9.531		12.824		2 × $\underline{1}$	[64]
$(NbSe_4)_{10}I_3$	tetragonal	$P4/mcc$	9.461		31.91		2 × $\underline{1}$	[65]
$(NbSe_4)_3I$	tetragonal	$P4/mnc$	9.489		19.13		2 × $\underline{1}$	[63]
$TaTe_4$	tetragonal	$P4cc$	2 × 6.513		3 × 6.812			[68]
$NbTe_4$	tetragonal	$P4cc$ or $P4/mcc$	6.496		6.823			[68, 69]
$ZrTe_5$	orthorhombic	$Cmcm$	3.988	14.502	13.727		2 × $\underline{2}$	[70]

chain axis is the first experiment to be performed on every new compound. Precise and reliable determination of the absolute resistivity can be obtained when – the resistance of a sample being known – its cross-section is measured with a scanning electron microscope. The resistance of the samples is measured either from the dc voltage–current relationship, or with a low frequency bridge giving dV/dI.

MX_3 compounds formed with Group IVB transition metals (the prototype of which is $ZrSe_3$) are diamagnetic semiconductors [73] with the exception of $ZrTe_3$ which shows metallic properties and becomes superconducting below $T = 2.0$ K [74]. Optical absorption on ZrS_3 and HfS_3 has revealed gaps of 2.6 eV and 3.1 eV respectively [75].

(i) $NbSe_3$ and $TaSe_3$

$NbSe_3$ and $TaSe_3$ remain metallic at low temperatures. The resistivity variation of $NbSe_3$ reproduced in Figure 7 [76, 77] shows two strong anomalies which appear at $T_1 = 145$ K

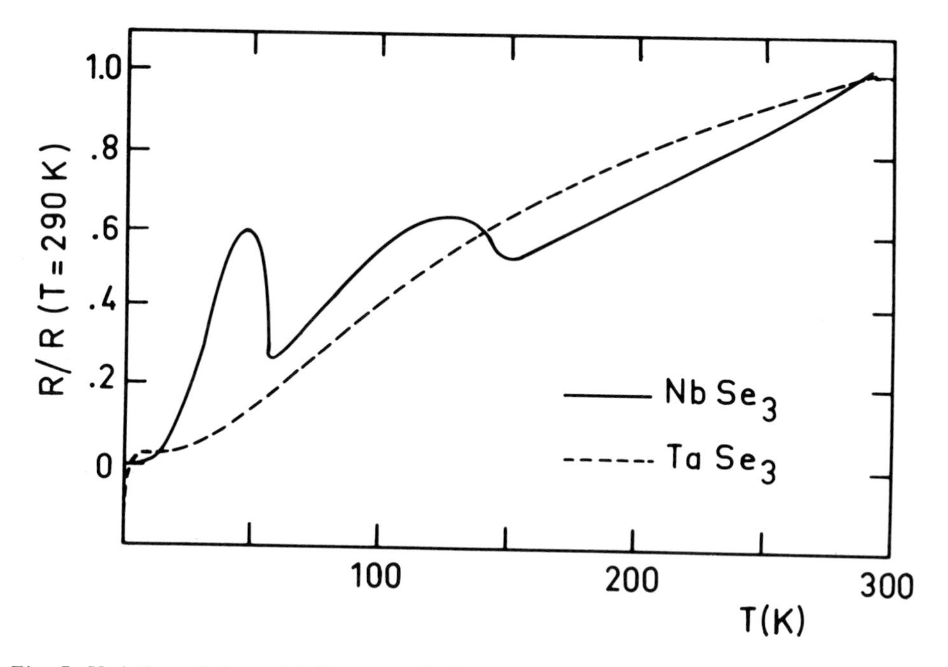

Fig. 7. Variation of the resistivity, normalized to room temperature, of $NbSe_3$ and $TaSe_3$ as a function of temperature (after [76, 200]).

and $T_2 = 59$ K. These giant anomalies can be explained in terms of the Fermi surface fraction destroyed by the formation of independent CDWs [78]. Approximately twenty per cent of the Fermi surface is destroyed by the transition at 145 K. When the temperature crosses T_2, a new gap appears over the remaining 80% of the Fermi surface and approximately 60% of this smaller Fermi surface is destroyed below 59 K. Conductivity anisotropy has been measured in a microwave cavity on a helicoidal coil made of $NbSe_3$ fibers leading to a transverse average anisotropy between 300 and 1200 [78]. Further measurements using the Montgomery method with the current flowing parallel to the

b monoclinic chain axis and parallel to the direction **c** gave a ratio of σ_b/σ_c between 10 to 20 only [79]. However, optical reflectance on a $NbSe_3$ single crystal reveals an optical behaviour characteristic of a semiconductor when the light is polarized perpendicular to the chain axis whereas, in the case of the light parallel to the **b** axis, a strongly damped plasma edge is observed below 1 eV [80]. From these optical reflectance spectra the dc conductivity at room temperature is estimated to be 2.6×10^3 $(\Omega\text{cm})^{-1}$, the band mass, $m^* = 0.4m$, the collision time for free carriers, 10^{-15} s, and the conduction band width 3 eV. The temperature variation of the resistance of $TaSe_3$, as shown in Figure 7, does not show any anomaly which might indicate a phase transition. A phase transition may, however, occur when $TaSe_3$ is intercalated with ammonia molecules [81]. $TaSe_3$ has been shown to become superconducting below $T \sim 2.2$ K [82].

All other transition metal tri- and tetrachalcogenides undergo a transition between a a high temperature state and a less conducting low temperature state. The variation of the resistivity, ρ, normalized to absolute resistivity at room temperature is plotted in Figure 8

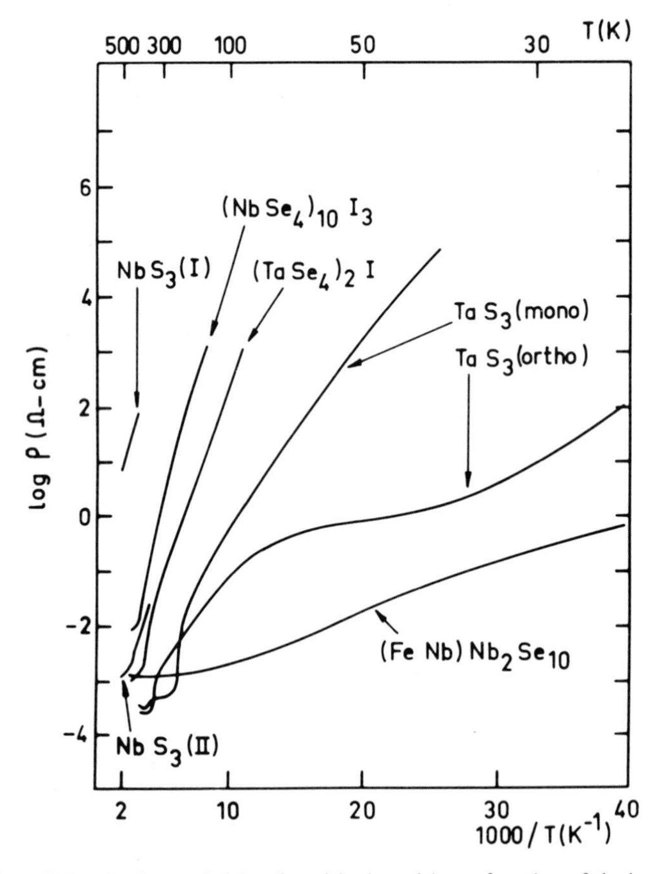

Fig. 8. Variation of the absolute resistivity (logarithmic scale) as a function of the inverse of temperature for transition metal trichalcogenides: NbS_3 type I and type II, TaS_3 with the monoclinic and the orthorhombic unit cell and $(Fe_{1+x}Nb_{1-x})Nb_2Se_{10}$ and for transition metal halogenated tetrachalcogenides $(TaSe_4)_2I$ and $(NbSe_4)_{10}I_3$.

on a logarithmic scale as a function of $10^3/T$ for both forms of TaS_3, $(Fe_{1+x}Nb_{1-x})Nb_2Se_{10}$, $(TaSe_4)_2I$, $(NbSe_4)_{10}I_3$ and both forms of NbS_3. In the insulating state, if there is a unique conduction mechanism, the temperature variation of the resistivity is expected to follow the activation law corresponding to single excitations through the gap, $\Delta'(T)$, such as:

$$\rho(T) = \rho_0 \exp\left[-\frac{\Delta'(T)}{2kT}\right] \quad (4\text{-}1)$$

It is commonly admitted that the metal–insulator Peierls transition is detected by a strong peak in the variation of the logarithmic derivative as a function of the inverse of the temperature [83] such as:

$$\frac{\mathrm{d}\log\rho}{\mathrm{d}T^{-1}} = A\left[\Delta'(T) + T\frac{\mathrm{d}\Delta'(T)}{\mathrm{d}T}\right] \quad (4\text{-}2)$$

When T is largely reduced below the Peierls transition, (4-2) allows us to measure the gap, $\Delta'(0)$, at saturation. In the mean field theory and in the limit of weak coupling in the electron–phonon interaction, the ratio between the CDW gap at $T = 0$ and the critical temperature is similar to the BCS theory of superconductivity [8]:

$$2\Delta = 3.52\, kT^{MF} \quad (4\text{-}3)$$

For a strictly one-dimensional material there is no phase transition and therefore $T_P = 0$. When the interchain coupling is taken into account a three-dimensional phase transition occurs at $T_P < T^{MF}$. Lee and Rice have calculated $T_P \sim T^{MF}/4$ [84]. Large one-dimensional fluctuations develop in the temperature range between T^{MF} and T_P. It should also be noted that large ratios between 2Δ and the Peierls transition have been measured in transition metal dichalcogenides: $2\Delta/T_c \sim 28$ in $2H$–$TaSe_2$, which has been explained by Varma and Simons by developing a microscopic strong-coupling theory [85].

(ii) TaS_3

For TaS_3 with the monoclinic unit cell, two transitions occur at the temperatures at which the logarithmic derivative shows peaks, namely at $T_1 = 240$ K and $T_2 = 160$ K [86, 87]. The increase in the resistance between T_1 and T_2 is approximately 50%. For samples of good quality, after the transition at T_1, the resistance decreases slowly before increasing at the onset of the T_2 transition. Below T_2, a gap of $\Delta' = 1900$ K is deduced from the activation energy.

For the orthorhombic TaS_3, the sharp peak in $[\mathrm{d}\log\rho/\mathrm{d}T^{-1}]$ as shown in Figure 9 corresponds to the Peierls transition at $T_0 = 215$ K as originally reported by Sambongi *et al.* [88]. However a broad maximum in the logarithmic derivative is also observed at $T_0' = 135$ K. As is shown in the following section, at T_0', the wavelength of the CDW in the chain direction becomes locked to the commensurate value of four lattice distances. From the plateau in Figure 9 and following (4-2) the gap, Δ', = 1600 K = $7.44\, kT_c$ is deduced, in keeping with results in [88, 89] but at variance with the value $2\Delta(0) \sim$ 770 K obtained by Raman spectroscopy [90]. Higgs and Gill have measured a thermal

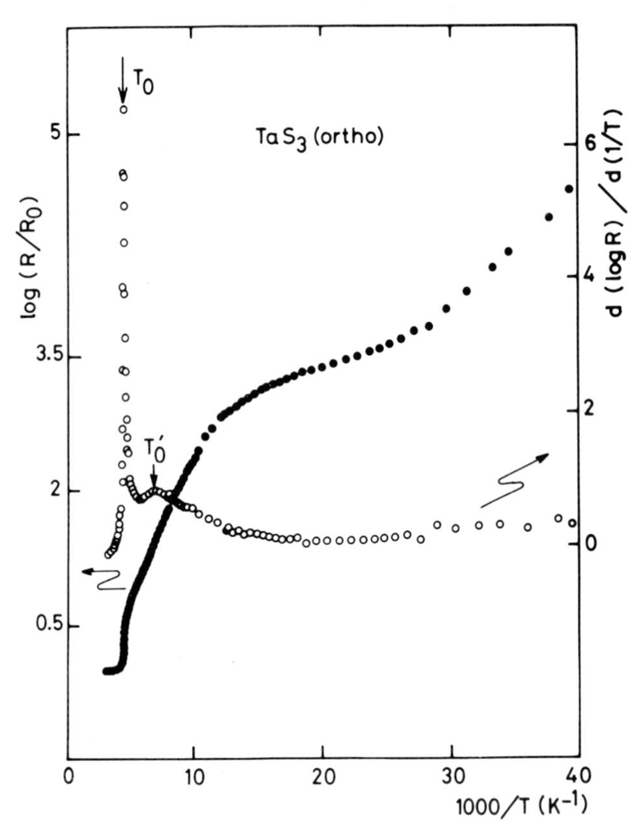

Fig. 9. Variation of the resistivity and of its logarithmic derivative as a function of inverse temperature for TaS_3 with the orthorhombic unit cell. The resistivity is normalized to its value at room temperature and is plotted on a logarithmic scale. The Peierls transition T_0 is defined as the temperature at which $d \log R/d(1/T)$ shows a sharp peak. The bump at T_0' in the variation of $d \log R/d(1/T)$ is associated with the locking of the wavelength of the CDW along the chain at the commensurate value of four lattice distances.

hysteresis below T_0 and down to 55 K in their electrical measurements [91]. They attribute this hysteresis to a non-equilibrium distribution of discommensurations [92] interacting with the defects in the crystal. An important difference in the temperature variation of the resistivity of both TaS_3 concerns the low temperature variation: as a function of $1/T$, $\log \rho$ keeps a linear variation for monoclinic TaS_3 whereas a curvature is observed for the orthorhombic form. This curvature, which indicates an increase of the conductivity at low temperature, might be due to another source of carriers different from normal electron excitations through the CDW gap. Similarly to KCP [34] (see Section 2.3), solitons have been supposed [93] to give a contribution to the conductivity at low temperature.

(iii) NbS_3

As mentioned above in Section 3.1, several polytypes of NbS_3 can be synthesized. From the variation of log ρ *vs* $10^3/T$ for the dimerized type I variety an energy gap of 0.44 eV is deduced. Transport measurements have been performed on the same fibers, defined as type II, as those studied in electronic microscopy by Roucau [42]. A phase transition occurs above room temperature at T_c = 323 K but beyond this temperature the resistivity remains activated. The room temperature resistivity for type II NbS_3 is 10^{-3} times lower than for type I NbS_3, which is a consequence of the absence of Nb—Nb pairing along the chains. Another polytype, called type III, shows a transition near 140 K [43]. Finally Izumi *et al.* [44] have reported that some sulfur-rich samples showed two successive resistivity anomalies around 235 K and 140 K, their room temperature resistivity being around 40–50 Ωcm. After annealing sulfur-rich crystals, a metallic NbS_3 is claimed to have been obtained with a resistivity at room temperature of 3×10^{-3} Ωcm and a superconducting transition between 1.65 and 2.15 K depending on the crystals [44]. If the real structure of this compound has to be studied, it might be worthwhile to note that a parent compound, Nb_3S_4, is superconducting at T_c = 3.65 K [94] and does not show any anomaly in its temperature resistivity variation [95].

(iv) $(MSe_4)_nI$

As explained in Section 3.3, $(MSe_4)_nI$ compounds are characterized by bond alternation along the chain axis. $(TaSe_4)_2I$ has the highest conductivity at room temperature, ρ = 1.5×10^{-3} Ωcm, because in this compound all Ta—Ta distances are identical. For $(NbSe_4)_{10}I_3$ in which the pairing between Nb is weak, $\rho = 10^{-2}$ Ωcm but for $(NbSe_4)_3I$ ρ = 1 Ωcm. The latter compound is semiconducting above room temperature with a semiconducting gap of 0.19 eV [61]. Around 275 K a sharp decrease of the activation energy is measured, which has been shown to be due to a space group change [96] and not to a Peierls transition. $(TaSe_4)_3I$, the structure of which has not yet been determined, shows a metallic state at room temperature and becomes semiconducting approximately below 70 K. The origin of this metal–insulator transition is again a space group change [96]. By contrast, $(TaSe_4)_2I$ and $(NbSe_4)_{10}I_3$ clearly show a Peierls transition respectively at 263 K [97, 98] and 283 K [99], detected by a sharp peak in the logarithmic derivative. The gap deduced from the activation energy below the transition is 2Δ = 3000 K = 11.4 kT_c for $(TaSe_4)_2I$, and 3900 K = 13.7 kT_c for $(NbSe_4)_{10}I_3$. The two latter compounds do not behave as metals at high temperature because their resistance decreases when T is increased above the Peierls transition. Another compound of the same family, $(NbSe_4)_2I$, has been reported to undergo a Peierls transition at T = 210 K [100]. Polytypes with the same chemical formula may exist: the existence of a $(NbSe_4)_xI$ compound with $2 < x < 3$ has been reported; it has the same unit cell parameters as $(NbSe_4)_3I$ but different transport properties [101].

(v) $(Fe_{1+x}Nb_{1-x})Nb_2Se_{10}$

Hillenius *et al.* [55] have measured the resistivity of $(Fe_{1+x}Nb_{1-x})Nb_2Se_{10}$ and they have shown that log ρ increases by a factor of 10^9 between $T_c \simeq 140$ K and down to 4 K. This transition temperature is roughly the same as for the upper CDW in $NbSe_3$. The activation energy below T_c gives Δ' = 900 K = 2.55 kT_c. Metal–insulator transitions have also been measured in (Fe, Ta)Nb_2Se_{10}, (Cr, Nb)Nb_2Se_{10} at approximately the same critical temperature with values for the gap of Δ' = 1.9 kT_c, 3.58 kT_c, respectively.

The conductivity variation in these compounds cannot be explained only by the occurrence of a Peierls transition. Localization has been suggested [55]. CDW and localization will be discussed below in Section 4.5.

(vi) $HfTe_5$ and $ZrTe_5$

These compounds are metallic at room temperature. They exhibit strong resistivity anomalies at low temperature with a peak around 150 K for $ZrTe_5$ [103] and 75 K for $HfTe_5$ [104], which is reminiscent of the resistivity anomalies in $NbSe_3$. No sharp peak in the logarithmic derivative suggesting a particular phase transition has been observed, however [105].

$NbTe_4$ and $TaTe_4$ are metallic with a regular decrease of the resistivity down to 20 mK without any phase transition or a superconducting transition [106].

Bronze $K_{0.30}MoO_3$ undergoes a metal–insulator transition around 180 K [107, 108]. The temperature variation of the resistivity does not follow the simple activation law (4-1). Below 50 K a value of 0.025 eV was found for the activation energy. From optical reflectivity measurements it was shown that the Peierls gap which appears at 180 K has an energy of 0.15 eV [109] corresponding to $2\Delta' = 9.7\ kT_c$.

4.1.2. *Superlattice Structure*

To ascertain that the transitions detected by transport properties are Peierls transitions, elastic and inelastic X-ray or neutron scattering studies are necessary, but the size of the samples has precluded such experiments up to now. The tendency of the lattice to distort is revealed dynamically in the phonon dispersion by a giant Kohn anomaly (or soft mode) which condenses below the Peierls transition in a three-dimensional modulated structure [17]. At high temperature, X-ray diffuse scattering due to phonon modulation leads to continuous satellite diffuse lines. These diffuse lines correspond to planes in the reciprocal space perpendicular to the chain direction.

(i) $NbSe_3$

These precursor effects have been reported by Tsutsumi *et al.* [110] in the first experiment which showed superlattice spots associated with the T_1 transition in $NbSe_3$ but they were not detected in further studies [47, 111]. However a more recent X-ray scattering study [112] reveals two kinds of weak diffuse sheets considered as precursor effects of the two phase transitions. The correlation length along the chain corresponding to the upper transition was found to increase from 10–15 Å at 295 K to 85 Å near 150 K, the correlation line for the lower transition, at the latter temperature, being about 10 Å. Below the Peierls transition these diffuse lines condense in well-defined satellite spots which accompany the main Bragg spots. Measurements of the superstructure components on the reciprocal axes $\mathbf{a}^*$, $\mathbf{b}^*$, $\mathbf{c}^*$ have given the scattering vector components for the T_1 and T_2 transition. Recent experiments performed with the synchrotron X-ray source of the Stanford Synchrotron Radiation Laboratory have led to a better resolution. At 10 K, the wave vectors were determined to be $\mathbf{Q}_1$ ($0 \times \mathbf{a}^*$, $0.24117 \times \mathbf{b}^*$, $0 \times \mathbf{c}^*$) for the T_1 transition and $\mathbf{Q}_2$ ($0.5 \times \mathbf{a}^*$, $0.26038 \times \mathbf{b}^*$, $0.5 \times \mathbf{c}^*$) for the T_2 transition [113]. The coherence length of the lower temperature CDW was estimated to exceed 5000 Å but its determination was limited by the instrumental resolution. The two superlattice spot intensity has been measured as a function of the temperature. Near T_1 and T_2 the intensity decreases to zero continuously indicating a second order phase transition. A very slight change in the intensity of the $\mathbf{Q}_1$ spots occurs when the T_2 transition appears [113]. Measurements of the distortion component along the chains in electronic diffraction patterns have shown that the $\mathbf{Q}_1$ component was $(0.243 \pm 0.003)\mathbf{b}^*$ above T_2, and $(0.241 \pm 0.003)\mathbf{b}^*$ below T_2 [87]. The exact mean values, 0.243 and 0.241, have been obtained from different pictures corresponding to different samples. Although it can be argued that the two values are within the range of uncertainty it cannot be excluded that, when the second transition occurs below 59 K, the component of the first transition changes from 0.243 $\mathbf{b}^*$ to 0.241 $\mathbf{b}^*$, indicating a slight coupling between both CDWs. Measurements in the vicinity of T_2 with the synchrotron X-ray source are needed in order to detect a possible slight temperature variation of $\mathbf{Q}_1$ and $\mathbf{Q}_2$. Below T_1 or T_2, however, there is no tendency for $\mathbf{Q}_1$ or $\mathbf{Q}_2$ to lock at the commensurate value of 0.25 $\mathbf{b}^*$. It can also be noted that the sum of the two wave vectors $\mathbf{Q}_1 + \mathbf{Q}_2$ is not the commensurate value $(0.5 \times \mathbf{a}^*, 0.5 \times \mathbf{b}^*, 0.5 \times \mathbf{c}^*)$. It had been suggested that such a coupling would lead to commensurability pinning to the lattice [114–116].

(ii) TaS_3

Similarly to $NbSe_3$, within the accuracy of electronic diffraction patterns, the $\mathbf{Q}$ vectors of the two distortions in monoclinic TaS_3 are temperature independent and respectively, $\mathbf{Q}_1$ $(0 \times \mathbf{a}^*, 0.253 \times \mathbf{b}^*, 0 \times \mathbf{c}^*)$ and $\mathbf{Q}_2$ $(0.5 \times \mathbf{a}^*, 0.247 \times \mathbf{b}^*, 0.5 \times \mathbf{c}^*)$. But in contrast with $NbSe_3$, intense diffuse lines are visible above T_1. Unfortunately, electron diffraction techniques preclude precise intensity measurements which would give an estimation of the coherence lengths. Careful electron microscopic diffraction measurements on orthorhombic TaS_3 have shown that the Peierls transition which appears at T_0 = 215 K is weakly incommensurate with the lattice [42, 117] at variance with earlier results [88, 118]. Just below T_0, the components of $\mathbf{Q}$ along $\mathbf{b}^*$ and $\mathbf{c}^*$ are: $0.1 \times \mathbf{b}^*$ and $0.255 \times \mathbf{c}^*$ (no measurement was made in this temperature range for the component along $\mathbf{a}^*$). The distortion components are continuously temperature dependent between T_0 and a temperature, T_0', at which they are locked to the values $0.5 \times \mathbf{a}^*$, $0.125 \times \mathbf{b}^*$, $0.250 \times \mathbf{c}^*$, without further variation down to 4 K [119]. Therefore, below T_0', the wavelength of the distortion is four lattice distances. T_0' is the temperature at which the logarithmic derivative of the resistivity shows a broad peak, as shown in Figure 9. It has to be noted that, because the distortion wavelength in the transverse direction comprises eight unit cells, the transition towards a commensurate component along the chain axis does not lead to commensurability pinning [120]. Following (2-11) the commensurability energy is $W_0 \cos 4\phi$. Each adjacent chain is out of phase by $\pi/4$. If the phase is $\phi = 0$ on a given chain, the energy gain from commensurability is $-W_0$. But on the adjacent chain, $\phi = \pi/4$, the energy gain is $+W_0$ and the net energy gain vanishes. Therefore, if the chains are arranged with the following sequence of phases: 0, $\pi/4$, $\pi/2$, $3\pi/4$, ... there is no commensurability pinning. It can be argued, however, that another sequence of phases is possible namely: 0, 0, $\pi/2$, $\pi/2$, π, π, $3\pi/2$ .... In this case the energy gain per chain is $-W_0$ and commensurability pinning occurs [121].

Higher order X-ray reflections might lead to the determination of this precise sequence of phases.

(iii) NbS_3

It was suggested that the Nb—Nb pairing in type I NbS_3 was the result of a Peierls transition [40]. However by heating the specimens, they are irreversibly damaged before entering a distortion-free phase. Type II NbS_3 does not show a Nb—Nb pairing. Superlattice spots appear as doublets near $\pm \mathbf{b}^*/3$ [40, 122]. Recent measurements of these superlattice spots have led to different components than those reported previously [40, 122]. The components of the wave vectors are found as: $\mathbf{Q}_1 = (0.5 \times \mathbf{a}^*, 0.298 \times \mathbf{b}^*, 0 \times \mathbf{c}^*)$ and $\mathbf{Q}_2 = (0.5 \times \mathbf{a}^*, 0.352 \times \mathbf{b}^*, 0 \times \mathbf{c}^*)$ [123]. Furthermore, by heating above room temperature, the intensity of the $\mathbf{Q}_1$ superlattice set decreases sharply and disappears around 77°C. The disappearance of the $\mathbf{Q}_1$ spots can be associated with the phase transition in resistivity measurements as shown in Figure 8. The doublet would be due to two different transitions.

(iv) $(Fe_{1+x}Nb_{1-x})Nb_2Se_{10}$

Two series of superlattice scattering have been observed. The first distortion has the wave vector, $\mathbf{Q}_1 = (0 \times \mathbf{a}^*, 0.27 \times \mathbf{b}^*, 0 \times \mathbf{c}^*)$ and it appears below the temperature at which the resistivity shown in Figure 8 starts to increase sharply [55, 124]. $\mathbf{Q}_1$ is slightly temperature dependent, its value being equal to 0.258 around 10 K. This CDW has been attributed to a distortion which occurs in the trigonal chains. However, if the composition permits a range of stoichiometry with $0.25 < x < 0.40$, a variation of $\mathbf{Q}_1$, according to different compositions, should be found. A second set of diffuse scattering has recently been detected [124] in the form of rods with wave vectors $(0.5 \times \mathbf{a}^*, 0.33 \times \mathbf{b}^*, l \times \mathbf{c}^*)$ probably related to partial ordering of the iron and niobium atoms in the octahedral disorder chains.

(v) $(MSe_4)_nI$

Superlattice spots have been observed in $(TaSe_4)_2I$ and $(NbSe_4)_{10}I_3$ [96]. No evidence of strong precursor effects has been detected above the Peierls temperature. The wave vector components for $(NbSe_4)_{10}I_3$ are found to be $(0 \times \mathbf{a}^*, 0 \times \mathbf{b}^*, 0.487 \times \mathbf{c}^*)$.

For $(TaSe_4)_2I$ the distortion vector which connects the satellite to the nearest Bragg peak has the components $0 \times \mathbf{a}^*, 0 \times \mathbf{b}^*, 0.057 \times \mathbf{c}^*$ in [96] and $0.05 \times \mathbf{a}^*, 0.05 \times \mathbf{b}^*, 0.084 \times \mathbf{c}^*$ in [125]. The distortion $\mathbf{Q}$ is defined modulo a reciprocal lattice vector, $G = h \times \mathbf{a}^* + k \times \mathbf{b}^* + l \times \mathbf{c}^*$ with $h + k + l = 2n$ with regard to the crystal symmetry. In order to be consistent with band calculations and the electronic concentration $\mathbf{Q}$ has been chosen as $1 \times \mathbf{a}^*, 0 \times \mathbf{b}^*$ and $0.943 \times \mathbf{c}^*$ [96]. For this compound, intense higher harmonic spots have also been detected.

$TaTe_4$ and $NbTe_4$ were shown to exhibit a superstructure, the unit cell being a $2 \times \mathbf{a}^* \times 2 \times \mathbf{b}^* \times 3 \times \mathbf{c}^*$ supercell. Recent electronic diffraction experiments have shown that, at room temperature, three incommensurate distortions occur in $NbTe_4$ [126, 127]: a distortion $\mathbf{Q}_1 = (0.5 \times \mathbf{a}^*, 0.5 \times \mathbf{b}^*, 0.344 \times \mathbf{c}^*)$, a distortion $\mathbf{Q}_2$ with weaker intensity, $\mathbf{Q}_2 = (0 \times \mathbf{a}^*, 0 \times \mathbf{b}^*, 0.311 \times \mathbf{c}^*)$ and diffuse spots which condensate below 60 K in commensurate spots $\mathbf{Q}_3 = (0.5 \times \mathbf{a}^*, 0 \times \mathbf{b}^*, 0.33 \times \mathbf{c}^*)$. For $TaTe_4$ the same satellites have been observed but with the commensurability 1/3 on $\mathbf{c}^*$. These satellite spots have been related to the occurrence of three independent CDW distortions on different types of chains. It has, however, to be noted that, although the relative intensity of the superlattice spots decreases when T is increased, these spots are still observed at $T = 625$ K when the crystals begin to loose their stability.

(vi) $K_{0.30}MoO_3$

The Peierls nature of the metal–semiconductor transition in $K_{0.30}MoO_3$ has been demonstrated by X-ray studies. The components of the superlattice are ($0 \times \mathbf{a}^*$, $0.74 \times \mathbf{b}^*$, $0.5 \times \mathbf{c}^*$) [128]. An incommensurate–commensurate transition at 110 K has also been reported [129]. Finally, no trace of extra diffraction spots has been shown in $ZrTe_5$ and $HfTe_5$ by X-ray diffraction [105] or by electronic diffraction [130] techniques.

In Table II we have gathered data concerning Peierls transitions for transition metal tri- and tetrachalcogenides such as: the Peierls transition temperature, the amplitude of the gap deduced from resistivity measurements, the ratio between 2Δ and the temperature kT_P, the components of the superstructure and the nature of the ground state. The absolute value of the resistivity at room temperature as well as the superconducting transition temperature are also listed. The value of 700 K for the lower CDW gap in $NbSe_3$ is obtained from preliminary tunneling experiments [131]. However far-infrared reflectance studies of $NbSe_3$ have shown the existence of a CDW gap the value of which lies between 180 K and 280 K [132].

4.1.3. *Domain Structure of the CDW State in $NbSe_3$ and TaS_3*

Superlattice dark field electron microscope images on $NbSe_3$ and TaS_3 have revealed interesting properties [113, 133–135] which can be summarized as follows:

(i) Strands

The bend contours formed from satellite reflections are irregular with discontinuities of contrast along lines approximately parallel to the **b** axis. The contrast of these strand domains was found to be time dependent ('twinkling effect') with a time-scale of several tens of seconds. The size of these strand domains is typically 1 μ long and 200 Å wide for $NbSe_3$ [134] and only 0.3 μ long for TaS_3 [135]. These strands have been interpreted as local changes of the modulation vector, **Q** [134]. The sample under examination in the microscope is divided into strand domains in which **Q** deviates from the **b*** direction: its amplitude is constant but it points at different orientations (up to 15° away from the **b*** axis) between strands. The time dependence is supposed to be a change of the mutual orientation of **Q** inside these domains resulting from the electron beam radiation.

(ii) Broad fringes

On some images, broad fringes were seen along the strands. Depending on the thickness of the whisker, one or several strands can be accomodated. Broad fringes are Moiré fringes between the two slightly different **Q** values within the thickness of the sample.

(iii) Fine fringes

On some, but not on all images, fine and weak-intensity fringes separated by 100 to 200 Å have been observed [133–134]. The weakness of these fringes has led Steeds

TABLE II

Data concerning Peierls or superconducting transitions in transition metal tri- and tetrachalcogenides. Listed are: the resistivity at room temperature, the Peierls or the superconducting temperature, the amplitude of the Peierls gap and its ratio with the Peierls temperature, the nature of the ground state at low temperatures and the components of the superlattice structure on the reciprocal axes.

	$\rho(\Omega \times \text{cm})$ at room temperature	Peierls temperature (K)	2Δ (K)	$\frac{2\Delta}{kT_c}$	Surstructure a*	Surstructure b*	Surstructure c*		Superconducting temperature	Ground state at low temperature
NbS_3 type I	80 [124]									Insulating
NbS_3 type II	8×10^{-2} [124]	330 [124]	4400	13.3	0.5 0.5	0.298 0.352	0 0	[42]		Semiconducting
$TaSe_3$	6×10^{-4} [200]								2.0 [82]	Superconducting
$NbSe_3$	2.5×10^{-4} [238]	145 59 [77]	700 [132]	11.9	0 0.5	0.24117 0.26038	0 0.5	[113]	3.5 under 5.5 kbar [183]	Metallic
TaS_3 orthorhombic	3.2×10^{-4} [120]	215 [88]	1600	7.44	? 0.5	0.1 0.125 (T < 130 K)	0.255 0.250	[117]		Semiconducting
TaS_3 monoclinic	3×10^{-4} [120]	240 160 [86]	1900	11.9	0 0.5	0.253 0.247	0 0.5	[87]		Semiconducting
$(Fe_{1+x}Nb_{1-x})Nb_2Se_{10}$	10^{-3} [99]	~140 [55]	360	2.55	0 0.5	0.27 0.33	0 l	[55] [124]		Semiconducting
$(TaSe_4)_2I$	1.5×10^{-3} [97]	263 [97]	3000	11.4	0.05 1	0.05 0	0.084 0.943	[125] [96]		Semiconducting
$(NbSe_4)_{10}I_3$	1.5×10^{-2} [99]	285 [99]	3900	13.7	0	0	0.487	[96]		Semiconducting
$NbTe_4$	1.2×10^{-4} [127]				0.5 0 0.5	0.5 0 0	0.344 0.311 0.333	[127]		Metallic
$TaTe_4$	1.2×10^{-4} [127]				0.5 0 0.5	0.5 0 0	0.333 0.333 0.333	[127]		Metallic

et al. [134] to wonder "how fundamental the fine fringes really are". By contrast, broad fringes do not seem to have been observed in TaS_3 [135], but the latter crystals are very sensitive to radiation damage, which reduces the longitudinal strand length from 0.3 μ to 0.1 μ; it reduces the perpendicular coherence much more, however (less than 50 Å). These dark field experiments clearly demonstrate the CDW domain structure with a maximum coherence for $NbSe_3$ on a length of 1 to 2 μ and a width of around 200 Å. This domain structure is strongly dependent on the CDW nucleation process, especially the cooling speed which can induce metastable states in the CDW superstructure the consequences of which will be discussed in Section 5.5. The microstructure is the consequence of non-equilibrium CDW condensation and it is not surprising that these domains have been observed in the 'commensurate' phase of orthorhombic TaS_3. If discommensurations are relevant, they are fine fringes sometimes observed as discussed in the following section.

4.1.4. *Periodic Modulation or Discommensurations*

Although the periodic lattice distortions described in 4.1. seem to result from an instability induced by the strong one-dimensional character in the structure of the compounds investigated, the origin of the modulated structure has been questionned. It was firstly noted that for many of the compounds the periodicity of the modulation is close to four atomic distances. Secondly, discommensurations have been observed in the incommensurate–commensurate transition of the transition metal dichalcogenide $2H$–$TaSe_2$ [136, 137] and TaS_3 was reported to have a temperature dependence of the CDW wavelength along the chains and a locking to four atomic distances at $T_0' = 140$ K [117]. Finally metal–metal bonding leading to a cation–cation pairing occurs in some of the MX_3 compounds such as type I NbS_3 and some of the MX_4 compounds such as $(NbSe_4)_3I$. The modulated structure may result from the clustering of cations along the chains coming from the correlation effects due to electron–electron interaction [138]. A suggestion has been made to describe the distortions in $NbSe_3$ by a discommensurate array of cation pair bonding [41, 139]. The precise $\mathbf{Q}$ determination for $NbSe_3$ [113] allows one to calculate the periodicity of these defects. Wilson [140] has remarked that the $\mathbf{Q}_1$ component along the chain can be represented by $1/4 \times 26/27$ (0.2407_5 as compared to the measured value of 0.2408_5) and $\mathbf{Q}_2$ by $1/4 \times 26/25$ (0.2600 as compared with 0.2599). The basic span is formed with 13 wavelengths of four atoms. The discommensuration unit for the T_1 transition is contracted to 50 atoms from 52 atoms while for the T_2 transition it is expanded from 52 atoms to 54 atoms. This compression and extension is brought about inside a discommensuration width of two waves. With this model, the distance between discommensurations is respectively 188 Å for the T_1 transition and 174 Å for the T_2 one, these values being very similar to distances between the fine fringes observed by Steeds *et al.* [133, 134] in dark field microscope imaging. With this nonsinusoidal distortion, however, strong harmonics are expected in the diffraction experiments. The intensity of the second harmonic in $NbSe_3$ was shown to be only 1% of the primary one [111], much weaker than in a discommensurate system such as $2H$–$TaSe_2$. Wilson accounts for this small intensity by adequate phasings of the distortions between adjacent chains. As already explained above, any similarity between the incommensurate–commensurate transition in $2H$–$TaSe_2$ and in orthorhombic TaS_3 should not be unduly stressed, as far as discommensurations are concerned. The commensurate transition in TaS_3 is provided by a transverse coupling between two types of chains [117]. The question of the shape of the modulation has, however, a primordial importance because it could be one of the clues for the explanation of non-linear properties described in Section 5. A proof for the discommensuration model would be the study of the variation of **Q** under pressure: in the case of a sinusoidal distortion one would expect a continuous variation of **Q** whereas in the nonsinusoidal case, **Q** must jump discontinuously to values corresponding to a different number of wavelengths of four atoms between discommensurations. Finally S. Barišić in Chapter I, Part I of this work, theoretically treats the question of anharmonic interactions of Peierls distortions.

4.2. Electronic Structure

4.2.1. *Semi-Metals or Quarter-Filled Bands*

It has been shown that the strong pairing between Se in the trigonal chains of $NbSe_3$ induces a variable covalent–ionic bond. Wilson [139] was the first to point out a very simple electron count as follows: there are six $NbSe_3$ units in the unit cell. The six cations introduce 30 electrons. Each chain (X4) with strong Se—Se interaction (chains III and I) accounts for two electrons for Se_2^{2-} and two electrons for Se^{2-}. Chain III (X2) with the weakest Se—Se interaction accounts for 2 X 3 electrons for each Se^{2-}. Two electrons remain to be shared between only four niobium atoms because the energy band of the chains with an equilateral basis is much higher than the Fermi level. Thus, there are 0.5 electrons per niobium atom. With the spin degeneracy, the Fermi level in a one-dimensional picture is exactly located at 0.25 $\mathbf{b}^*$. In fact, the four niobium atoms are not exactly equivalent and there are not exactly 0.5 electron per niobium. Experimentally, the superstructures appear at 0.243 $\mathbf{b}^*$ for the T_1 transition and at 0.259 $\mathbf{b}^*$ for the T_2 one. Wilson [139], moreover, considering the phasing between chains of the same type coordinated by inversion centers, has concluded that the upper transition, T_1, appears on type III chains and the lower one, T_2, on type I chains, type II chains with nearly equilateral basis being insulating.

From the number of electrons per unit cell the electron concentration at room temperature is found to be 3.9×10^{21} cm^{-3}. Unfortunately, an incorrect deduction of the electronic concentration from early Hall effect measurements performed at room temperature [141] gave a value of 5×10^{19} electrons cm^{-3} which led Wilson to propose a semi-metal description for the band structure. This band structure, resulting from hybridization between crossing bands from type III and type I chains was recognized to be incorrect [140] as shown by the calculations of Bullett [142, 143] and Hoffmann *et al.* [144]. Furthermore, the Hall effect measurements have been remeasured [145] and an electronic concentration at room temperature of 1.4×10^{21} cm^{-3} deduced. It has to be noted that the deduction of carrier density from Hall constant measurements may not be as straightforward in complicated band structures.

The same calculations can be made for monoclinic TaS_3 because of the similarity of its structure with $NbSe_3$. The room temperature electron concentration is found to be 4.5×10^{21} electrons cm^{-3}. XPS measurements on TaS_3 orthorhombic [146] show that there are two oxydation states of tantalum: Ta^{+4} and Ta^{+5} in a ratio of 1 : 1 and also S^{2-} and S^- in the same proportion. The formal ionic description of orthorhombic TaS_3 should be: Ta^{+4} Ta^{+5} 3 $S^{2-}(3/2)S_2^{2-}$. Although the structure is unknown, adopting the hypothesis of trigonal chains, one can imagine that half the chains in orthorhombic TaS_3 have a cross-section with an S—S pairing and do not participate in the electrical conductivity such as type II chains in $NbSe_3$; lattice distortions would then only appear on half the chains which would have an S—S pairing. Hall effect measurements at room temperature have given an electronic concentration of 1.0×10^{22} carriers cm^{-3} [147].

With the same covalent–ionic bond description, $TaSe_3$ should be semiconducting. Among four chains in the unit cell, two have no Se—Se pairing. Therefore, the four cations introduce 20 electrons. The chain (×2) with uncorrelated anions takes 6 electrons and the chain (×2) with anion pairing takes 4 electrons which leads to an empty d band for tantalum. In fact, there is some overlap between p and d bands and $TaSe_3$ is a semimetal undergoing no transition when the temperature is lowered. At low temperatures Shubnikov–de Haas oscillations have been recorded in magnetoresistance measurements with many orbits which reveal a rather complicated Fermi surface [148].

This electronic count can also be made for $(MSe_4)_nI$ compounds. $(TaSe_4)_2I$ can be formulated as Ta^{+4} Ta^{+5} $4(Se_2)^{2-}$ I^- if the iodine atoms between chains are considered as isolated, which is a reasonable assumption because of the large distance between them. There are two electrons per chain, which lead to an electronic concentration above the Peierls transition of 3.2×10^{21} cm^{-3}. Again the number of d electrons on each tantalum is 1/2 and the d_z bands are 1/4 filled. With such an electronic concentration one expects the Fermi level, $2k_F$, to be near **c*** [97] and, therefore, the superlattice spots in the vicinity of the main Bragg spots, as is found experimentally [96]. For $(NbSe_4)_{10}I_3$ there are seven electrons per chain in the unit cell. Six electrons fill up three full bands and one conduction electron per chain remains which leads to an electronic concentration of 0.7×10^{21} electrons cm^{-3}.

The electron count has been made for $(Fe_{1+x}Nb_{1-x})Nb_2Se_{10}$ in [149]. The Fermi level depends on stoichiometry x. For $x = 0.25$ there are three conduction electrons to fill and k_F must be at 0.20 **b*** whereas for $x = 0.40$ k_F is at 0.13 **b***.

4.2.2. *Band Dispersion*

(i) MX_3

Band calculations have been performed on $ZrSe_3$, $TaSe_3$ and $NbSe_3$. Several methods have been used for calculating the band structure of $NbSe_3$: a chemical pseudopotential method using neutral atomic potentials of niobium and selenium atoms by Bullett [142, 143], an extended Buckel method with semi-empirical parameters by Hoffmann *et al.* [144] assuming identical chains (a refinement of the calculation has appeared in [150] by considering nonequivalent pairs of chains), a LCAO-local exchange correlation potential-self-consistent field method by Shima [151]. Although there seems to be agreement about the results obtained, they do, however, differ with regard to some details concerning the assignment of a band to a well-defined type of chain. There are six broad conduction bands which rise rapidly along the chain direction. Dispersion in orientations perpendicular to the chain axis reveals a very weak dependence, which would indicate a pronounced one-dimensional character for $NbSe_3$, in keeping with the optical reflectance measurements [80]. All the models attribute two of these six bands to chains III because of the predominant character of the niobium orbitals of these chains at the Fermi level. It was also recognized that the two conducting bands with higher energy especially concern chains II. If the Fermi level is below these two bands, they are empty; the Fermi level crosses only four bands corresponding to two types of chains, III and I, on which two CDW transitions occur which corresponds to the model given by Wilson [139]. However, Bullett [143] and Shima [151] show that the Fermi level crosses five bands and that chain II is not completely empty of *d* electrons. Shima has modelled the Fermi surface as being made up of five pieces: two plane-like flat surfaces connected to chain III, two other flat surfaces connected to chains I and II and a small warped and partially closed Fermi surface also assigned to chains I and II. The flat surfaces can nest independently when translated by some vector equal to a fraction of **b*** giving rise to two phase transitions. The small non planar Fermi surface survives after these transitions and its cross-section is thought to have been deduced from Shubnikov–de Haas oscillations observed in magnetoresistance measurements [148, 152, 153]. This band scheme does not separate the individual contributions of chains I and II which are considered as a unit. Such a model is also in keeping with the structural description made by Hodeau *et al.* [47] in which, considering the Nb and Se bonds, $NbSe_3$ is described as being formed by two rigid units: one with two chains III and one with four chains I and II, the chains of each unit being related to each other by a centre of symmetry. Figure 10 reproduces the electron band structure along the chain axis calculated by Bullett [143], Whangbo and Gressier [150].

Wada *et al.* [154], by NMR study and ^{93}Nb relaxation time T_1 measurements, have also concluded that chains II have a small electron concentration. T_1 was shown to

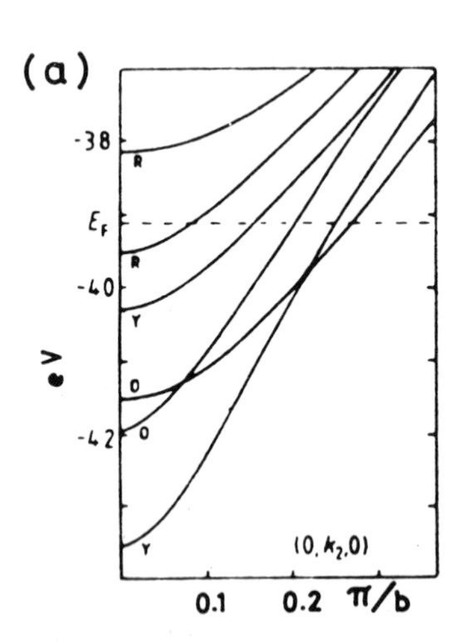

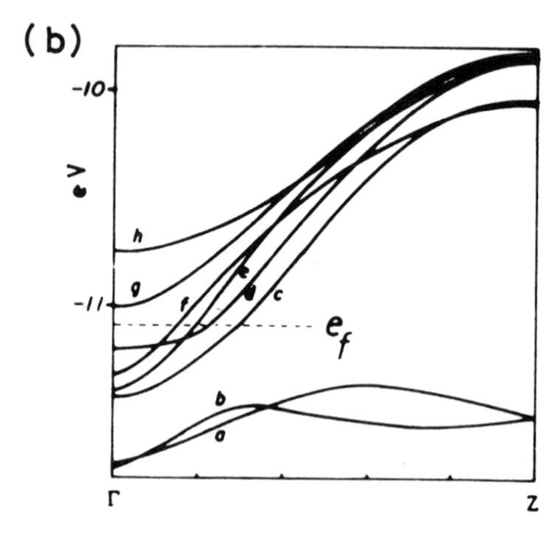

Fig. 10. Electron-band structure of $NbSe_3$ along the chains in fraction of the reciprocal vector **b*** calculated by (a) Bullett [143] and (b) Whangbo and Gressier [150]. Bands are assigned to a given type of chain in the $NbSe_3$ structure when their wavefunctions are predominant on them. In (a) *R* is predominently type II chains, *Y*: type III chains and 0: type I chains, and in (b) bands *g* and *h* are assigned to type II chains, *d* and *e* to type III chains, *c* and *f* to type I chains.

increase at each Peierls temperature, T_1 and T_2, indicating the growth of a CDW gap at the Fermi level. A frequency dependent T_1 has been measured at 4.2 K and attributed to the nuclear relaxation of nuclei of chains II, indicating that the electron motion is a one-dimensional diffusion like. Furthermore, the relaxation time rate has been shown to be very anisotropic which seems to be at variance with the mapping of the Fermi surface deduced from Shubnikov–de Haas oscillations.

Contrarily to $NbSe_3$, $TaSe_3$ shows a larger dispersion perpendicular to the chain axis [142]. The trigonal prismatic chain unit is again met in $(Fe_{1+x}Nb_{1-x})Nb_2Se_{10}$. Band calculations lead to a one-dimensional d band based on Nb atoms from the trigonal chain [143, 149].

(ii) $(MSe_4)_nI$

Band dispersion has also been calculated for some $(MSe_4)_nI$ compounds [155]. For a unique chain made up of $NbSe_4$ units, calculations lead to a d_{z^2} band the width of which is 3 eV. Sub-bands of a single chain of $(NbSe_4)_3I$ and of $(TaSe_4)_2I$ have been estimated. The same calculations made on $TaTe_4$ and $NbTe_4$ [156] indicate that the d band of the transition metal is half-filled, giving rise to a Fermi surface formed of two nearly flat parts. It has been expected that the lattice distortions take place from different nestings between these surfaces.

(iii) $ZrTe_5$

Finally, band calculations on $ZrTe_5$ show a semi-metallic band overlap between valence and conduction bands arising from coupling between adjacent $ZrTe_5$ layers [157, 158]. There is, therefore, no reason to expect any CDW transition. The anomalies observed in transport properties may be related to temperature dependence of the carrier concentration and of the carrier mobility [157].

4.2.3. *Unequivalent Chains in* $NbSe_3$

Unequivalent niobium sites have been clearly shown by an NMR study performed by Devreux [159] on a pseudo-monocrystal of aligned fibers of $NbSe_3$. Figure 11 shows his results. At room temperature, with the magnetic field aligned parallel to the chain axis, the spectra allow us to distinguish three unequivalent sites with different Knight shifts and quadrupolar frequencies. At 4.2 K, one site remains unaffected by the two CDW transitions. The quadrupolar structures for the other two sites disappear successively as a consequence of the modulation of the local electric field by the CDW whereas the Knight shift changes because of the partial suppression of the Fermi surface. While in the model proposed by Wilson [139] one expects an insulating chain at low temperature, consequently undergoing no Knight shift, the unaffected chain is the chain with the largest quadrupolar interaction and the largest Knight shift. To take into account this result, Devreux has made the hypothesis that the d_{z^2} bands of chains II are completely filled, the CDW occuring on chains I and III. The refinement of band calculations made by Whangbo and Gressier [150] taking into account inequivalent chains gives rise to two filled bands below the Fermi level but not with type II chain character. Therefore, if the result of three different types of niobium sites is unquestionnable, the exact assignment of a CDW on a specified chain is still ambiguous.

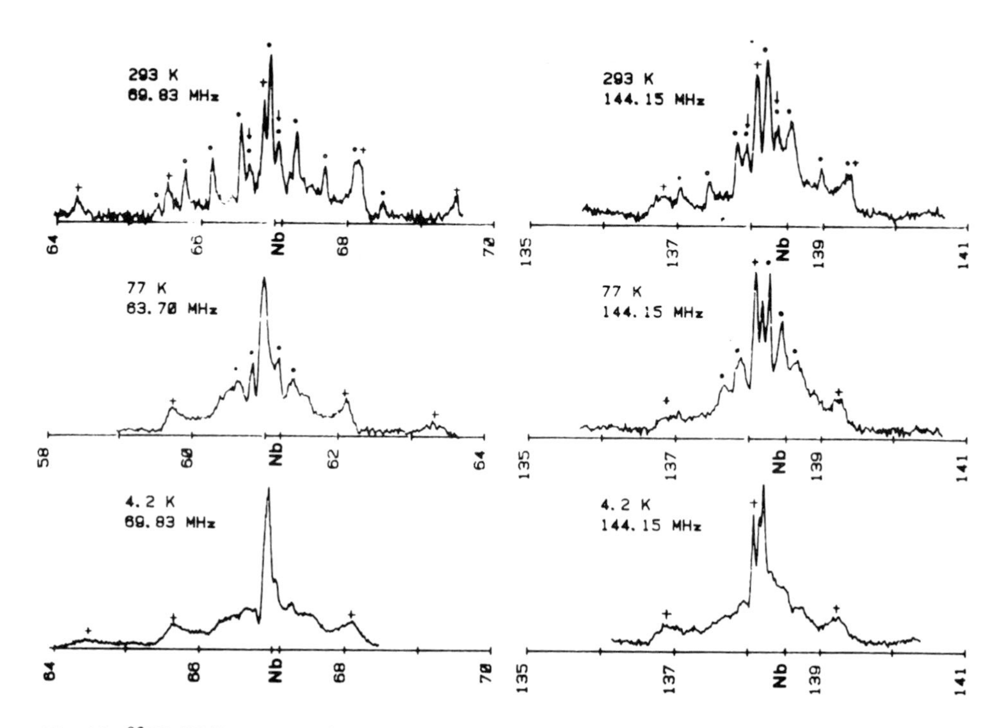

Fig. 11. ^{93}Nb NMR spectra of a pseudo-monocrystal of $NbSe_3$ made up of aligned fibers with the magnetic field H parallel to the **b** axis at two frequencies and three temperatures. X axis is labelled in kG. The position of the resonance of the diamagnetic Nb (γ = 1.0407 MHz/kG) is shown. At room temperature three sets of lines can be indexed; they correspond to three different niobium sites marked ○ for site 1, + for site 2 and • for site 3. At 77 K, the well-defined structure corresponding to site 1 has disappeared. At 4.2 K the Knight shift and quadrupolar interactions of site 2 remain unchanged whereas the quadrupolar structure of site 3 is suppressed (after [159]).

4.2.4. *Summary*

(i) Band calculations allow us to distinguish compounds which show a semi-metal band overlap such as $TaSe_3$, probably $ZrTe_3$, $ZrTe_5$, $HfTe_5$ from compounds characterized by broad pseudo-one-dimensional d_{z^2} bands with a band filling approximately 1/4 for most of them: CDW transitions only occur in the latter compounds.

(ii) Although superstructures in type I NbS_3, in $NbTe_4$ and in $TaTe_4$ have been explained as resulting from CDW instabilities, however, a transition temperature has never been measured below a temperature at which the crystals start to be deteriorated. Being perhaps restrictive we disregard these compounds as far as CDW properties are concerned.

(iii) In $NbSe_3$ and TaS_3 the modulation resulting from the lattice distortion is only homogeneous in domains with a maximum length of 2 μ and a maximum width of 200 Å. The misalignment between **Q** vectors of different domains can relax with time according to external conditions.

4.3. FURTHER CONSEQUENCES OF THE PEIERLS INSTABILITY ON THE PHYSICAL PROPERTIES OF $NbSe_3$ AND TaS_3

Among transition metal trichalcogenides, $NbSe_3$ and TaS_3 have mostly been investigated. Experiments on such small crystals, however, raise difficulties when a finite volume, for instance for heat capacity, magnetic susceptibility, NMR measurements . . . , is required because of a possible contamination, inhomogeneity or mixing between different polytypes as in the case of TaS_3. The sample has to be prepared fiber-by-fiber. $(MSe_4)_nI$ compounds are more suitable for such experiments.

4.3.1. *Other Transport Properties*

A review of concepts used for transport properties in one-dimensional compounds has been given by Chaikin [160]. The nature of charge carriers can be deduced from the sign of the thermoelectric and of the Hall voltage.

(i) The thermopower for a metal follows a linear dependence with the temperature and decreases with T. Its sign is positive (negative) if the transport of energy is carried by electrons (holes) which are located above the Fermi level. The thermopower derived from the Boltzmann equations in the tight binding approximation, valid for one-dimensional compounds, is given by [160]:

$$S = -\frac{\pi^2 k_B^2 T}{3|e|}\left[\frac{\cos \pi \frac{\rho}{2}}{2|t| \sin^2 \frac{\pi\rho}{2}} + \frac{\tau'(E_F)}{\tau(E_F)}\right] \tag{4-4}$$

where t is the transfer integral along the chains (1/4 of the electron bandwidth), ρ the number of conduction electron per transition atom and τ the scattering time for electrons of energy E. For a semiconductor with an energy gap, E_g, the thermopower is given by [161]:

$$S = -\frac{k_B^2}{|e|}\left[\frac{b-1}{b+1}\,\frac{E_g}{kT} + \ln \frac{m_h}{m_e}\right] \tag{4-5}$$

where b is the ratio of the electron to hole mobility, m_h and m_e the hole and electron effective mass. In this case, S follows a T^{-1} dependence. Figure 12 shows the thermopower of $NbSe_3$ and of both forms of TaS_3 (with a different scale for orthorhombic TaS_3).

Above the Peierls transition, the thermopower of $NbSe_3$ and monoclinic TaS_3 is negative (n-type) and weak indicating that the major carriers at room temperature are electrons: $-15\ \mu V/K$ for $NbSe_3$ [162–164], $-1\ \mu V/K$ [167] for monoclinic TaS_3. For $NbSe_3$, S is approximately constant with T between 300 K and T_1 [162] or decreases very slowly with T down to T_1 [164]. At T_1 the thermopower in $NbSe_3$ increases and becomes more negative which is interpreted as the formation of a CDW gap at the Fermi surface with hole like carriers. At T_2 another gap occurs and S becomes more positive. A change in the sign of S is also observed in monoclinic TaS_3: type n above T_1, type p between T_1 and T_2 are type n again below T_2 [167]. For orthorhombic TaS_3

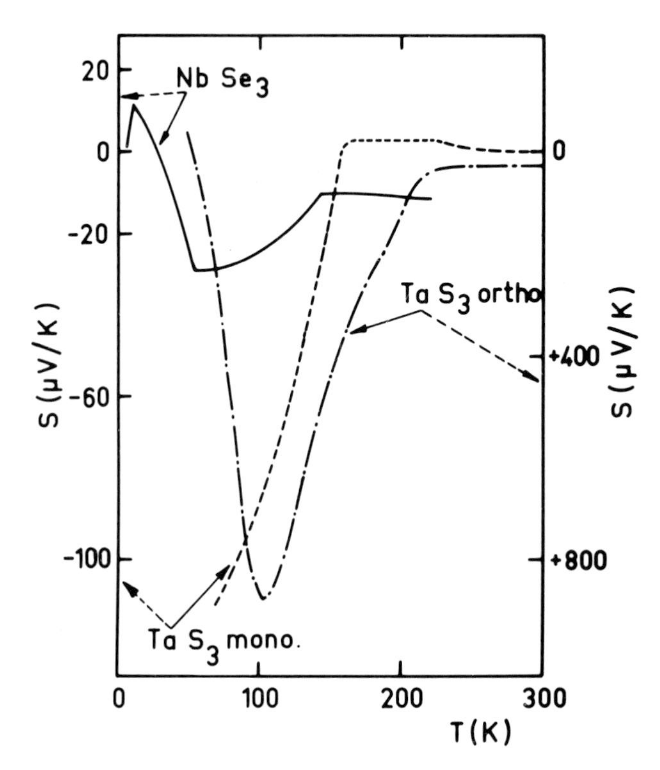

Fig. 12. Variation of the thermopower, S, as a function of temperature for $NbSe_3$ [162] and monoclinic TaS_3 [167] (left scale) and for orthorhombic TaS_3 [166, 168, 293] (right scale).

contradictory results on the sign of S have been reported [166, 168, 293]. It is now recognized that the thermopower is positive at all temperatures [293]; it increases strongly at T_0 and is at a maximum around 100 K. S does not behave like a metal for both TaS_3 because it stays constant or increases when T is decreased [166] which may indicate strong precursor effects above the Peierls transitions. Below T_0, Fisher has measured the T^{-1} dependence of S in orthorhombic TaS_3, leading to a "gap" of 0.2 eV. But as relative values of m_n, m_e and b in (4–5) are unknown, a value of the CDW gap cannot be deduced from the temperature dependence of S.

(ii) The Hall effect in $NbSe_3$ at room temperature is negative [141], which indicates that the conductivity is mainly due to electrons, in keeping with thermopower measurements. As has already been mentioned, the Hall constant, (-1.0×10^{-2} cm^3/c), derived in [141] has been corrected and found to be $-(4.5 \pm 0.5) \times 10^{-3}$ cm^3/c [145]. At T_1 and T_2, R_H shows an abrupt change in magnitude but not in sign. Below 15 K, however, the sign of R_H becomes positive. If, at T_1, the sign of carriers involved in the transition deduced from thermopower and Hall effect measurements is the same, at T_2, S becomes positive whereas R_H becomes more negative. In order to explain this inconsistency, Chaikin *et al.* [164] have suggested that the two transport coefficients measure different averages of the Fermi surface parameter, which may indicate a rather complicated Fermi surface. These results have led Ong [170] to present a two-band model for transport properties in $NbSe_3$ below T_2. The magnitude, but not the sign of R_H, $|R_H| = 6.3 \times 10^{-4}$ c cm^{-3}, has also been measured for orthorhombic TaS_3 [147].

4.3.2. *Thermodynamics of CDW Phase Transitions*

(i) Magnetic Susceptibility

The magnetic susceptibility of $NbSe_3$ is diamagnetic at all temperatures and has been measured to be: -43×10^{-6} emu/mole at 4.2 K [76]; and -55×10^{-6} emu/mole [171] or -70×10^{-6} emu/mole [172] at room temperature. The core diamagnetism largely exceeds the diamagnetic value; it is not very well known because it depends on the real ionicity of niobium atoms and of the bonded selenium. The core diamagnetism has been estimated to be: -110×10^{-6} emu/mole in [171]; and -84×10^{-6} emu/mole in [172]. Therefore, Kulick and Scott give a paramagnetic susceptibility of $NbSe_3$ at 300 K of 55×10^{-6} emu/mole and Di Salvo *et al.* a value of 14×10^{-6} emu/mole. At each CDW transition, Kulick and Scott have measured a decrease of χ. This is interpreted as resulting from the decrease of the density of states due to the gap formation at the Fermi surface. χ decreases between 25% and 13% at T_1 and 13% and 5% at T_2 according to the core diamagnetism correction. Estimation of the spin and orbital contribution to the paramagnetic susceptibility has been performed by Devreux from Knight shift data [159]. He found that less than half the paramagnetic susceptibility is due to spin contribution ($\sim 20 \times 10^{-6}$ emu/mole) which explains the small relative change of the total χ found in [171]. The room temperature spin susceptibility corresponds to a density of states of 0.31 states per eV-spin-Nb atom, in keeping with band calculations [142]. Spin susceptibility has also been shown to decrease between 300 K and 4.2 K by measurements of $(T_1 T)^{-1}$ (where T_1 is the nuclear relaxation time) [154, 159], which is proportional to the square of the density of states at the Fermi level. Following this analysis, 50% of the Fermi surface is destroyed by both CDW transitions and at 4.2 K the spin susceptibility has a value of 9×10^{-6} emu/mole. The conduction electron susceptibility of orthorhombic TaS_3 has been found to be approximately 10×10^{-6} emu/mole (although the core contribution was not specified) [166].

Below T_0 in TaS_3 and T_1 in $NbSe_3$ the CDW gap can be deduced from the temperature variation of χ: $2\Delta = 1300$ K is found for orthorhombic TaS_3 [166], in relatively good agreement with the gap value obtained from conductivity measurements (see Table II) and $2\Delta = 500 \pm 50$ K for the T_1 transition in $NbSe_3$. Between room temperature and the Peierls transition, T_0 in TaS_3 and T_1 in $NbSe_3$, χ does not behave as expected for a Pauli susceptibility: χ decreases with T which suggests a pseudo-gap formation resulting from precursor effects due to one-dimensional fluctuations. In orthorhombic TaS_3 because of these fluctuations the density of states is reduced by 50% between 300 K and T_0.

(ii) Thermal Conductivity

Thermal conductivity, K, of $NbSe_3$ has been measured between 35 K and room temperature [173] and between 2 K and 70 K [174]. In this temperature range the main energy carriers are phonons. Below 10 K, the thermal conductivity can be separated into two parts:

the electronic contribution [0.15 W m^{-1} K^{-2}] × T in keeping with the value deduced from the Wiedeman–Franz law, and the phonon contribution represented by [0.39 W m^{-1} K^{-3}] × T^2. At room temperature, the electronic thermal conductivity is approximately a quarter of the total value. A small anomaly of K has been detected at T_2 [174] and a sharp one with an amplitude of 0.7 W m^{-1} K^{-2} at T_1 [175]. The increase in K is caused by the loss of the electronic contribution in the thermal conductivity. The size of the anomaly corresponds to approximately 25% of the total electronic thermal conductivity at this temperature.

(iii) Heat Capacity

An anomaly in heat capacity measurements of $NbSe_3$ has been found at T_1 and at T_2 [176] (the large anomaly at T_2 reported in [77] has not been reproduced in further experiments). Data for both transitions are shown in Figure 13. Precursor effects are

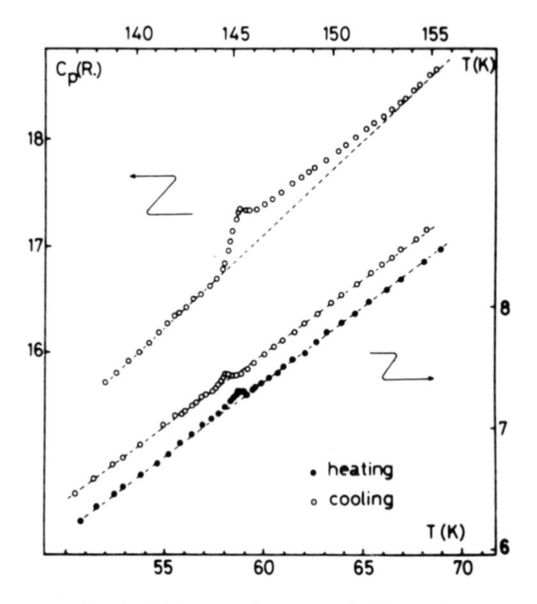

Fig. 13. Heat capacity anomalies in $NbSe_3$ at the upper CDW transition T_1 (upper and left scales) and at the lower CDW transition T_2 (lower and right scales). Pretransitional effects are visible 7 K above the transition at T_1 = 145 K (after [176]).

visible 7 K above T_1, but no pretransitional effect has been found at T_2. From the area under the specific heat anomaly, the change in entropy is estimated to be 0.01 R_0 at the T_1 transition and 0.005 R_0 at the T_2 one. When compared to the total conduction electron entropy at the phase transitions (neglecting the phonon entropy), these numbers allow one to estimate that 25% of the conduction electrons are condensed at T_1 and a further 30% at T_2 [176, 177]. When the sample was cooled extremely slowly (1 K per hour), the phase transitions at T_1 and T_2 have also been found to have a first order character with a latent heat of approximately 0.4 $R_0 T_1$ at the T_1 transition and 0.02 $R_0 T_2$ at the T_2 transition. Moreover, an hysteresis of 1 K is detected at the lower CDW transition. This experiment, unique as far as the time scale is concerned, clearly indicates the strong dependence of physical properties of $NbSe_3$ on the thermal history of the sample. Specific heat measurements have also been performed in $NbSe_3$ at low temperatures between 0.05 K and 6.5 K [178]. Below 1 K the electronic linear contribution, γT, to the heat capacity has been found to be 24.5 erg/g K^2. The lattice contribution deviates from a Debye variation; it obeys a $T^{2.8}$ power law which can be separated into a normal cubic term, due to longitudinal modes, and a $T^{2.5}$ term attributed to a bending force constant dominating transverse modes. The latter term, important down to 0.5 K, reveals a large anisotropy in the lattice.

(iv) Elastic Anomalies at the CDW Phase Transition

Elasticity and internal friction have been measured in $NbSe_3$ and TaS_3 [179, 180] using a vibrating reed technique [181]. The Young's modulus at room temperature is $(2.5 \pm 0.5) \times 10^{12}$ dynes/cm² for $NbSe_3$ and $(3.5 \pm 0.5) \times 10^{12}$ dynes/cm² for orthorhombic TaS_3, indicating that these compounds are very stiff. A strong anomaly in the Young's modulus ($\Delta E/E \sim 2\%$) is detected in orthorhombic TaS_3 at the Peierls transition as shown in Figure 14; on the other hand, $\Delta E/E$ is only 6×10^{-4} at T_1 and less than

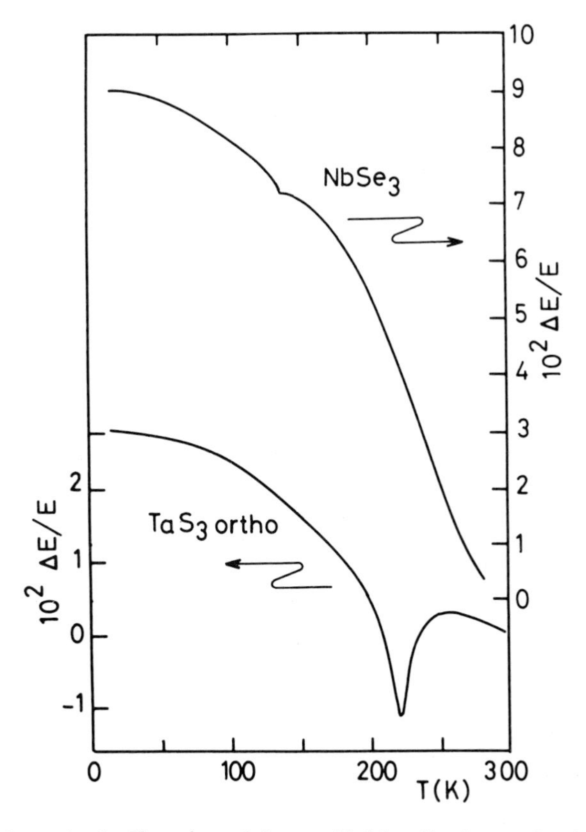

Fig. 14. Relative change in the Young's modulus provided by vibrating reed techniques with respect to room temperature value as a function of temperature for $NbSe_3$ (upper curve and right scale after [179]) and for orthorhombic TaS_3 (lower curve and left scale after [180]).

3×10^{-5} at T_2 in $NbSe_3$. At a second order phase transition, the Young's modulus in the ith direction has an anomaly obtained from the Clausius–Clapeyron equation such as:

$$\frac{\Delta E_i}{E_i} = -E_i \frac{\Delta C_p}{T_c} \left(\frac{\partial T_c}{\partial \sigma_i} \right)^2 \tag{4-6}$$

where σ_i is the ith component of longitudinal stress, ΔC_p the specific heat jump at the transition temperature, T_c. As mentioned above, ΔC_p has only been measured in $NbSe_3$. From (4-6), with the values of $\Delta C_p = 0.45\ R_0$ at T_1 and $0.1\ R_0$ at T_2 [176] it has been deduced that $|\partial T_1/\partial \sigma_\| | \sim 0.2$ K/kbar and $|\partial T_2/\partial \sigma_\| | < 0.07$ K/kbar in $NbSe_3$. Recently, the stress dependence of the CDW transition temperatures has been measured in $NbSe_3$ and TaS_3 [182]. In $NbSe_3$, T varies linearly as a function of $\sigma_\|$ such as: $\partial T_1/\partial \sigma_\| = -0.44$ K/kbar and T_2 has a quadratic variation as a function of $\sigma_\|$ with an initial slope $\partial T_2/\partial \sigma_\| = -0.04$ K/kbar. These results confirm the validity of (4-6) and the relationship between thermodynamic quantities. It will be shown in a following section that T_1 and T_2 decrease under hydrostatic pressure at the rate of $\mathrm{d}T_1/\mathrm{d}P = -4$ K/kbar and $\mathrm{d}T_2/\mathrm{d}P = -6.3$ K/kbar [183]. Therefore, uniaxial stress has the same effect as pressure, which has led Lear *et al.* [182] to conclude that the Peierls transition temperatures are predominantly affected by contractions transverse to the chain axis.

The stress dependence of T_0 for orthorhombic TaS_3 has been found to have a linear variation with $\sigma_\|$ such as $\partial T_0/\partial \sigma_\| = -1.5$ K/kbar (this figure is the same value as for $\mathrm{d}T_0/\mathrm{d}P$ [184]). With the experimental values of $\partial T_0/\partial \sigma_\|$, ΔE, E, and assuming the validity of (4-6) one expects an anomaly of $0.62\ R_0$ in the heat capacity of orthorhombic TaS_3, similar to that detected in $NbSe_3$. It has also been noted that the large amplitude of the Young's modulus anomaly is comparable to that measured at the incommensurate–commensurate transition in $2H$–$TaSe_2$ the amplitude of which is much larger than the anomaly at the normal–incommensurate transition in the same compound [181]. The internal friction (measured as the inverse of the resonance width of the reed) increases ($\Delta Q^{-1} \sim 5 \times 10^{-4}$) at the Peierls transition in orthorhombic TaS_3 which has been interpreted as the growth of three dimensional domains taking place at T_0 and saturating a few degrees below T_0. In the modulated phase these domains have a large pinning which inhibits their motion. As explained in [180] this domain motion is not related to the depinning of the CDW (as will be explained in Section 5), but it is more probably a consequence of the reorganization of the domain wall distribution (as observed by Steeds *et al.* in dark field electron microscope imaging [134]) under the applied stress.

4.3.3. *Critical Resistive Fluctuations*

When heated up to 400 K, the resistance of orthorhombic TaS_3 still shows an upward curvature as a function of temperature, as is shown in Figure 15. This rounding is interpreted as the signature of a pseudo-gap due to large CDW fluctuations. Electronic diffraction measurements clearly reveal these fluctuations and, as has already been mentioned in the preceding sections, they influence all the physical properties. These structural deformations provide a periodic potential at the Fermi level; electrons are scattered by this potential and suffer large changes in momentum. Electronic scattering from these structural deformations is resistive. The contribution of this critical scattering

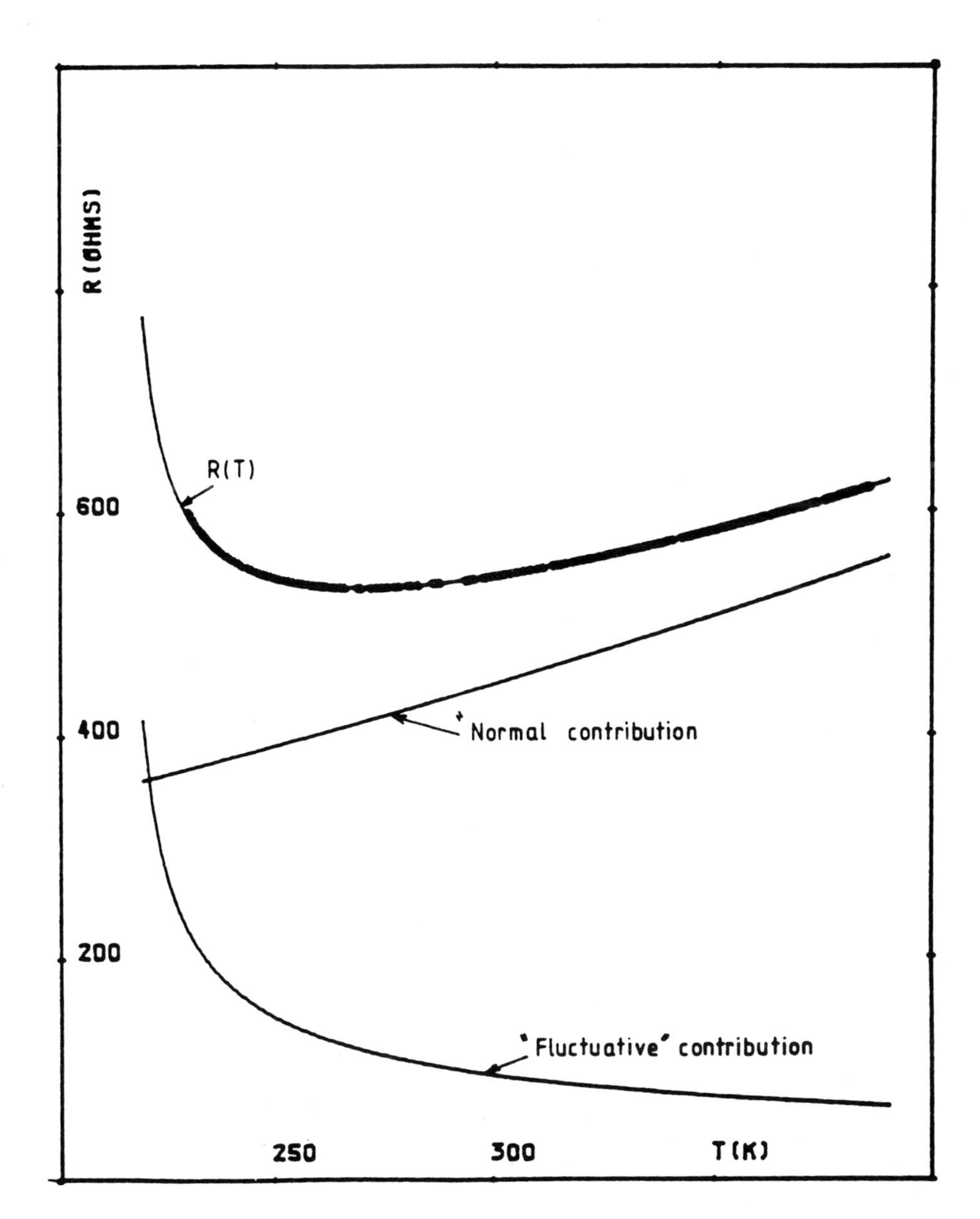

Fig. 15. Variation of the resistance, R, of an orthorhombic TaS_3 sample as a function of temperature, T, for T above the Peierls transition, T_0. The rounding in the variation of $R(T)$ is attributed to CDW fluctuations. Following (4-8) the resistance is divided in two contributions: one due to normal electrons, R_n, the other due to resistive fluctuation, R_f, (after [186]).

to the electrical resistivity in anisotropic metals has been calculated by Horn and Guiddoti [185] in the case of TTF–TCNQ and TSF–TCNQ. They have found that the critical behaviour of the resistivity follows a power law which depends on the dimensionality,

d, of the allowed final states for scattering (although the fluctuations are assumed to be three dimensional) such as:

$$\frac{d\rho_f}{dt} \alpha t^{-1.5} \quad \text{for} \quad d = 1 \tag{4-7a}$$

$$\frac{d\rho_f}{dt} \alpha t^{-1.0} \quad \text{for} \quad d = 2 \tag{4-7b}$$

$$\frac{d\rho_f}{dt} \alpha t^{-0.5} \quad \text{for} \quad d = 3 \tag{4-7c}$$

with $t = (T - T_0)/T_0$. The resistance of TaS_3 can be separated into two parts such as:

$$R = R_n + R_f \tag{4-8a}$$

$$= A + BT + CT^2 + D/(T - T_0)^\alpha \tag{4-8b}$$

where the three first terms are the normal resistance and the last one the resistance due to fluctuations. The relative contribution of R_n and R_f is shown in Figure 15 resulting from the best fit of (4-8b) with the experimental data. It can be seen that the pseudo-gap contribution to the resistivity is approximately 50% at T_0; the same figure has been obtained in susceptibility measurements [166]. The variation of dR_f/dt as a function of the reduced temperature, t, is drawn in Figure 16 in a log–log scale. The critical exponent for the variation of dR_f/dt near T_0 is -1.5 which indicates that the dimensionality of states for scattering are one-dimensional near the Peierls transition [186]. Corroboration with the Horn and Guiddoti model has been found in the two-dimensional dichalcogenide $2H$–$TaSe_2$ [187]; dR_f/dt varies as $t^{-1.03}$, indication that the scattering from periodic distortion is strongly limited to the hexagonal plane of the crystal.

Critical fluctuations have also been investigated near T_1 and T_2 in $NbSe_3$ [188, 189] but dR_f/dt seems to vary as t^{-2} for the T_1 transition and t^{-1} for the T_2 transition.

4.4. Superconductivity of $NbSe_3$ and $TaSe_3$

4.4.1. *CDW and Superconductivity in $NbSe_3$*

$NbSe_3$ becomes superconducting when the CDW transitions are weakened, which can be achieved by applying pressure or by doping.

The CDW transition temperatures and the amplitudes of the resistivity anomalies are affected by pressure. T_1 and T_2 decrease with an initial slope of $dT_1/dP = -4$ K/kbar [77] and $dT_2/dP = -6.25$ K/kbar [183]. The resistive anomaly, the signature of the lower CDW transition, is barely visible at 5.5 kbar. T_2 decreases to zero around 5.5–6 kbar. Beyond the pressure which suppresses T_2, $NbSe_3$ is superconducting at approximately $T_c \sim 3.5$ K with a sharp resistive transition (less than 0.1 K). The superconducting transition temperature decreases when P is further increased. Variation of T_2 and T_c as a function of pressure is drawn in Figure 17. On the other hand, initial susceptibility measurements under pressure have also been performed on a sample made up of many fibers (0.5 g) [190]. The superconducting temperature derived from the variation of the

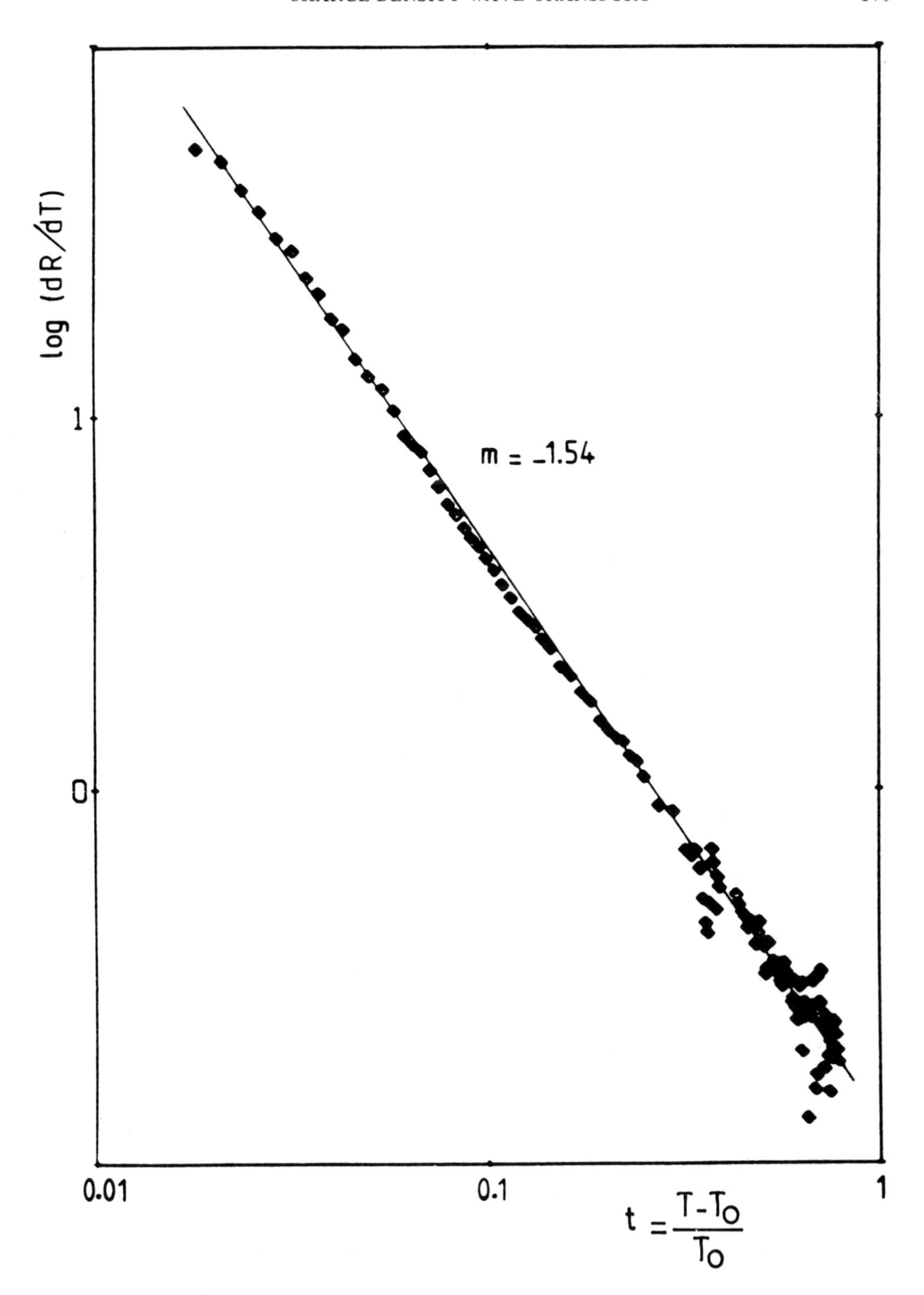

Fig. 16. Log–log plot of the derivative of the fluctuation resistance, R_f, as a function of the reduced temperature, $(T - T_0)/T_0$, above the Peierls transition T_0 for a TaS_3 sample (after [186]).

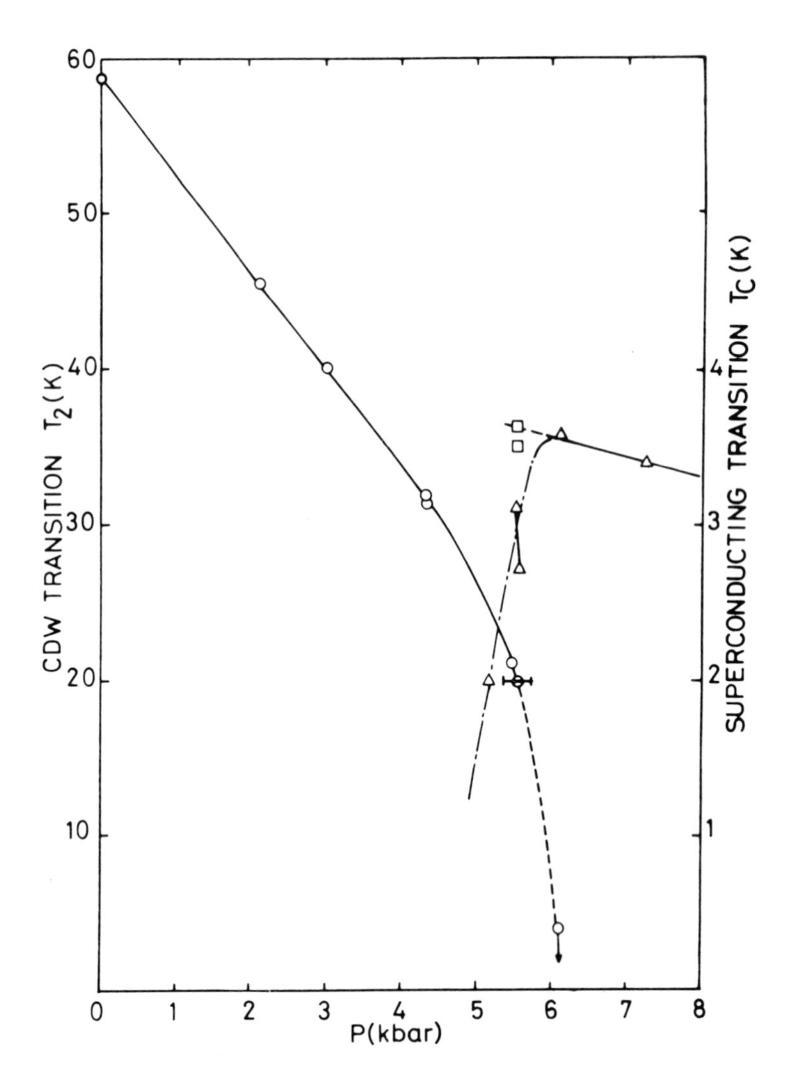

Fig. 17. Variation of the CDW transition T_2 (○) and of the superconducting transition (Δ, □) of $NbSe_3$ as a function of pressure (after [183]).

susceptibility as a function of T at several pressures increases linearly with P and saturates around 2.5 K at a pressure of 5–6 kbar. The magnitude of the susceptibility indicates that the volume of the sample is at least 75% superconducting.

Zero resistivity has also been measured on pure $NbSe_3$ sintered powder [191], on $NbSe_3$ crystals doped with impurities of tantalum [192, 193], titanium [192, 194] and zirconium [195]. The critical superconducting temperature is approximately 2–2.5 K. No magnetization on these doped samples has yet been reported. As shown in the subsequent section, T_i and T_a impurities strongly affect the T_2 Peierls transition.

A decrease of the CDW formation temperature with pressure has already been observed for many transition metal dichalcogenides with trigonal prismatic coordination [196]. This decrease is interpreted as the result of the competition between lattice strain energy which increases under pressure and the gain of the electronic energy given by the opening of a gap at the Fermi surface. The variation of T_c with pressure can be estimated from the pressure and temperature dependence of the electron–phonon coupling constant:

$$\lambda = \frac{N(E_F)\langle I^2 \rangle}{M\langle \omega^2 \rangle} \tag{4-9}$$

where $\langle I^2 \rangle$ is the Fermi surface average of the electron–phonon matrix element, $\langle \omega^2 \rangle$ an average of phonon frequencies, $N(E_F)$ the density of states at the Fermi level. As mentioned in Section 4.3.2, the density of states is strongly reduced by the CDW gap formation. The pressure reduces the Fermi surface nesting and restores areas of the Fermi surface lost by the gap formation (see the band structure calculated by Shima in Section 4.2.2 [151]) and, therefore, increases $N(E_F)$, which is favourable to the onset of superconductivity [197]. The partial gapping of the Fermi surface resulting from the competition between CDW and superconductivity has been studied by Bilbro and McMillan [198]. They have derived an equation between the superconducting transition temperature, T_c, in the presence of partial gaps on a fraction N_1/N of the Fermi surface ($N_1 + N_2 = N$), the transition temperature, T_0, at which these gaps appear and the superconducting transition, T_{c0}, in the absence of any CDW gap, as follows:

$$T_c^{N_2/N} T_0^{N_1/N} = T_{c0} \tag{4-10}$$

Fuller *et al.* [194] have shown that this theory is appropriate to $NbSe_3$, the fraction of the Fermi surface affected by the CDW gap being estimated from the fractional increase of the resistivity at T_2. Therefore, the reduction of the density of states due to the CDW gap accounts for the variation of T_c without large changes in the electron–phonon interaction. However, the superconducting transition of $NbSe_3$ [199] decreases with pressure which may be explained by a lattice stiffening. The occurrence of superconductivity in doped samples also requires a modification of the band structure. Probably, the change in electron density introduced by doping affects the nesting properties (like pressure) which favours the onset of superconductivity.

4.4.2. *'Filamentary' or 'Bulk' Superconductivity in $TaSe_3$ and $NbSe_3$ at Ambient Pressure*

The resistivity of $NbSe_3$ at ambient pressure has been found to drop below $T \sim 2.5$ K by 30 to 75% (depending on the samples) of its residual value at 4.2 K and to reach a plateau approximately at 1.5 K [194, 200]. Measurements have been performed at lower temperatures down to 7 mK. Some samples have shown another drop around 0.4 K [201]. The resistivity below 2.2 K is strongly current-dependent for current densities higher than 0.1 A/cm^2; the drop in resistivity is completely suppressed for a current density of approximately 50 A/cm^2 (see Figure 20 in Chapter 5, Part I). These measurements can be interpreted as due to superconducting filaments becoming normal when the current density increases. Similar results have been obtained on $TaSe_3$ crystals with a superconducting transition at 1.7 K [200]. It has recently been reported, however, that a giant resistivity anomaly appears just in the temperature range where $TaSe_3$ becomes superconducting when measurements are performed with a very small current density [202]. With a current density of 0.4 A/cm^2 the resistance increases at $T_c \sim 2.3$ K, has a peak and decreases to zero at a temperature below T_c. The maximum amplitude of the peak is four times the residual value above T_c. Its amplitude is strongly current-dependent; with a current density of a few A/cm^2, the resistance of $TaSe_3$ drops below T_c and reaches zero at 1.7 K.

Superconductivity is a bulk effect if flux expulsion (Meissner effect) occurs and, therefore, if the sample under test shows a diamagnetic susceptibility of $-1/4\pi$ per unit volume below the superconducting transition. Burhmann *et al.* [203] have performed very precise measurements of the magnetization of $TaSe_3$ and $NbSe_3$. They have measured a diamagnetic susceptibility and flux trapping which clearly prove a superconducting behaviour in these compounds. However, χ is very small at T_c (even when measured in magnetic fields as low as 0.1 Oe). χ increases sharply when T is reduced below T_c but its magnitude is never larger than 1% of the value expected for a total Meissner effect. Therefore, superconductivity in $NbSe_3$ and $TaSe_3$ cannot be a bulk homogeneous superconductivity. Furthermore, Kawabata and Ido [204] have performed combined resistivity measurements at low temperatures in $NbSe_3$ with optical and scanning electron microscope observations. They have found that a drop in resistivity below 2 K is measured only if striations are visible on the crystal surface. They conclude that single crystals without striation do not show superconductivity at ambient pressure, at least above 0.38 K. Strain between neighbouring crystals, the presence of another phase at the crystal boundaries, impurity migration at the boundaries may be possible factors which induce this filamentary superconductivity. If the same conclusions can be applied to $TaSe_3$ remains an open question. The nature of the giant anomaly in $TaSe_3$ at low-current density has also to be investigated. As mentioned already, $NbSe_3$ under pressure is a bulk superconductor with an almost total Meissner effect [190, 191].

4.4.3. *Anisotropy of the Critical Magnetic Fields*

The critical magnetic fields have been measured for $TaSe_3$ at ambient pressure [205] and for $NbSe_3$ at several pressures above 5.5 kbar when $NbSe_3$ is fully superconducting [191]. A precise determination of the angular orientations needs to be made. $TaSe_3$ and $NbSe_3$ samples are long ribbons with a large width compared to the thickness. The ribbon plane is the plane $(\bar{2}01)$ for $TaSe_3$ and the plane (**b**, **c**) for $NbSe_3$. The magnetic field can be rotated in two independent crystallographic planes: one is the plane including the **b** axis and the axis normal to the ribbon; in this plane the polar angle is θ with $\theta = 0$ when H is parallel to **b**. The second plane is the plane (010) perpendicular to the **b** axis; the azimuthal angle, ϕ, is zero when H is parallel to **c** in both samples (θ and ϕ are inverted in [205]). H_{c_2} is defined either as the midpoint of the transition curves for $TaSe_3$ or as the field at which R is 90% of the resistance in the normal state. H_{c_2} strongly varies with the polar angle, θ: $H_{c_2}(\theta = 0)/H_{c_2}(\theta = \pi/2) = 27$ for $TaSe_3$ at 1.53 K, 5.2 at $T = 2$ K and $P = 7.4$ kbar for $NbSe_3$. H_{c_2} is also angular-dependent in the plane perpendicular to the **b** axis: $H_{c_2}(\phi = 0)/H_{c_2}(\phi = \pi/2) \sim 24$ for $TaSe_3$ at $T = 1.53$ K, but only 2 for $NbSe_3$ at $T = 2$ K and $P = 7.4$ kbar. Figure 18 shows the temperature variation of H_{c_2} for $TaSe_3$ and for $NbSe_3$ at 7.4 kbar with H parallel to the chain axis **b** ($H_{\parallel}$) and H perpendicular to the ribbon plane. When H is parallel to **b**, H_{c_2} shows an upward curvature characteristic of low-dimensional conductors [206].

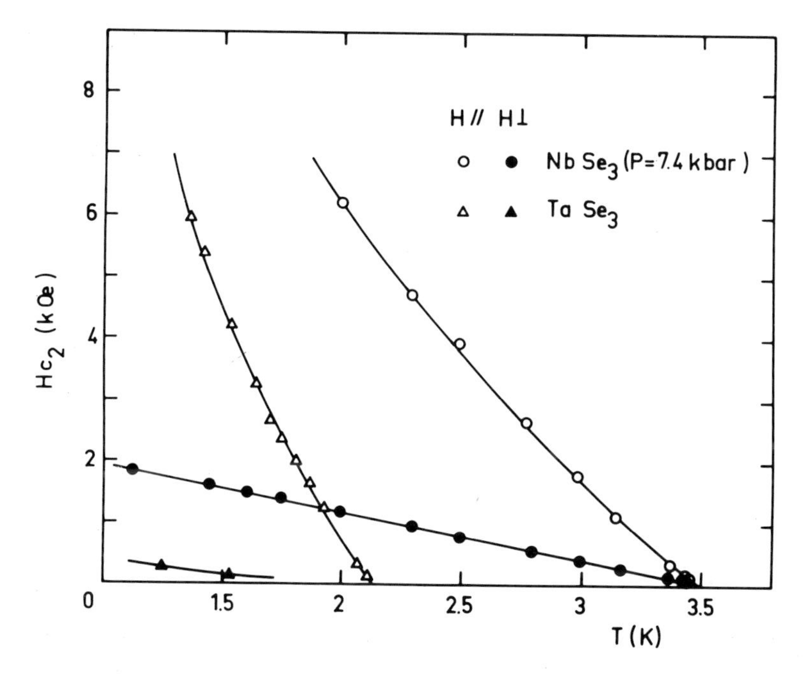

Fig. 18. Variation of the critical magnetic fields of $TaSe_3$ and of $NbSe_3$ under a pressure of 7.4 kbar as a function of temperature for two orientations: H parallel ($\theta = 0$) and H perpendicular ($\theta = \pi/2, \phi = \pi/2$) to the chain axis (after [191, 205]).

Critical fields of filamentary superconductors have been calculated by Turkevitch and Klemm [207] (a review is given by R. A. Klemm, Chapter 5 of Part I). They have extended the model of Lawrence and Doniach [208] of Josephson-coupled planes, valid for multilayered superconducting compounds such as 2H transition metal dichalcogenides. They consider an assembly of filaments parallel to the z direction which form a lattice with dimensions a and b. When H is applied perpendicular to the chains, the vortex current follows the chain and tunnels by Josephson effect from one chain to another. When H is parallel to the chains, the vortex current is entirely a Josephson tunnel current and the system behaves as a Josephson grid. Near T_c, the coherence lengths are much larger than the interchain distance, a or b, and a generalized effective mass model can be used. With H perpendicular to **b** the conductivity is anisotropic and the angular azimuthal dependence of H_{c_2} is:

$$H_{c_{2\perp}}(\phi) = \frac{\phi_0}{2\pi\xi^2} \frac{1}{[\epsilon_x^2 \cos^2 \phi + \epsilon_y^2 \sin^2 \phi]^{1/2}} \tag{4-11}$$

where ϕ_0 is the flux quantum, ξ the coherence length, $\epsilon_x = (m/m_x)^{1/2}$ and $\epsilon_y = (m/m_y)^{1/2}$, m_x and m_y being the effective mass in the x and y direction, respectively. The polar dependence of H_{c_2} when H is rotated from being parallel to perpendicular to the chains is:

$$H_{c_2}(\theta) = \frac{\phi_0}{2\pi\xi^2} \frac{1}{\epsilon_y} \frac{1}{[\cos^2 \theta + \epsilon_x^2 \sin^2 \theta]^{1/2}} \quad (4\text{-}12)$$

At low temperatures the vortex core can fit between chains. Below a temperature, T^*, the chains decouple leading to a divergence of the critical field but T^* is not the same when H is applied parallel or perpendicular to the chains. For $TaSe_3$ and $NbSe_3$ the coherence lengths when T is extrapolated to zero are much larger than the chain separation and consequently, this effective mass model can be used. The polar dependence of H_{c_2} for $TaSe_3$ at 1.53 K (see Figure 5 in Chapter 5, Part I) and for $NbSe_3$ at 2 K and 7.4 kbar fit expression (4-12) with $\epsilon_x = 1/27$ for $TaSe_3$ [205] and $\epsilon_x = 0.1820$ for $NbSe_3$ [191]. Therefore, the anisotropy of the critical fields leads to an effective mass anisotropy of ~30 for $NbSe_3$ when the electrons are travelling parallel to **b** or parallel to **c**. In the plane perpendicular to the chain axis, the effective mass ratio is four, the lower effective mass being with the current along **c**. This anisotropy is in keeping with cyclotron effective masses derived from Shubnikov–de Haas oscillations in magnetoresistance measurements [152]. For $TaSe_3$, the polar anisotropy leads to an effective mass anisotropy of 700. Furthermore, in the plane perpendicular to **b** the azimuthal dependence of H_{c_2} does not follow (4-11): an effective mass of several hundreds is found while the anisotropy deduced from magnetoresistance gives an anisotropy of band masses only of 3. Similar to the polymeric sulfur nitride $(SN)_x$ [209], (a review of $(SN)_x$ is given by K. Kaneto *et al.* in Chapter 2 of this volume) size effects are expected when superconducting filaments have a thickness d less than the coherence length. Size effects may explain such a large anisotropy in $TaSe_3$. It is also worthwhile to recall the absence of a bulk diamagnetism as mentionned above. Although the upward curvature of $H_{c_{2\parallel}}$ as a function of T seems an indication, these pseudo-one-dimensional conductors do not clearly show the dimensional crossover between an anisotropic three-dimensional behaviour near T_c and a behaviour of Josephson-coupled chains at low temperatures. This crossover between the two regimes is now observed in two-dimensional layered composites [210].

4.5. Disorder

Disorder lowers and eventually suppresses Peierls transition temperatures. It was shown by Bulaevskii and Sadovskii [211] that the effect of disorder on T_p is qualitatively similar to the effect of magnetic impurities on a superconducting transition. Disorder in $TaSe_3$, $NbSe_3$ and both forms of TaS_3 has been created either by doping or by irradiation with electrons, protons and neutrons.

4.5.1. *Doping*

Doping can be achieved by replacing the transition metal in the chain by another transition metal either from the same column or from another column, or by chalcogen (selenium or sulfur) mixing. Cation mixing of V—V elements is thought to be isoelectronic. However, it has to be noted that the elementary prisms $NbSe_6$, $TaSe_6$, TaS_6 have different chalcogen—chalcogen bonding lengths and as mentionned in Section 2 this distance leads directly to the effective charge on the transition metal atom. Therefore, the substitution of a Nb atom by a Ta atom, like the substitution of a selenium by a sulfur atom, produces a small local change of charge. Compounds $Nb_{1-x}Ta_xSe_3$ [162, 212–214], $Ta_{1-x}Nb_xS_3$ [215], $TaS_{3-x}Se_x$ [215] or $TaSe_{3-x}S_x$ [216] have been synthesized. Cation mixing of IV—V elements is non-isoelectronic and compounds such as $Nb_{1-x}Ti_xSe_3$ [162, 212–214], $Nb_{1-x}Zr_xSe_3$ [193, 213, 214] have been prepared. Doping with iron [55, 217] with a concentration much less than that used for the synthesis of $(Fe_{1+x}Nb_{1-x})Nb_2Se_{10}$, with cobalt, and with molybdenum [218a] has also been investigated. One of the difficulties with doping is the determination of the real concentration of dopant. In contrast to transition metal dichalcogenides, the concentration of impurities which can be introduced into MX_3 compounds is extremely weak. The distribution of impurities is inhomogeneous for crystals from the same batch and probably also inside each monocrystal.

The resistivity of $Ta(S_xSe_{1-x})_3$ samples shows a semiconducting behaviour when $x > 0.06$. One-dimensional diffuse lines observed in some compounds with $x = 0.15$ may indicate a Peierls transition at approximately 70 K [216]. Doping with Group IV cations strongly affects the Peierls transitions of $NbSe_3$: 500 ppm of Zr(Ti) decrease T_1 and T_2 by 1 K (9 K) and 4 K (11 K) respectively [213]. By contrast, T_1 and T_2 are weakly decreased with tantalum impurities: a few degrees with 5% of tantalum. Figure 19 shows the temperature variation of $NbSe_3$ samples doped with titanium [164].

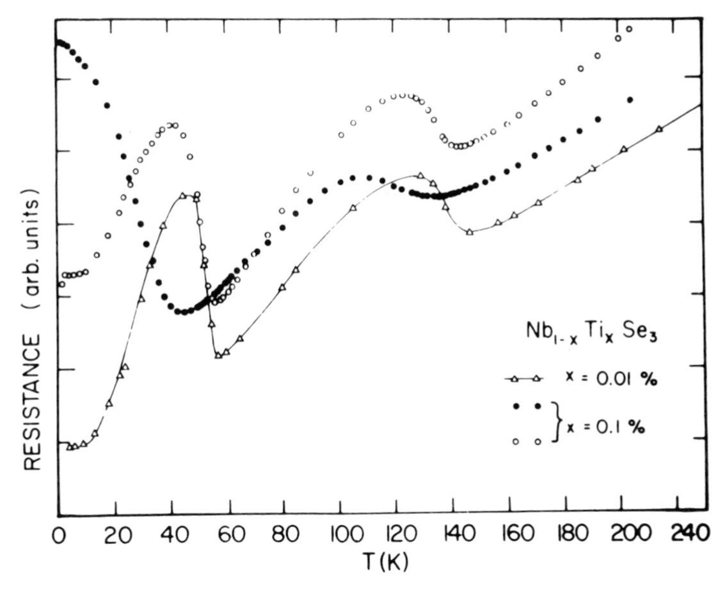

Fig. 19. Variation of the resistance of $NbSe_3$ samples doped with titanium as a function of temperature. Inhomogeneity in titanium doping is revealed by a different resistance variation between samples from the same growth batch (after [164]).

The resistivity increases strongly at low temperatures, similarly to a semiconductor, although the phase transitions at T_1 and T_2 are still visible. The Peierls transitions in $Ta_{1-x}Nb_xS_3$ and $TaS_{3-x}Se_x$ crystals decrease linearly with x at the rate: $dT_0/dc \sim$ 70 K at %. Cation and anion disorder seems to be similar, although for a d band phenomenon one expects cation disorder to be stronger.

Zero resistivity has been measured below 2 K in $NbSe_3$ samples doped with tantalum, titanium or zirconium. Therefore, superconductivity in $NbSe_3$ is induced either by doping or by pressure.

4.5.2. *Irradiation*

Irradiation with high energy electrons (3 MeV) creates stable defects which are displaced transition metal atoms and the corresponding vacancies. From the displacement threshold energy it is possible to calculate the fraction of displaced atoms. Although this calculation may be imprecise, the relative increase in defect concentration between two successive doses is known very accurately because the number of defects created by the small electron flux used in the experiments is proportional to the dose. Figure 20 shows the

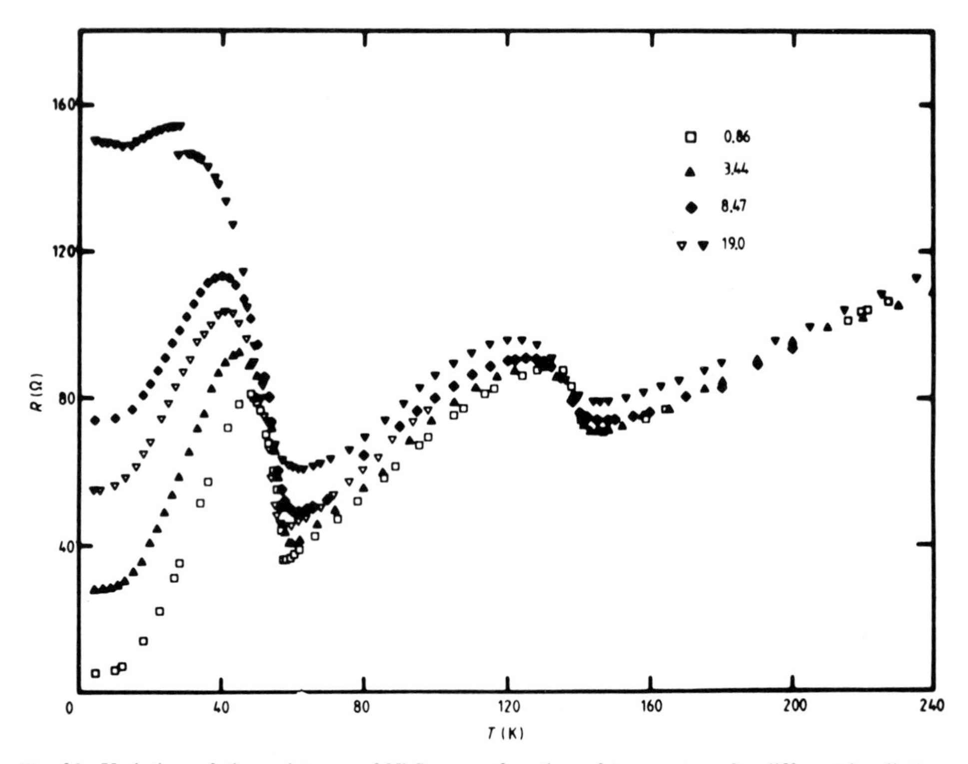

Fig. 20. Variation of the resistance of $NbSe_3$ as a function of temperature for different irradiation doses (given by symbols in 10^{17} electrons cm^{-2}). The filled symbols are for increasing temperature and the open symbols for decreasing temperature. The drop in resistance at 32 K for the dose of 19×10^{17} electrons cm^{-2} is due to the annealing of the defects at this temperature during the time taken to study nonlinear properties.

resistance variation of a $NbSe_3$ sample as a function of temperature when irradiated with doses up to 19×10^{17} e cm^{-2} [218b]. T_1 and T_2 are not affected but the transitions are softened. The increase of the resistivity at low temperatures is all the more large as the electron dose is higher. Beyond a dose of 10^{18} e cm^{-2} the resistivity at 4.2 K exceeds that at room temperature. Irradiation of $NbSe_3$ with 2.5 MeV protons leads to similar features as far as the low electric field resistivity is concerned [219]. Irradiated $NbSe_3$, however, does not exhibit superconductivity, in contrast to doped samples. When monoclinic TaS_3 is irradiated with electrons, the plateau in the resistivity variation between the two Peierls transitions is suppressed and R increases monotically when T is decreased [220, 221]. Below T_2, the activation energy seems to be independent of the irradiation dose up to 10^{-2} displacement per atom. The Peierls transition in orthorhombic TaS_3 decreases with the electron dose and the peak in $d \log R/\mathrm{d}(1/T)$ at the transition is smeared out. T_0 varies non-linearly with the defect concentration, a decrease of 15 K being measured for a dose of 3×10^{-3} displacement per atom [222]. The coherence of the CDW is very sensitive to irradiation for both TaS_3 [220, 222]. Superlattice spots broaden in diffuse rings. The CDW domain structure observed in dark field imaging is strongly affected by the irradiation beam, principally the transverse size of the domains [135].

4.5.3. *Carrier Localization*

(i) The resistivity of disordered (doped or irradiated) $NbSe_3$ samples has been shown to increase strongly at low temperatures below T_2. It may be thought that disorder leads to localization of conduction electrons which are not affected by the two CDW transitions. A large increase in the resistivity has also been measured in doped $1T\text{–}TaSe_2$ [223] and $1T\text{–}TaS_2$ [224] with a resistivity variation as a function of temperature as follows:

$$\rho = \rho_0 \exp(T_0/T)^n \tag{4-13}$$

with $n \sim 1/3$ between $0.03 \text{ K} < T < 2.2 \text{ K}$. Impurities can be considered as defects interrupting the conduction along the chains. Such an interrupted strand model has largely been developed for transport properties of KCP [225]: the electrons jump from one chain to an adjacent chain instead of crossing the interruption caused by the defect and of staying in the same strand. In this model, impurities act as impenetrable barriers between metallic one-dimensional boxes along a given strand. In order to sustain such a description, resistivity measurements have been performed on $NbSe_3$ samples made up of: (a) a great number of filaments randomly orientated and gently pressed; (b) a sieved powder with diameter, $\emptyset$, such that: $100 < \emptyset < 200\ \mu$, and $40 < \emptyset < 63\ \mu$. The powder was compacted under a press. Results of the resistivity measurements are shown in Figure 21. The similarity with Figure 19 and 20 is clear. The Peierls transition temperatures are identical to those of monocrystal $NbSe_3$ samples, indicating no effect of the pressure used to agglomerate the powder. The increase of the resistance at low temperatures is larger when the powder diameter is reduced. Even for grains with a small diameter, however, the CDW anomalies are still visible. These results lead to the conclusion that another resistivity is superposed on the resistivity variation due to $NbSe_3$, which precludes the use of the Matthiesen rule for calculating the impurity concentration, even for small concentrations. In Figure 22, the extra conductivity, σ, for the powder with the smallest diameter is plotted as a function of the temperature: σ is defined as $\sigma = 1/(\rho - \rho_0)$, which ρ is the resistivity of the compacted $NbSe_3$ sample, as shown in the insert of Figure 22, and ρ_0 is the resistivity for a $NbSe_3$ monocrystal. σ follows a linear variation with T. This dependence can be accounted for by a photon-assisted tunneling

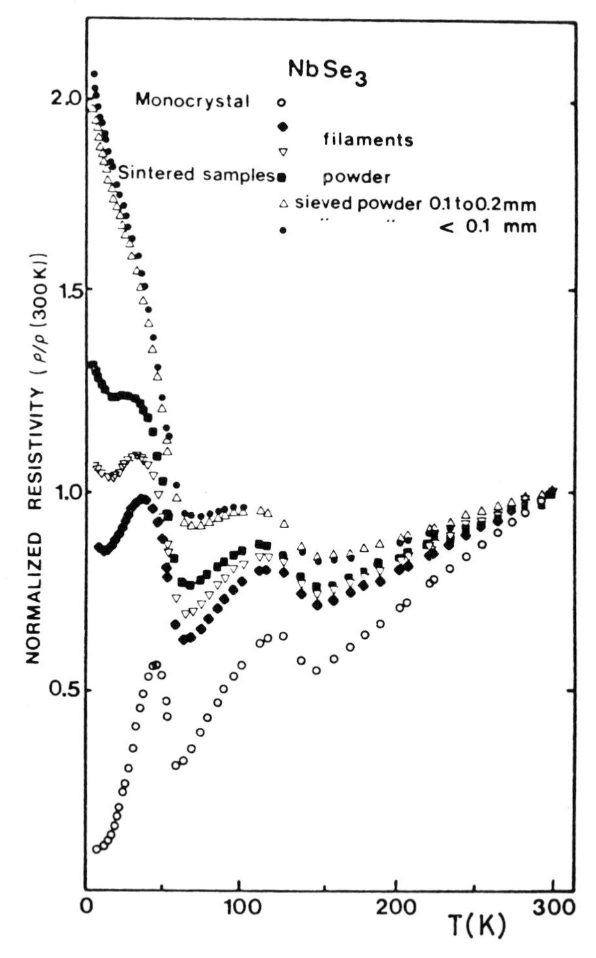

Fig. 21. Temperature variation of the resistance of $NbSe_3$ samples made up of (i) numerous fibers randomly orientated and gently pressed; (ii) a compacted powder sifted through a filter with a diameter $\emptyset$ with 100 μm $< \emptyset <$ 200 μm and with $\emptyset <$ 100 μm.

model between semimetallic $NbSe_3$ grains separated by a barrier. In each grain the electronic levels are quantified in $k = n\pi/L$ where L is the grain size [213].

(ii) Among all the crystals the properties of which have been reviewed in the preceding sections, compounds with the (Fe, Nb)Nb_2Se_{10} structural type seem to be the best suitable for studying Anderson localization. As has been mentioned in Section 3.2, disorder is structurally present in octahedral chains in which iron and niobium are randomly distributed. Hillenius *et al.* [55] have reported that the resistivity of $(Fe_{1+x}Nb_{1-x})Nb_2Se_{10}$ does not follow the activation law (4-1) at low temperatures. But, below T = 19 K, the resistivity follows the functional law (4-13) on five orders of magnitude, with n fairly well approximated by the value 1/4. The ratio, $2\Delta/kT_c$, in this compound is extremely low (see Table II) when compared to other compounds exhibiting Peierls transitions.

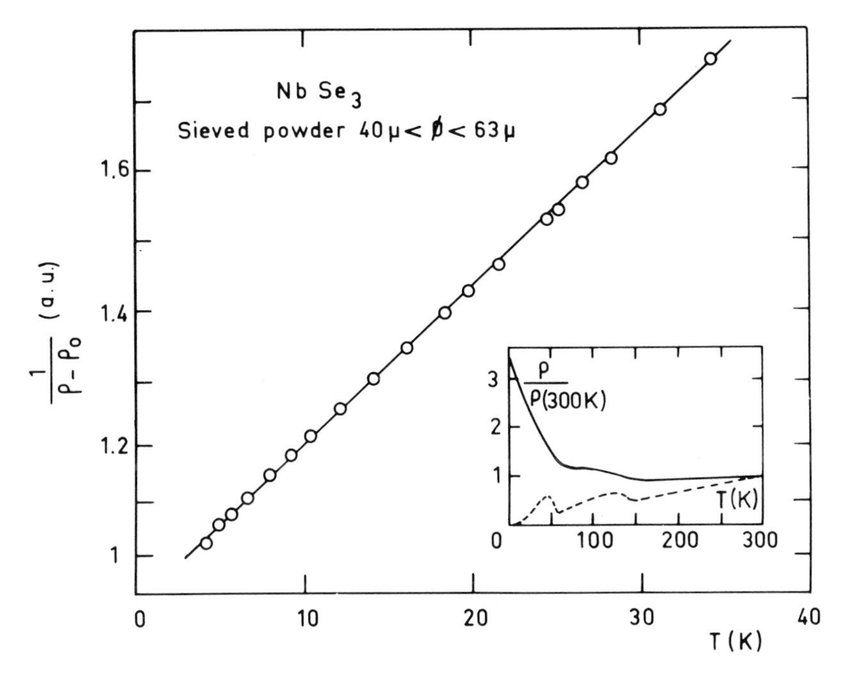

Fig. 22. Variation of the conductivity $\sigma = 1/\rho - \rho_0$ as a function of temperature; the variation of ρ, the resistivity of a compacted $NbSe_3$ powder with 40 μm < ∅ < 63 μm and of ρ_0 the resistivity of a $NbSe_3$ monocrystal as a function of temperature is shown in the insert.

The electronic conduction in (Fe, Nb)Nb_2Se_{10} type systems can be described as follows [226]: a random potential dominates on the octahedral chains due to disorder and, at high temperatures, the conductivity is provided by the pure $NbSe_3$ chains. Localization takes place on these trigonal chains because, after the CDW formation, the conduction electron density is reduced which enhances the random potential. This structural type seems to open large possibilities for studying combined effects between CDW and localization in chain structures.

4.6. Summary

We have reviewed the physical properties of transition metal tri- and tetrachalcogenides. Many of them undergo a Peierls transition. The ratio between the amplitude of the CDW gap and the Peierls transition temperature indicates strong-coupling phenomena. Non-linear transport properties have been measured in type II NbS_3, $NbSe_3$, both forms of TaS_3, $(TaSe_4)_2I$ and $(NbSe_4)_{10}I_3$. These properties are now described in the following section.

5. Properties of the CDW Current-Carrying State

Whereas the consequences of a CDW formation on the physical properties of many one- or two-dimensional compounds such as transition metal dichalcogenides or organic compounds had been very well established, it was shown for the first time in 1976 [227] that the resistivity anomalies associated with the CDW formation in $NbSe_3$ are affected by the dc current density applied to the crystal under investigation and by a microwave field [78]. Figure 23a shows the temperature variation of the resistivity anomaly of $NbSe_3$ for the lower transition as a function of the current density, and Figure 23b the temperature variation of the real part of the resistivity at 9.3 GHz. These measurements clearly showed that:

(i) The CDW transition temperature is not affected by the current density or by the frequency. A small resistivity anomaly is still detectable at T_1 and T_2 at the frequency of 9.3 GHz as in the case where the applied dc current density is 100 A/mm^2 (400 A/mm^2 in later experiments [228]).

(ii) The dc conductivity increases as a function of the electric field, E, following a law in $\exp(-E_0/E)$.

(iii) The dc conductivity saturates at high electric field but not above the value which would have existed if the CDW gap had not been formed.

Bardeen [229, 230] was the first to interpret these experimental results as the Fröhlich conduction induced by the CDW motion as explained in Section 2. X-ray measurements performed on a $NbSe_3$ monocrystal in the nonlinear state have clearly indicated that the CDW is not destroyed by the electric field [111]. The intensity of the superlattice spots and, therefore, the amplitude of the CDW has been shown to be electric field independent. This result rules out the possibility that the electric field creates normal domains – similarly to the current applied to a type I superconducting wire – which grow when the field is increased up to fill the whole volume of the sample. Later, Fleming and Grimes [231] showed that the nonlinearity in transport properties starts only above a threshold field, E_c. Beyond E_c, noise is generated through the voltage leads of the sample which consists of a broad band noise and a periodic voltage the fundamental frequency of which increases with the electric field. The threshold field has been calculated by Lee and Rice [232]. Typical curves giving a clear indication of non-linear transport properties are shown in Figure 24 for an orthorhombic TaS_3 sample: the voltage–current characteristic, $V(I)$, deviates from Ohm's law above the field E_c (= $R \times I_c/l$ where l is the length of the sample); a better determination of the threshold field is obtained from the differential resistance dV/dI: E_c is the field at which dV/dI starts to decrease (Figure 24b). Figure 24c shows the noise power measured at a frequency of 100 kHz as a function of the applied current. Below the threshold, the noise is essentially instrument noise but it increases sharply at E_c. We now describe the dynamical properties of the CDW state namely the conductivity variation as a function of the electric field and the frequency, the noise analysis and the metastable properties of the CDW. Reviews on the nonlinear CDW state have already been published by Fleming [233], Grüner [234, 235], Ong [236] and in [214, 237].

5.1. Conductivity as a Function of the DC Electric Field – Threshold Field

5.1.1. *Temperature Dependence*

Figure 25 shows the electrical conductivity normalized to its ohmic value, σ_a, as a function

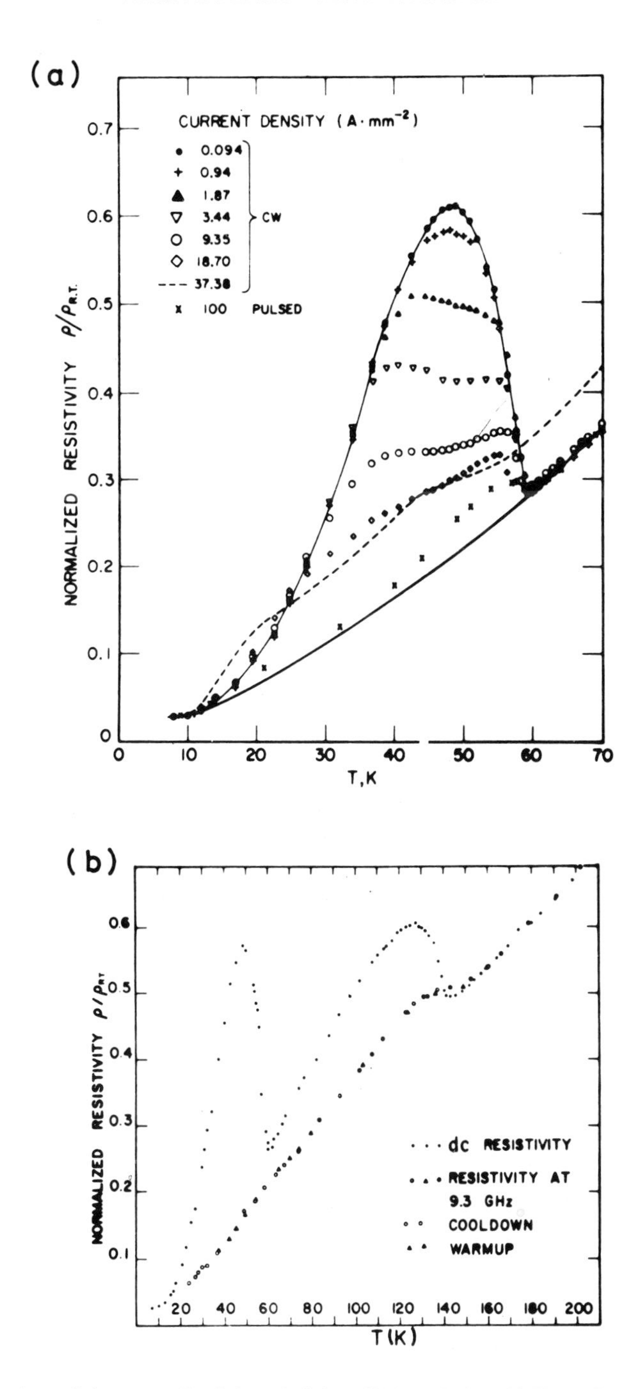

Fig. 23. (a) Variation of the normalized dc resistivity of $NbSe_3$ below 59 K as a function of temperature at several current densities applied to the sample (after [227]). (b) Real part of the resistivity at 9.3 GHz with regard to the dc resistivity of $NbSe_3$ as a function of temperature (after [78]).

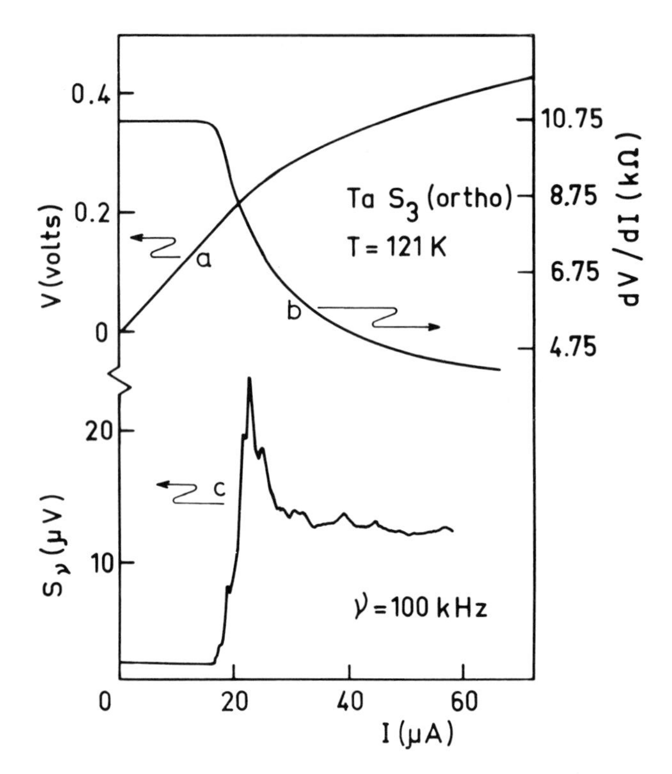

Fig. 24. Nonlinear transport properties in TaS_3 as a function of the applied current (a) the dc voltage–current characteristics, (b) the differential resistance dV/dI. The threshold field is defined when dV/dI starts to decrease, (c) the noise power through the voltage leads measured at 100 kHz.

of the reduced electric field, E/E_c, for different compounds which have a semiconducting ground state below their Peierls transitions. In the case of $(NbSe_4)_{10}I_3$, however, between the Peierls transition temperature at 285 K and approximately 180 K, the resistance on increasing the electric field has a discontinuous decrease at a field E_c' whereas the resistance makes a smooth transition back to ohmic behaviour at a field $E_c < E_c'$. E_c', but not E_c, depends on the external conditions such as the maximum field applied, the rate of variation of the electric field . . . [99]. If the field is reduced from a value above E_c' to a value between E_c' and E_c, the resistance of the sample becomes time-dependent and increases slowly towards a less conducting state with a time constant of several hours.

As far as $NbSe_3$ is concerned, it can be seen in Figure 23(a) that the conductivity never increases beyond a value approximately twice the ohmic value. Whereas the variation of the differential resistance for the upper CDW as a function of the applied current is typically represented by the curve shown in Figure 24(b), a different behaviour occurs for the lower CDW as indicated in Figure 26. Near T_2, dV/dI has a smooth decrease above the critical electric field. At lower temperature, however, dV/dI shows a sharp drop with a deep minimum. At low temperature, T = 24.3 K in Figure 26, two critical

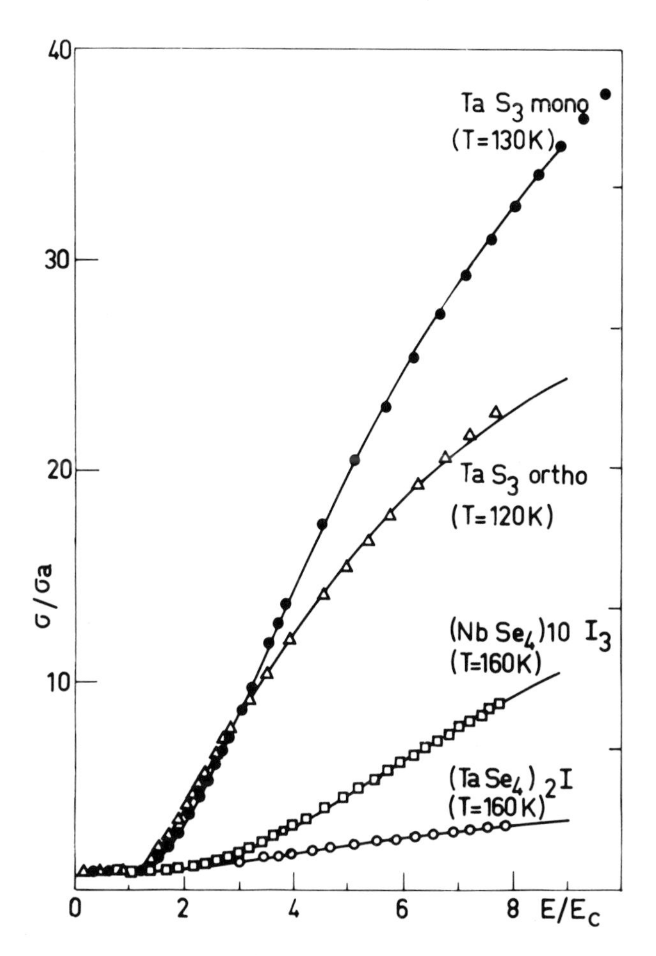

Fig. 25. Variation of the electrical conductivity in the nonlinear state (normalized to the Ohmic value, σ_a, when $E \ll E_c$) as a function of electric field (normalized to the threshold field) for both forms of TaS_3, $(TaSe_4)_2I$, $(NbSe_4)_{10}I_3$ samples, at a given temperature. The solid lines are fits to the theoretical expressions (6-41) and (6-42) obtained by Bardeen in his tunneling theory. The parameters used for the best fit are given in Section 8-6.

electric fields can be defined: E_c when low-frequency noise (in this experiment at 33 Hz) appears without any decrease in dV/dI and E_{cr} when dV/dI starts to decrease [238, 239]. The ratio E_{cr}/E_c increases when the temperature is lowered: its value is 2.6 at $T \sim 25$ K but ~ 1 at the temperature at which the resistance of $NbSe_3$ below T_2 has its maximum ($T \sim 47$ K). This sharp drop in dV/dI corresponds to a jump or a switching in the resistance between a higher resistance state to a lower resistance state [240].

The temperature variation of the threshold fields is drawn in Figure 27 for all the compounds exhibiting nonlinear properties. The data correspond to samples in each family which, when measured, have shown the lowest threshold fields. When two threshold fields are detected in some temperature range such as E_c and E_c' in $(NbSe_4)_{10}I_3$ or E_c

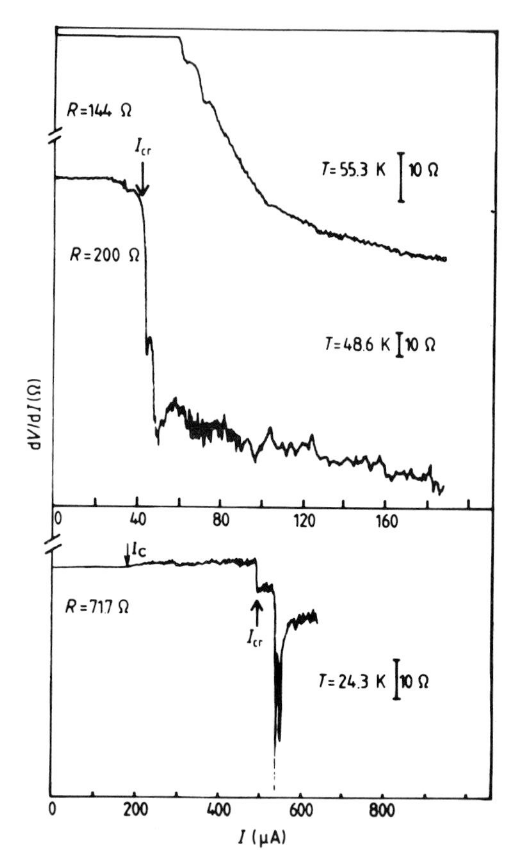

Fig. 26. Variation of the differential resistance, dV/dI, as a function of the applied current at several temperatures below the onset of the lower CDW in $NbSe_3$. At low temperatures two threshold fields can be defined: one at I_c at which low frequency noise is generated but without variation in dV/dI, the second one at I_{cr} at which dV/dI jumps discontinuously ([238, 239]).

and E_{cr} in $NbSe_3$, only the variation of the lower one, i.e., E_c, is plotted. The following general features can be deduced from Figure 27.

(i) For all the compounds, E_c decreases in a small temperature range in the vicinity of the Peierls transition and goes through a minimum. The temperature dependence of E_c in the extreme vicinity of T_p is not precisely known because the threshold field is not well defined in this temperature range; this probably is the consequence of the lack of three-dimensional ordering in the CDW state in the vicinity of the transition. It will be seen in Section 5.1.3 that a different behaviour occurs at T_2 for $NbSe_3$.

(ii) The minimum value of E_c is in the range of 0.1–1 V cm^{-1}. Its value is all the higher as the Peierls transition occurs at a higher temperature. It can be roughly estimated that log E_c varies linearly with the Peierls transition temperature. The lowest value for $NbSe_3$ is ~6 mV/cm approximately at 47 K.

(iii) E_c increases sharply at low temperatures. The slope of the log E_c as a function of T is roughly the same for each compound except for orthorhombic TaS_3 which

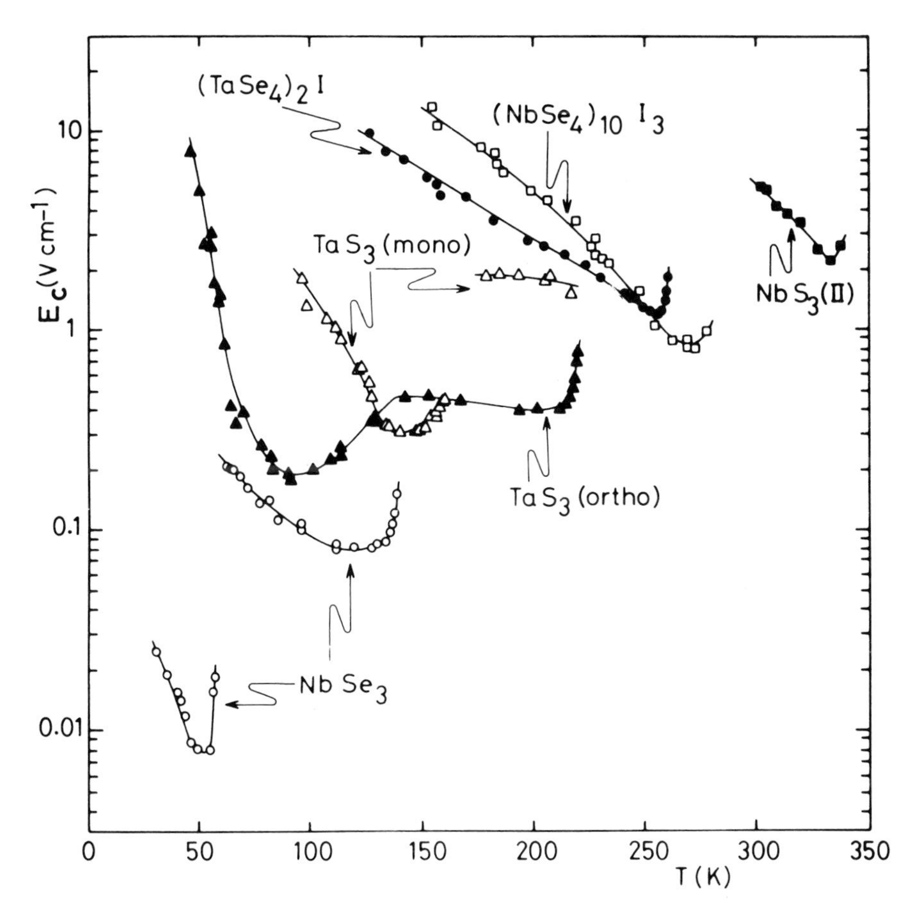

Fig. 27. Variation of the threshold field E_c (logarithmic scale) as a function of temperature for NbS_3 [123], $NbSe_3$ [247, 248] both forms of TaS_3 [221], $(TaSe_4)_2I$ [97] and $(NbSe_4)_{10}I_3$ [99].

shows a larger variation [91, 241–244]. Moreover, for the latter compound, E_c decreases in the vicinity of $T_0' = 130$ K down to 90 K. It should be recalled that T_0' is the temperature at which the wavelength of the CDW along the chain axis becomes exactly four lattice distances. The behaviour of E_c at low temperatures is not understood but: (1) the exponential increase of E_c when T is reduced; and (2) the close relationship between the minimum value of E_c and the Peierls temperature transition clearly indicate the importance of thermal activation in the CDW dynamics.

5.1.2. *Impurity Dependence*

Threshold field has been measured on $NbSe_3$ and TaS_3 samples doped with iso- or non-isoelectronic impurities and samples deteriorated by electron or proton irradiation. When $NbSe_3$ is doped with titanium or irradiated with protons, E_c increases sharply with the impurity content [212, 219]. For the lower CDW in $NbSe_3$, 0.01% of damage increases E_c up to a few V cm^{-1}. Moreover, E_c, below T_2, becomes approximately temperature independent. When $NbSe_3$ [218b] and monoclinic TaS_3 [221] are irradiated with electrons of 2.5 or 3 MeV, E_c does not increase but the saturation value of the conductivity at high electric fields is strongly reduced (20 times lower after an irradiation of 10^{18} electron cm^{-2} in the case of monoclinic TaS_3). By contrast, other experiments on orthorhombic TaS_3 indicate that E_c increases linearly with the electron flux up to 10^{-4} displacement per tantalum atom – a dose which does not affect the Peierls transition – but much more slowly for higher doses [222]. Further experiments are needed to clarify these contradictory experimental results. E_c has been reported to vary as the square of the impurity concentration in the case where impurities are isoelectronic – tantalum for $NbSe_3$ [212], niobium for TaS_3 [215] – and linearly with the impurity concentration when the defects are created by irradiation with protons [219]. Lee and Rice [232] have calculated an increase of E_c with a c^2 dependence in the case of weak pinning and a c dependence in the case of strong pinning (see Section 6.1). Isoelectronic impurities are thought to be weak pinning impurities whereas charged impurities such as non-isoelectronic atoms in substitution or interstitial-vacancy pairs created by irradiation should be strong pinning impurities. The latter impurities, because of their charge, strongly interact with the CDW and locally fix its phase. Some criticism, however, has to be made concerning the experimental results which show a c^2 impurity dependence for E_c. For $NbSe_3$ doped with tantalum, the concentration of impurities has been determined by the resistance ratio between room temperature and helium temperature [212] but it was shown in Section 4.5.3 that the Matthiessen law is not followed in impure $NbSe_3$ samples. For TaS_3 doped with niobium, the Peierls transition, T_P, decreases linearly with an effective impurity concentration; subsequently, E_c varies quadratically as a function of T_P [215]. Hseih *et al.*, however, report that T_P of pure TaS_3 samples is spread between 210 and 223.5 K, probably because of crystal imperfections. Therefore, it may be thought that the variation of E_c with T_P is not only due to impurities but involves the crystal perfection of doped samples. Finally, the distinction between charged (non-isoelectronic) or uncharged (isoelectronic) impurities may not be relevant in transition metal trichalcogenides. It was explained in Section 3 that the exact charge on tantalum or niobium atoms is provided by the strength of the bond between chalcogen atoms; when a tantalum atom takes the place of a niobium one, it has to accomodate the $TaSe_6$ structure which cannot be made without a local distortion in the selenium framework and therefore, without a change in the local charge. In conclusion, impurities increase the threshold field and strongly reduce the extraconductivity in the limit of an infinite electric field.

5.1.3. *Depinning of the CDW Fluctuations in TaS_3 and $NbSe_3$*

It was shown in Section 4.3.3 that the critical fluctuation contribution is approximately 50% of the total resistivity in the vicinity of the Peierls transitions T_0 and T_1 in both forms of TaS_3. These resistive fluctuations can also be depinned by an electric field. Because fluctuations induce a non-uniform electric field, the well-defined threshold is suppressed. Figure 28(a) shows the variation of the resistance $\Delta R = R(E) - R_N$ normalized to the fluctuation contribution, $R - R_N = R_f$, as defined in (4-8(a)), as a function of the electric field. Below $T_0 = 215$ K nonlinearity occurs at a sharp threshold in the temperature range where the CDW is tri-dimensional. Nonlinearity is observed

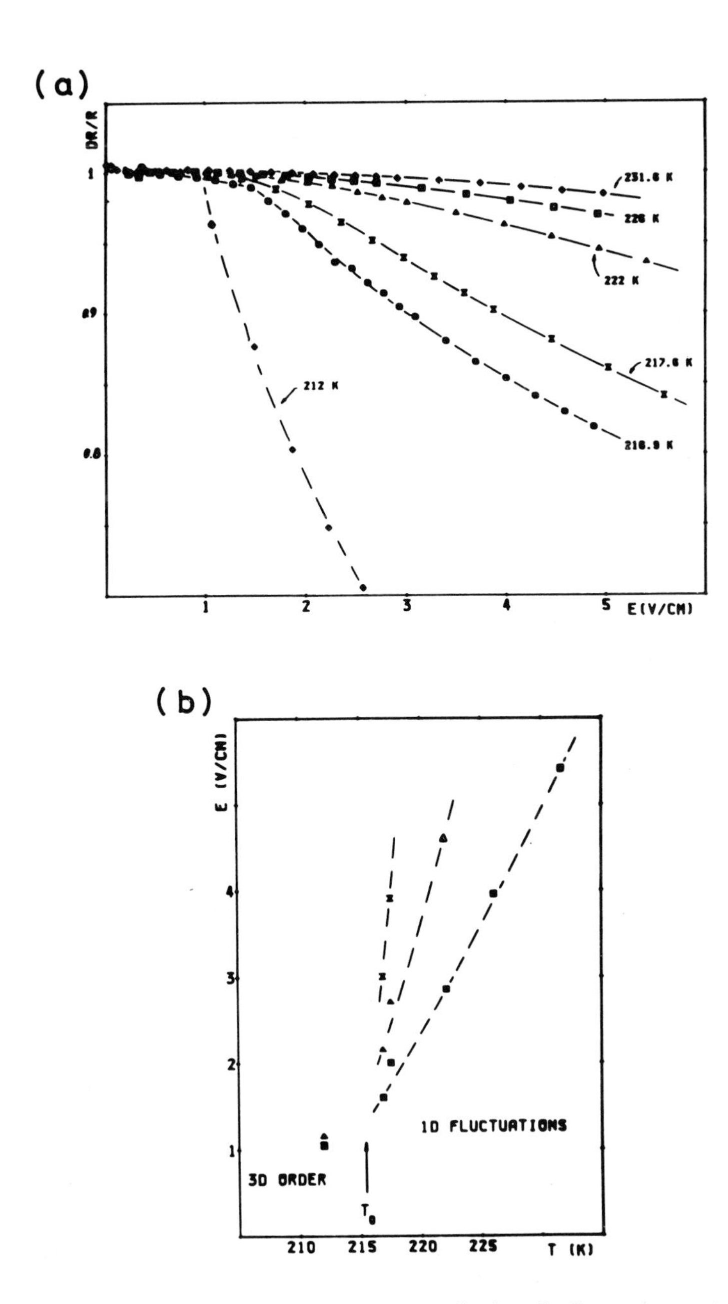

Fig. 28. (a) Variation of the nonlinear resistance (normalized to the fluctuation contribution) as a function of electric field at different temperatures below and above the Peierls transition T_0 = 215 K of orthorhombic TaS_3. (b) Variation of the electric field at which the conductivity of orthorhombic TaS_3 is $x\%$ up as a function temperature. Below the Peierls transition, T_0, the CDW is in a three-dimensional ordered state; CDW fluctuations contribute to the conductivity but all the less as T is increased above T_0 (after [186]).

at least 20 K above T_0. One can define a field for which the resistance due to fluctuations is $x\%$ down. In Figure 28(b), such an electric field with $x = 0.90, 0.95$ and 0.98 is shown to increase sharply when $T - T_0$ becomes larger [186]. In monoclinic TaS_3 the same behaviour occurs above the upper CDW transition, T_1. Moreover, below T_1, the fluctuations associated with the lower CDW transition at T_2 suppress the sharp E_c corresponding to the upper CDW [221]. CDW fluctuations are weaker in $NbSe_3$. Nonlinearity has only been measured 2 K above T_1. The peak in the logarithmic derivative defines the lower CDW at $T_2 = 57.90$ K. A well-defined threshold field, however, has been measured at 61 K. These experiments have been performed in liquid oxygen with a temperature stability of a few mK for several hours. The threshold above T_2 has been interpreted as the depinning of three-dimensional fluctuations. Figure 29 shows that the threshold field does not diverge at T_2 but has only a small bump when T crosses T_2 [186].

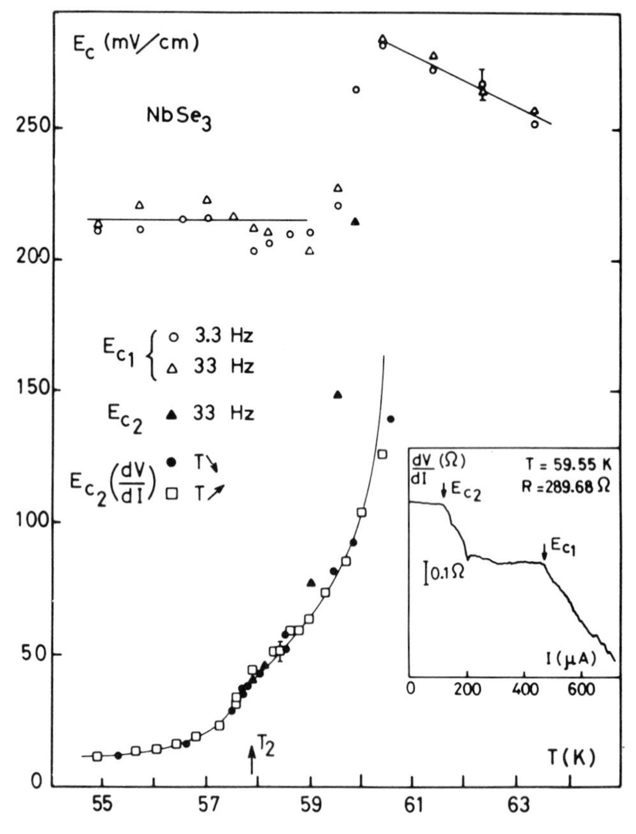

Fig. 29. Variation of the threshold fields E_{c_1} and E_{c_2} at which the upper and lower CDW in $NbSe_3$ are depinned, respectively, as a function of temperature in the vicinity of the Peierls transition temperature T_2: ○, △ and ▲ correspond to the fields at which noise (at the given frequency 3.3 Hz or 33 Hz) starts to increase; • □, are the data deduced from the sharp decrease in dV/dI. The insert shows the variation of dV/dI as a function of the applied current at $T = 59.55$ K. At this temperature the upper CDW is depinned at E_{c_1} and the three-dimensional fluctuations of the lower CDW are depinned at E_{c_2}.

Critical fluctuations in transition metal trichalcogenides behave differently from those in TTF–TCNQ (see Section 2.2). In the latter compound, fluctuations are thought to be unpinned and therefore conductive. On the contrary, in TaS_3 and $NbSe_3$ they are pinned and resistive.

5.1.4. *Interaction Between Both CDWs in $NbSe_3$ and TaS_3*

A possible locking between both CDWs in monoclinic $NbSe_3$ and TaS_3 has been considered when it was noted that twice the sum of both wave vectors is in the vicinity of a lattice reciprocal wave vector: $2(\mathbf{Q}_1 + \mathbf{Q}_2) \sim (\mathbf{a}^*, \mathbf{b}^*, \mathbf{c}^*)$ [114–116]. It has been calculated that the phase locking, described by a fourth-order term in the CDW amplitudes, may lead to a very small perturbation such as a small gap enhancement at T_2. But, Fleming has clearly shown that $\mathbf{Q}_1 + \mathbf{Q}_2$ is not a commensurate value [113]. Such a locking between both CDWs has been suggested to explain some nonlinear transport studies in $NbSe_3$ [245] but this has not been confirmed in further experiments [246]. Nevertheless, measurements of $\sigma(E)$ below T_2, besides the depinning of the lower CDW, have shown a discontinuity at higher electric fields which has been attributed to the depinning of the upper CDW [228]. Successive depinning of both CDWs is clearly shown in the insert of Figure 29. The experiment has been performed at a temperature for which the respective variation of dV/dI for each CDW has approximately the same magnitude. At this temperature the threshold E_{c_2} concerns the depinning of the three-dimensional fluctuations of the lower CDW. But in the temperature range below T_2, in which the resistivity anomaly resulting from the lower CDW transition takes place, the extra conductivity associated with the motion of the upper CDW is very small in comparison to that for the lower CDW. Consequently, Fleming has measured the onset of the depinning of the upper CDW below T_2 by the field at which noise increases [247]. In the vicinity of T_2, however, the onset of noise is a function of the frequency at which it is measured: E_{c_1} is smaller at a frequency of 3.3 Hz than at 33 Hz. The variation of E_{c_1} and E_{c_2} in the vicinity of T_2 for $NbSe_3$ is plotted in Figure 29. E_{c_1} is shown to decrease in the temperature range where fluctuations of the lower CDW are important, and to be temperature independent below T_2, when the lower CDW is a real three-dimensional state [248]. The temperature independence of E_{c_1} below T_2 confirms the results in [228] but is at variance with the experiments performed by Fleming [247]. The same results have been obtained for monoclinic TaS_3 in the vicinity of T_2, but over a larger temperature range [249].

5.2. Frequency Dependent Conductivity

It was shown in Figure 23(b) that the resistivity anomalies associated with the CDW transitions at T_1 and T_2 in $NbSe_3$ are strongly suppressed by a microwave field at 9.3 GHz. The frequency dependence of the conductivity has been studied in $NbSe_3$ [246, 250–253] and in TaS_3 [147, 254–256]. Real and imaginary parts of the conductivity have been measured with a network analyser as a function of the frequency up to 1 GHz and using microwave cavity techniques for higher frequencies.

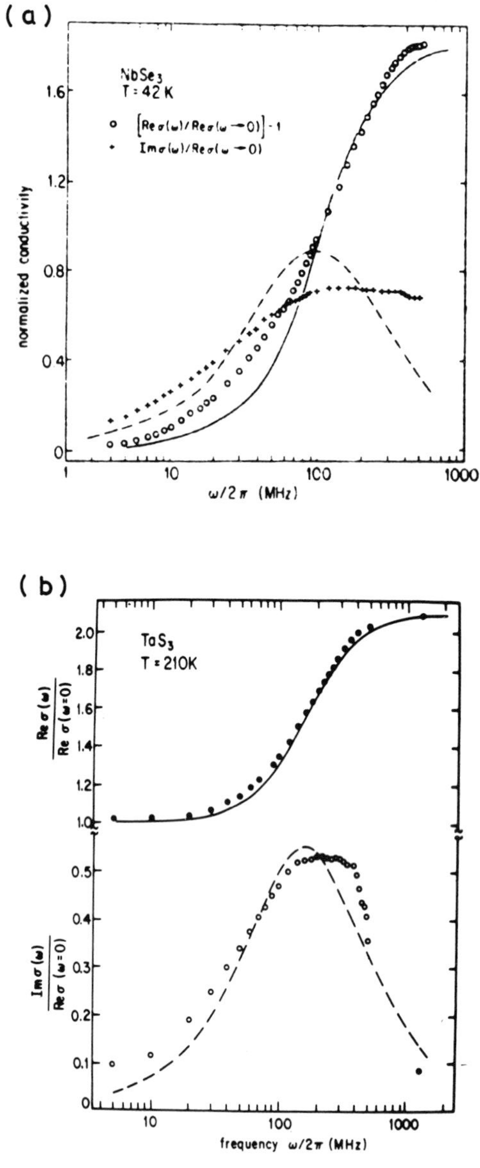

Fig. 30. Real, Re $\sigma(\omega)$, and Imaginary, Im $\sigma(\omega)$, parts of the conductivity (a) of $NbSe_3$ at 42 K (after [253]) as a function of frequency and (b) of orthorhombic TaS_3 at 210 K (after [255]). The solid lines are fits to the classical overdamped oscillator model – Equation (6-25) – with parameters given in Section 5.2.1.

5.2.1. *Overdamped Response*

Figure 30 shows the variation of the real, Re $\sigma(\omega)$, and imaginary, Im $\sigma(\omega)$, parts of the conductivity of a TaS_3 and a $NbSe_3$ sample at a given temperature as a function of frequency. In these experiments the amplitude of the ac voltage is small compared to the threshold voltage. These measurements allow us to conclude:

(i) Contrary to the response to a dc field, there is no threshold frequency beyond which Re $\sigma(\omega)$ starts to increase.

(ii) The frequency dependence of $\sigma(\omega)$ is that of an overdamped oscillator in which the damping term dominates the inertial term. The latter contribution would lead to a peak in Re $\sigma(\omega)$ at the frequency ω_0 and at a negative value of Im $\sigma(\omega)$ for $\omega > \omega_0$, as was measured in TTF–TCNQ or KCP at far-infrared frequencies (see Figure 7 of Chapter III). The curves drawn in Figure 30 are calculated from Equation (2-9b) obtained from the harmonic oscillator equation (2-8) without the inertial term. The parameters are $\omega_0^2/2\pi\Gamma$ = 100 MHz and $ne^2/\Gamma M^*$ = 0.85 σ_{dc} for $NbSe_3$ [253] and 160 MHz and 1.1 σ_{dc} for TaS_3 [255].

(iii) Re $\sigma(\omega)$ saturates at high frequencies at the same value as the dc conductivity in the limit of an infinite electric field.

This frequency dependence of $\sigma(\omega)$ can be described by a simple model as proposed in [250] and [252]. The CDW is represented by a R_cC series combination in parallel with a resistance, R_N, corresponding to the remaining normal electrons; R_c represents the friction of the unpinned CDW and C the pinning forces which act on the CDW. At frequencies smaller than $(R_cC)^{-1}$ the conductance is that due to normal carriers, R_N^{-1}. At frequencies beyond $(R_cC)^{-1}$ the impedance of the capacitance is negligible and the conductance increases smoothly towards the value $1/R_N + 1/R_c$. The crossover frequency, ω_p, between the low-frequency and the high-frequency limits can be chosen as the one at which the conductivity is half the value between high frequency and dc values. In the overdamped oscillator description, ω_p corresponds to ω_0^2/Γ.

5.2.2. *Dielectric Constant*

The dielectric constant is deduced from the imaginary part of $\sigma(\omega)$ by the relation, $\epsilon(\omega) = 4\pi$ Im $\sigma(\omega)/\omega$. A huge positive value in the range of 10^7–10^8 is obtained when the temperature crosses the Peierls transition. For $NbSe_3$, ϵ is independent of the frequency up to 40 MHz; it increases from approximately zero above T_1 to 0.5×10^8 below T_1, increases again below T_2 up to 2×10^8 [250]. For orthorhombic TaS_3 ϵ, measured at 10 MHz, increases to 10^7 just below T_0 and decreases slowly at lower temperatures [255]. At higher frequencies (9.6 GHz) ϵ has been measured to have a value of 7×10^3 near 100 K [254], but 10^5 in [256], which decreases approximately linearly with temperature when T is lowered further. The decrease of ϵ at low temperatures, in orthorhombic TaS_3, has been interpreted as resulting from a 'disordered' CDW state. The frequency response is expected to be the sum of the frequency response of many domains with a distribution of pinning energy and a distribution of relaxation times. In this case it has been shown that the frequency response of the CDW is similar to that of localized single-particle states [255].

5.2.3. *Fluctuations and Impurities*

In orthorhombic TaS_3, fluctuations in the vicinity of the Peierls transition dominate the frequency response. Re $\sigma(\omega)$ and the dielectric constant start to increase 20 K above T_0, indicating the response of the pinned fluctuations to the applied ac field [255, 264].

Frequency dependence conductivity measurements have also been performed on doped [212] and irradiated [219] $NbSe_3$ samples. The introduction of impurities shifts the frequency $\omega_p = \omega_0^2/\Gamma$ to higher frequencies; for instance, ω_p is approximately 9 GHz for $NbSe_3$ doped with 5000 ppm of tantalum, which has to be compared to the value of 50 MHz for a pure $NbSe_3$ sample at the same temperature. Defects increase both the free oscillation frequency, ω_0, and the damping frequency, Γ. Consequently, the dielectric constant decreases. It has also been shown, in the case of irradiated samples, that the inverse of the dielectric constant and the dc threshold field are proportional to each other and proportional to the defect concentration [219].

5.3. Time-Dependent Voltage

Fleming and Grimes [231] have shown that noise is generated through the voltage leads of the sample under investigation when the electric field exceeds the threshold one, E_c, as shown in Figure 24(c). Fourier analysis of this noise voltage reveals well-defined fundamental frequencies with their harmonics. Moreover, for pure samples and in some restricted temperature ranges, the waveform of the oscillations has been displayed directly on an oscilloscope.

5.3.1. *Periodic Voltage – Purity of Fourier Spectra*

In most of the experiments performed on $NbSe_3$, Fourier spectra reveal several frequencies, usually three, non-harmonically-related [231, 256, 259, 260]. A pure spectrum with one fundamental and six harmonics, however, has been recorded at one temperature, namely 42 K, in a two probe measurement [261]. The role of contacts on the quality of Fourier spectra has been studied by Ong and Verma [260]. They have shown that, the contacts acting as equipotentials, the electric field below the contact pads is smaller than the threshold one. The sample is truncated, therefore, in three parts, which oscillate independently. The four probe measurement would be the origin of the three fundamental frequencies detected in [257] and [258]. Such a breaking of the CDW in the neighbourhood of the contacts has also been suggested by Gill in order to explain the increase of the threshold field when the distance between contacts is reduced [262]. The large amplitude and the narrowness of the fundamental peaks in Fourier spectra of $NbSe_3$ by comparison with the poor quality spectra in orthorhombic TaS_3 [263, 264] has been ascribed to the remaining normal electrons which are thought to homogenize the electric field in the case of $NbSe_3$. Although possible, these arguments do not seem to be general. Indeed, Fourier voltage analysis performed on both TaS_3 with four probe measurements reveal a unique fundamental frequency and its harmonics in the temperature range well below the Peierls temperature where very few normal excitations remain. Figure 31(b) shows Fourier spectra for an orthorhombic TaS_3 sample at $T = 81$ K for several values of the current above the threshold. The spectrum is described with one frequency and in the vicinity of the threshold, 16 harmonics are detected. Therefore, only poor quality crystals lead to a CDW disordered state at temperatures below approximately 100 K as reported in [255, 264]. At higher temperature, spectra are less pure, as is shown in Figure 31(a) for a monoclinic TaS_3 sample at $T = 123$ K. When the field is increased above the threshold, the fundamental and the harmonic frequencies become broader,

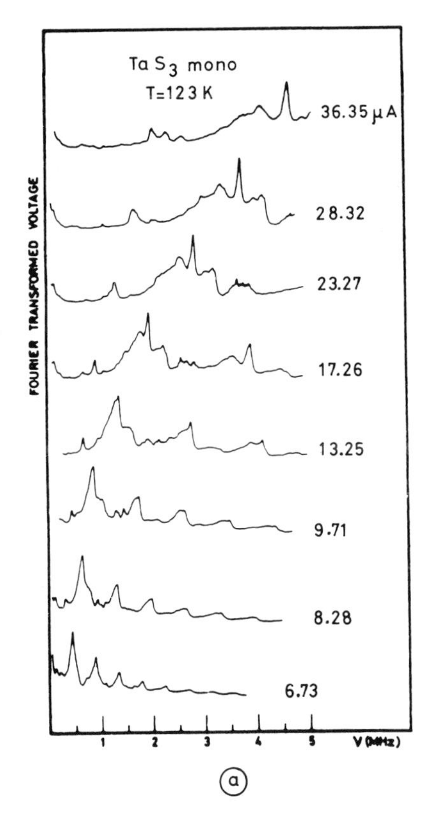

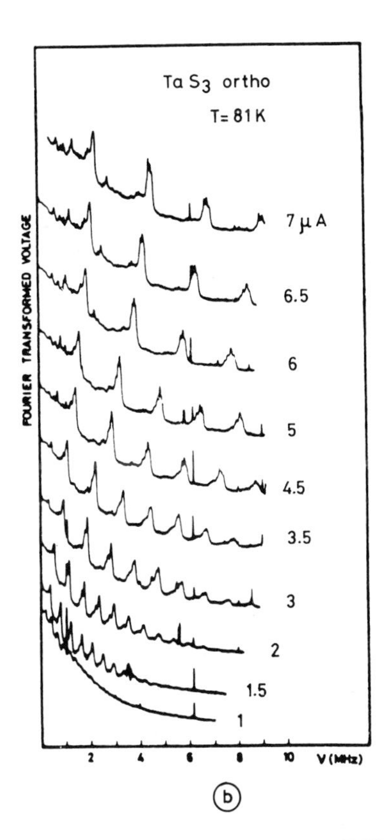

Fig. 31. Fourier spectra as a function of frequency for several values of the current (electric field) above the critical current (threshold field) for: (a) a monoclinic TaS_3 sample at $T = 123$ K which corresponds to the CDW motion of the lower CDW ($T_2 = 160$ K) (the length of the sample is 1.5 mm, its resistance 30.3 kΩ and the critical current 2.8 μA) (after [221]); (b) an orthorhombic TaS_3 sample (length 1.6 mm, resistance 1.18 MΩ and critical current 0.22 μA) at $T = 81$ K, a temperature at which the distortion component along the chain axis is commensurate with the lattice.

probably because of some inhomogeneity in the CDW motion. It can be seen, however, that sharp peaks emerge from the broad frequency distribution indicating that a part of the sample moves coherently.

In the case of some pure $NbSe_3$ samples, the voltage oscillations have directly been observed on an oscilloscope without Fourier analysis. In these experiments a current pulse is applied to the sample. The voltage oscillations clearly appear on the top of the corresponding voltage pulse. The periodic voltage is a significant fraction of the nonlinear response, a value which may approach 100% in the vicinity of the threshold [265, 266].

5.3.2. *Harmonic Content*

The amplitude of the harmonics in Fourier spectra of $NbSe_3$ [261] and orthorhombic TaS_3 [264] has been shown to decrease following an law intermediate between $1/n$

and $1/n^2$, which corresponds to a sawtooth or a triangular waveform, respectively. On the other hand, the amplitude of the harmonics shown in Figure 31(b) decreases exponentially. This exponential decay is followed up to the 14th harmonic. The same law has also been observed for monoclinic TaS_3 at low temperatures [221]. These contradictory results indicate either that the oscillation waveform is temperature dependent or that the experimental method, the quality of contacts, the purity of the sample . . . may affect the harmonic amplitude. It has to be noted that a novel feature appears at low temperatures in $(TaSe_4)_2I$ [97] and $(NbSe_4)_{10}I_3$ [99]: the odd harmonics have a smaller amplitude than the even ones, which may indicate the presence of two periodicities, one twice the other.

5.3.3. *Low-Frequency Oscillations Near the Threshold*

In their study of nonlinear properties of orthorhombic TaS_3 Gill and Higgs [267] have found at 77 K two regimes, as far as noise frequencies are concerned: (1) a 'high frequency' range for frequencies above 10^3 Hz which correspond to those measured in [221]; and (2) large amplitude voltage oscillations near the threshold with frequency as low as 1 Hz. The amplitude of these oscillations between 1 Hz and 10^3 Hz are typically 0.5% of the dc voltage value, which implies a modulation of the nonlinear current too great to be associated with the motion of the CDW. Therefore, they suggest that, near the threshold and at least at 77 K, the excess-current comes from an enhancement of the conduction by a single-electron process resulting from some spatial rearrangement of the CDW. This result has to be connected to low-frequency noise voltage pulses detected near the threshold in the blue bronze $K_{0.30}MoO_3$ [268]. It is worthwhile noting that these low-frequency oscillations occur at temperatures much lower than the Peierls temperature.

5.3.4. *Broad Band Noise*

Fleming and Grimes [231] have shown that the voltage noise, besides the periodic components, has a broad band frequency distribution. Extensive studies of this broad band noise have been performed at several temperatures and several electric fields for the lower CDW in $NbSe_3$ [239]. The results shown in Figure 32 can be summarized as follows:

(i) For any temperature below T_2, well beyond the threshold, the noise spectrum varies with a $1/f$ dependence.

(ii) For temperatures near T_2, namely T higher than approximately 47 K, and a field in the vicinity of the threshold, the noise spectrum follows a $1/f$ dependence, but is white above 1 kHz.

(iii) At low temperature, the noise power follows a $1/f^2$ dependence for electric fields between the two thresholds E_c and E_{cr}.

In fact, similarly to many other physical systems, the noise power variation as a function of frequency is a $f^{-\alpha}$ law, with $\alpha \sim 0.8$. The noise power has been found to be proportional to the length of the sample and, consequently, to be a volume property [239]. Finally, the amplitude of the broad band noise has to be noted: at 1 kHz, the noise temperature is typically 3×10^5 K. It is worthwhile pointing out the abrupt change in the noise variation at low temperatures: $1/f^2$ between E_c and E_{cr} and $1/f$ beyond

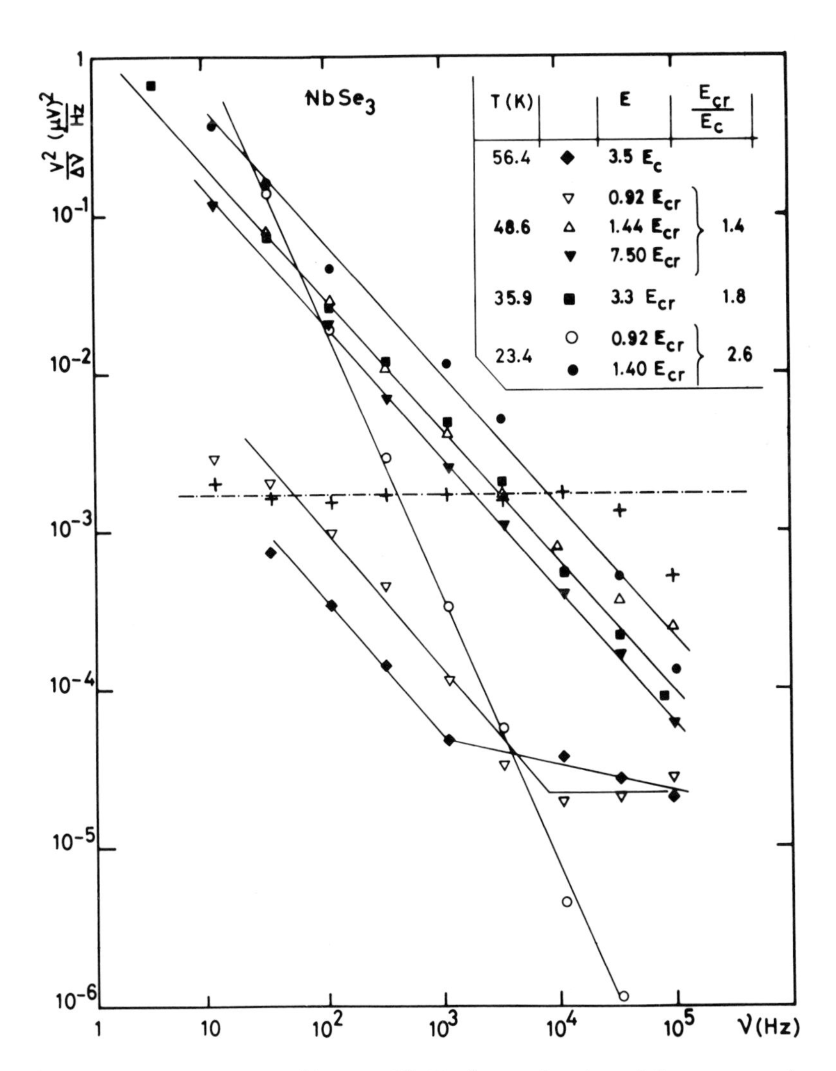

Fig. 32. Noise power spectrum $V^2/\Delta\upsilon$ in μV^2 Hz^{-1} as a function of frequency, υ, for a $NbSe_3$ sample at several temperatures below the onset of the lower CDW (T_2 = 59 K) and at different electric fields. Thermal noise at 300 K in a 100 kΩ resistor: +, experimental; —·— theory. The thresholds E_c and E_{cr} are defined in Figure 26.

E_{cr}, E_c being the field at which low frequency noise appears without change in dV/dI and E_{cr} the field at which dV/dI decreases with a discontinuity as already mentioned in Section 5.1.1 and shown in Figure 26. Papoular [116] has put forward an explanation for these two regimes; he has considered the motion of kinks–antikinks along a chain inside a domain: below E_{cr}, the motion of kinks is diffusive whereas, above E_{cr}, the kinks can propagate. These two channels for the decay of the current–current time correlation function can account for the two broad band noise regimes. The broad band noise has also been found to follow a $f^{-0.4}$ variation for orthorhombic TaS_3 [269]

and a $f^{-0.8}$ variation for monoclinic TaS_3 [270], the same as for $NbSe_3$. Qualitatively, when the quality of the sample is deteriorated, the amplitude of the broad band noise increases whereas that of the periodic voltage component is decreased. It has been reported, however, that the amplitude of the periodic voltage is independent of the purity of the sample [260]. A quantitative study in order to evaluate the ratios between the dc power dissipated in the sample, the power associated with the periodic component in the noise, and the broad band noise power remains to be made as a function of the purity of the samples, the temperature and electric fields.

5.4. AC–DC COUPLING EXPERIMENTS

As has been mentioned in Sections 5.1 and 5.2, the response of the CDW to a dc or to an rf excitation is highly nonlinear. Therefore, it is expected that the joint application of a dc and an rf field such as,

$$V = V_0 + V_1 \cos \omega t, \tag{5-1}$$

does strongly affect the CDW response. Such experiments are described in this section.

5.4.1. *Interference Between Voltage Oscillation and an External rf Field*

When an ac field is superposed on a dc field higher than the threshold, interferences occur between the frequency of this ac field and the voltage oscillation frequency induced by the CDW motion [257]. Figure 33 shows the differential resistance $\mathrm{d}V/\mathrm{d}I$ (measured with an ac modulation of 33 Hz) of a $NbSe_3$ sample as a function of the dc current swept through it when a rf field at a given frequency and a constant amplitude is superimposed. Without an rf field, the response is that shown in Figure 24b. When the rf is applied, peaks occur in $\mathrm{d}V/\mathrm{d}I$ which correspond to the frequency of the voltage oscillation. Experiments, however, are performed differently: the dc current, higher than the threshold, is kept constant, which fixes the electric field in the sample, and the frequency of the rf field, the amplitude of which is constant, is swept. The variation in the dc response is detected by a low-frequency, low-amplitude differential technique. Peaks appear at well-defined frequencies labelled F_0 in Figure 34. It has been shown that F_0 is identical to the fundamental frequency measured in the Fourier noise analysis at the same electric field [258], except in the vicinity of the threshold; indeed the latter is reduced by the ac field [270, 271] as will be shown in the following section. The amplitude of the rf field is kept as $I_{rf}/I_{dc} < 0.25$. It can be seen in Figure 34 that, besides F_0 and harmonics of F_0, subharmonics are also generated, as well as a second fundamental frequency, F_2. The peaks in the $\mathrm{d}V/\mathrm{d}I$ curves are steps in the $V(I)$ characteristics. These steps have been measured in the dc characteristics [253, 272]. The amplitude of the steps follows a periodic variation as a function of the ac amplitude which is reminiscent of the behaviour of the rf steps induced in the Josephson junction dc characteristics [273]. In the so-called resistively-shunted junction (RSJ) model [274, 275] the height of the dc steps induced by a microwave field have been found to vary as Bessel functions [276]. Several authors [269, 277, 278] have suggested this similarity between Josephson junctions and CDW transport, principally based on the similarity of their phenomenological

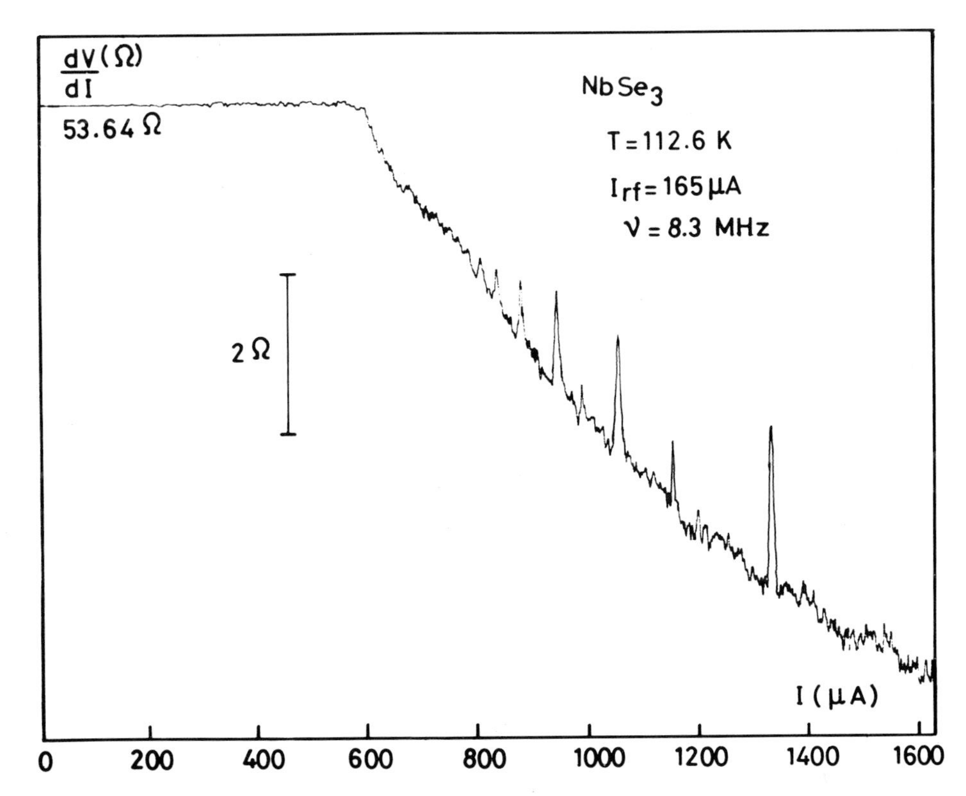

Fig. 33. Differential resistance dV/dI (at 33 Hz) of a $NbSe_3$ sample at T = 112.6 K as a function of the applied dc current on which is superimposed an rf field with a frequency of 8.3 MHz and an amplitude of 165 μA. Peaks occur when the voltage oscillation frequencies in the nonlinear state and the rf frequency are synchronized (after [258]).

non-linear equation of motion. The height of the steps is also strongly frequency dependent; its frequency variation is similar to that of Re $\sigma(\omega)$. The amplitude of the step $n = 1$ has been found to be 40 times larger at a frequency $\omega \gg \omega_0^2/\Gamma$ than at a frequency $\omega \ll \omega_0^2/\Gamma$ [272]. Interference effects obtained by Zettl and Grüner [272, 253] are larger than those measured in [257, 258] because the rf field used in their experiment had a frequency and an amplitude much higher than in [257, 258].

5.4.2. *Effect of an ac Field on the Threshold Field*

The variation of the differential resistance dV/dI as a function of the dc bias current has been measured in the presence of an ac field [271, 253]. It can be noted that the variation of dV/dI is smoother in the vicinity of the threshold and that the latter is decreased. When the rms amplitude of the ac current, I^{ac}, is not too large compared to the threshold current, I_c^{dc}, the threshold in the presence of the rf field $I_c^{dc,ac}$ is reduced to a value such as [271]:

$$I_c^{dc,ac} + I^{ac} = I_c^{dc} \tag{5-2}$$

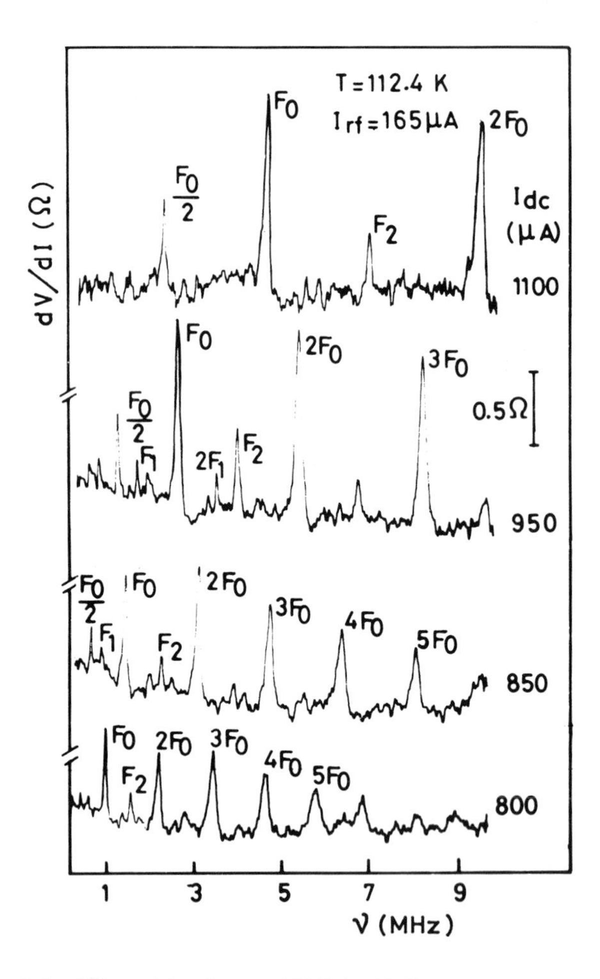

Fig. 34. Variation of the differential resistance dV/dI (at 33 Hz) of a $NbSe_3$ sample as a function of the frequency of a superimposed rf field (with a constant amplitude of 165 μA) at T = 112 K for different dc currents higher than the critical one (I_c = 600 μA).

If I^{ac} is increased, the dc threshold $I_c^{dc,ac}$ goes smoothly to zero [253]. Moreover, it has recently been shown that the ac dynamics is decoupled from the dc response for bias currents below the threshold I_c^{dc} [279].

In another experiment, I^{ac} was kept constant and much lower than I_c^{dc} and the frequency of the rf field was varied. The linear relation (5-2) is verified up to 4 MHz (for a $NbSe_3$ sample at T = 112 K). Above this frequency $I_c^{dc,ac}$ increases, and beyond 40 MHz the ac field has no more effect on the dc characteristic [271]. These experiments clearly show that at frequencies around several tens of MHz (which are function of the temperature and of the materials) the CDW cannot follow the ac excitation field; the contribution of the latter to the dc conductivity, possible at lower frequencies, is, therefore, suppressed.

5.4.3. *DC Conductivity Induced by a Large Amplitude ac Excitation*

The dc conductivity of a $NbSe_3$ sample has been measured as a function of a large amplitude ac signal at different frequencies. The conductivity value is measured with a dc field much smaller than the dc threshold field, E_c^{dc}, i.e., these experiments are without dc bias. An increase of the dc conductivity occurs for applied ac fields higher than E_c^{dc}. The variation of the ac amplitude which induces an extra dc conductivity of $x\%$ of the ohmic value, say 10%, increases strongly with the frequency above 5 MHz [280]. For a better precision, the dc characteristic has been measured with a low frequency–low amplitude ac modulation. A sharp threshold, I_c^{ac}, is observed, the value of which increases logarithmically with the frequency [271, 281]. Again, these experiments indicate that the overdamped CDW cannot follow the exciting field at high frequency and in order to increase the dc conductivity by the same amount, larger ac amplitudes have to be applied.

5.4.4. *Low Amplitude ac Conductivity in the Presence of a dc Bias Field*

These experiments concern the effect of a dc field on the low field ac conductivity. Figure 30 has shown the variation of Re $\sigma(\omega)$ and Im $\sigma(\omega)$ without dc bias. At a given frequency, Re $\sigma(\omega)$ and $\epsilon(\omega)$ have been measured as a function of a dc bias [253, 282]. There is no effect until the dc bias exceeds the threshold field. For $E > E_c^{dc}$, Re $\sigma(\omega)$ increases towards the conductivity value measured in the high frequency limit and ϵ decreases to approximately zero, which indicates the suppression of the ac polarization. Decrease of the dielectric constant with the electric field has also been reported by Djurek *et al.* [283]. Furthermore, steps in Re $\sigma(\omega)$ and large dips in $\epsilon(\omega)$ have been detected which correspond to interferences between the frequency at which the measurement is performed and the frequency of the oscillation voltage induced by the CDW motion, as has already been shown in Section 5.4.1.

5.4.5. *Harmonic Mixing*

Harmonic mixing between two frequencies ω and 2ω has been shown to be a powerful tool for the study of nonlinear transport properties in semiconductors [284]. The voltage

$$V(t) = V_0 + V_1 \cos(\omega_1 t + \phi) + V_2 \cos \omega_2 t \tag{5-3}$$

is applied to the sample; ϕ is the phase shift between ω_1 and ω_2 and $\omega_2 = 2\omega_1$. The mixing of ω_2 with the second harmonic of ω_1 generated by nonlinearity in the system leads to a rectified dc signal as follows:

$$\Delta U \sim V_1^2(\omega) V_2(2\omega) \cos(2\phi + \psi) \tag{5-4}$$

The additional phase shift, ψ, takes its origin from the characteristic time constant of the nonlinear mechanism. In the case of hot carriers in semiconductors, this time constant is the energy relaxation time which is determined by the carrier–lattice interaction. Harmonic mixing experiments have been performed on $NbSe_3$ [285, 286, 279] and TaS_3 [287].

(i) The first experiment was made on $NbSe_3$ in the microwave frequency range with $\omega \sim 9.5$ GHz and no dc bias, $V_0 \equiv 0$ [285, 286]. The harmonic mixing signal, ΔU, increases at both CDW transitions at T_1 and at T_2. The phase shift ψ is measured with an unknown additive value; however, ψ is temperature independent at all temperatures above 30 K.

(ii) A harmonic mixing signal generated in TaS_3 [287] and in $NbSe_3$ [279, 288] has been measured in the rf frequency range between 1 and 250 MHz as a function of the dc bias V_0. When $\omega_0/2\pi = (\omega_2 - 2\omega_1)/2\pi \sim 1$ kHz, a dc signal, ΔU, appears at all frequencies, only when V_0 is higher than the threshold field; moreover, the phase shift in TaS_3, ψ, is zero when the frequency is varied from 1 to 250 MHz. Finite phase shifts, however, have been measured in $NbSe_3$ when the experiments are performed at frequencies ω_1 higher than 50 MHz [288]. On the other hand, when ω_2 is not equal to $2\omega_1$, which is called a pseudo-harmonic mixing, a dc signal is detected at ω_0 for any dc bias below the threshold. In the experiments performed by Miller *et al.* [279], $\omega_0/2\pi$ was kept to a constant value of 12 MHz and ω_1 varied from $\omega_1/2\pi = 10$ MHz to 60 MHz. For a given ω_0, the amplitude of the dc signal below the threshold decreases sharply down to zero when ω_1 is increased, i.e., when the frequencies ω_1 and ω_2 tend towards the values corresponding to a real harmonic mixing situation.

5.5. Metastability – Memory Effects – Switching

At the onset of depinning, the CDW does not move as a rigid unit but suffers distortions. In this section, we discuss the influence of these CDW deformations on physical properties.

5.5.1. *Metastability – Memory Effects*

The ground state of the CDW below the Peierls transition contains many metastable states. The existence of these states has been shown by Gill [289]. He has applied a pulse with an amplitude larger than the threshold one to a $NbSe_3$ sample below T_2. The low field ($E < E_c$) resistance, R, increases by a fraction, δR, which can be greater than 2%; this δR persists for a period of time which may be longer than half an hour. If the sample is heated above T_2 and cooled down to the same temperature, the original resistance, R, is recovered. The pulse brings about a change in the CDW state from its equilibrium state to a metastable state.

Memory effects are related to the transient response of a sample in the nonlinear state to rectangular pulses. This response has been shown to be dependent on the electrical history of the specimen. When current pulses in excess of threshold with same polarity are applied, the voltage rise time is fast (at least for pure samples [266]); but if a current pulse has a polarity opposite to the preceding pulse, the voltage response is found to be initially fast and to become sluggish with a long tail towards its steady value. The voltage response for a $NbSe_3$ and a bronze $K_{0.30}MoO_3$ sample is shown in Figure 35 [290]. For $NbSe_3$, an active bridge circuit was used and the signal shown in Figure 35 is $U(t) = R_0 I(t) - V(t)$ where $I(t)$ is the current pulse, R_0 the Ohmic resistance and $V(t)$ the voltage response. In the nonlinear state, $V(t)$ is smaller than $R_0 I(t)$. Therefore, if the voltage response is initially less than the steady value, the bridge circuit will show a larger difference or an 'overshoot' [266, 289]. The time-constant of the voltage variation towards its steady value is typically, in the case of $NbSe_3$, 50 μs for $T_1 < T < T_2$

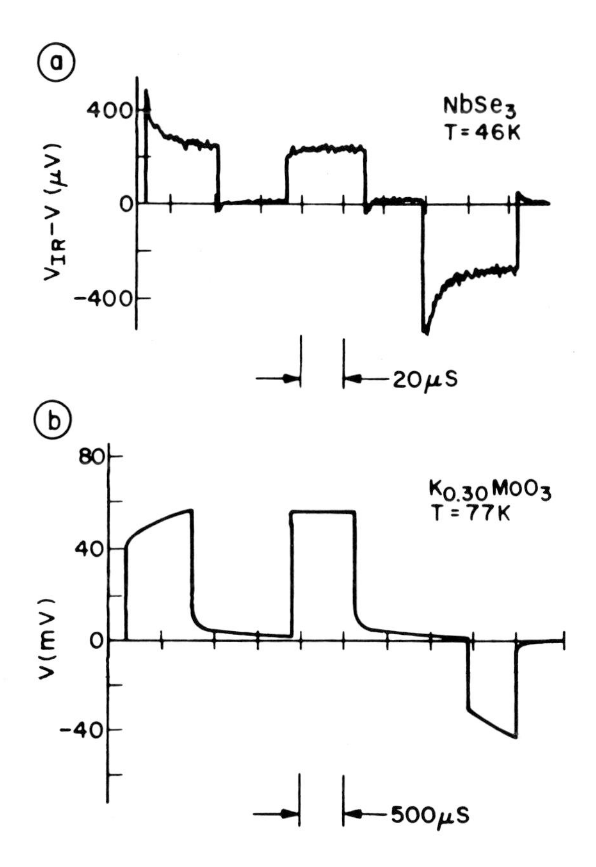

Fig. 35. Response of $NbSe_3$ and $K_{0.30}MoO_3$ to a repetitive series of constant-current pulses consisting of two positive, followed by one negative pulse. The amplitude of the pulses are in excess of threshold. When the current direction is reversed, the response becomes very slow. (a) for $NbSe_3$; the linear response is subtracted so that only $I\,R_{\text{linear}} - V(t)$ is plotted as a function of time; (b) the total signal $V(t)$ is plotted (after [290]).

and 15 μs for $T < T_2$ whereas the value is 500 μs for $K_{0.30}MoO_3$. This sluggish response is only found for pulses of which the polarity has been inverted. It seems, therefore, that the CDW retains a memory of the direction it last moved almost indefinitely. It has been shown by Gill [289] that the amplitude of this effect is independent of the time interval between pulses of opposite signs, at least up to 25 minutes. It is not necessary to apply pulses with inverse polarity in order to observe this sluggish response. Fleming and Schneemeyer [290] have shown that, when a pulse is superimposed on a dc bias, the slow response is measured as soon as the voltage pulse crosses the threshold field. Moreover, in the case of bronze, the slow variation of the voltage, $V(t) - V(\infty)$, is a function of the pulse width and is nearly proportional to $-\ln(t)$. These experiments clearly indicate that, after a current pulse, the CDW is left in a metastable state; a pulse with opposite polarity leads to some rearrangement of the CDW, which implies very long relaxation times.

The out-of-phase relationship between the voltage and the current when I is driven beyond the threshold current has allowed the study of hysteresis loops at given frequencies. Hysteresis has been recorded at a frequency of 100 Hz for $K_{0.30}MoO_3$ [290] and between 0.1 and 0.5 MHz for $NbSe_3$ [291]. For the latter sample, anticlockwise loops indicating an inductive behaviour have been found at frequencies much lower than $\omega_p = \omega_0^2/\Gamma$ ($\omega_p/2\pi$ is typically 50 MHz in $NbSe_3$). However, if the CDW is modelled by an harmonic oscillator, its dynamics at such low frequencies should be adiabatic, leading to recovery of the dc behaviour. The inertia term has been attributed to metastable states in the depinning process [291].

Finally another spectacular memory phase effect is shown in Figure 36. When a pulse

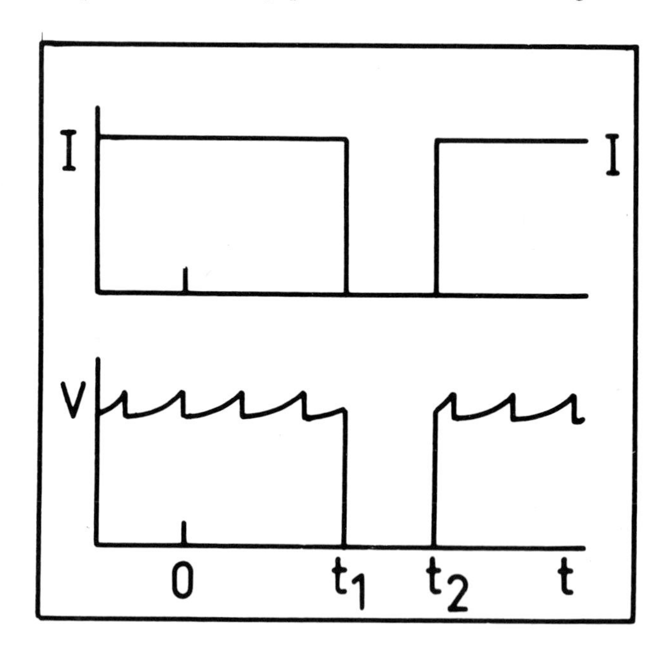

Fig. 36. Waveform of I and idealised response of V, demonstrating the phase memory effect (after [267]).

current is applied to a good quality sample, the voltage pulse displays the oscillatory behaviour due to voltage oscillations induced by the CDW motion. In the experiment performed by Gill and Higgs [267], the current pulse could be stopped at any part, t_1, of the cycle, $V(t)$; then the current pulse was restored at the time, t_2. Figure 36 shows that the waveform of V for $t > t_2$ continues from the point reached at $t = t_1$ when the CDW has stopped moving. This phase coherence memory at a temperature of 77 K for an orthorhombic TaS_3 sample has been shown to hold for at least 30 ms.

5.5.2. *Switching*

Typical $V(I)$ and dV/dI characteristics have been shown in Figures 24 and 26. Sharp jumps in $V(I)$, however, have been measured in some $NbSe_3$ samples [189, 248, 292] for temperatures below 49 K, in some bronzes $K_{0.30}MoO_3$ at 77 K [268], and in $(NbSe_4)_{10}I_3$ [99] between the Peierls transition 283 K and approximately 220 K. Figure 37 shows

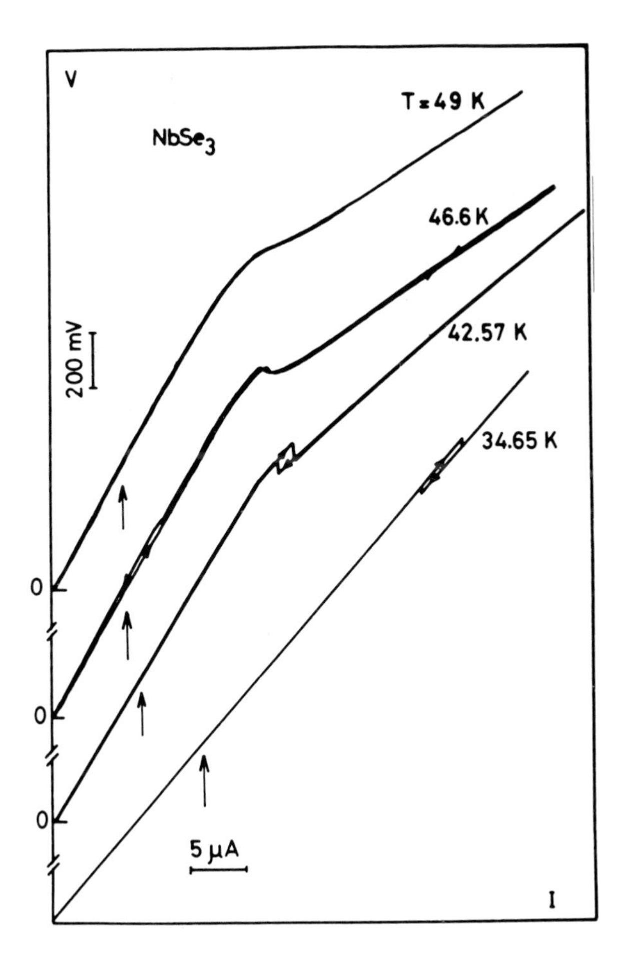

Fig. 37. $V(I)$ characteristics of a $NbSe_3$ sample at several temperatures below the onset of the CDW lower transition (T_2 = 59 K) showing a 'switching' effect. The threshold current, I_c, determined by the value at which dV/dI has slightly decreased is indicated by an arrow (after [189]).

the characteristics $V(I)$ for a $NbSe_3$ sample showing this effect at different temperatures [189]. For temperatures below approximately 45 K the sample jumps discontinuously from a state with higher resistance to a state with a lower resistance. This jump or 'switching' occurs at a higher value when increasing the current than when the current is decreased. The hysteresis loop enlarges when the temperature is decreased. This switching, however, does not correspond to the onset of nonlinearity in the sample: dV/dI shows very little decrease at a lower current, indicated by an arrow in Figure 37. Beyond this threshold, periodic noise is measured and new frequencies appear at the discontinuity in $V(I)$ [189, 248]. It has been shown by Zettl and Grüner [240] that the switching between the two conducting states, at a temperature of 28 K, requires a finite time that they associate with the coupling of different CDW domains in the sample. Moreover, they measured a switching in the conductivity only on samples in which the coherent voltage oscillations (as described in Section 5.3.1) were clearly observed. However, switching in the conductivity of iron-doped $NbSe_3$ samples have recently been reported [217]. The resistance ratio of samples with a concentration of iron between 1 to 7% is in the range of 6–35. The nonlinear properties of these doped samples are quite complicated. Two threshold fields are detected, however, each one with an hysteresis loop, as shown in Figure 38(a). The change to the nonlinear state occurs in a narrow range of multiple transitions. Figure 38(b) shows oscilloscope traces of hysteresis loops for a $Fe_{0.03}NbSe_3$ sample at 43.5 K, measured at a frequency of 10 Hz. The shape of multilevel hysteresis loops is a function of drive frequency. The depinning process and, consequently, the motion of the CDW is strongly affected by the Fe impurities. Analogy with domain wall motion in ferromagnetism has been drawn [217]: in the latter case hysteresis is also an impurity process which needs a substantial coercive field to initiate the domain wall motion. Therefore it can be concluded that the purity of the samples is not the essential feature which determines the switching in the conductivity, more important is the domain structure associated with the spatial distribution of impurities which depends on crystal growth. Depinning of a sample made up an array of $m \times n$ CDW domains has been simulated by Joos and Murray [292]. These multiple thresholds measured in the hysteresis loops of $Fe_{0.03}NbSe_3$ are also reminiscent of the voltage spikes detected at 77 K in $K_{0.30}MoO_3$ [268]. In increasing the current, the conductivity of $(NbSe_4)_{10}I_3$ makes an abrupt transition to the non-linear state whereas the conductivity makes a smooth transition back to the Ohmic behaviour when the current is decreased [99]. This hysteresis loop in the depinning process may be a consequence of the interaction between the CDW motion and regularly spaced planar faults seen in high resolution electron microscope pictures [96].

5.6. SUMMARY

The dc transport properties of NbS_3, $NbSe_3$, both forms of TaS_3, $(TaSe_4)_2I$, $(NbSe_4)_{10}I_3$, $K_{0.30}MoO_3$ are highly nonlinear beyond a very small threshold field. The energy per electron associated with this threshold is in the range of 10^{-2}–10^{-4} K, a value much smaller than the thermal energy, kT. Therefore, only a condensed state made up of many electrons can be stable against thermal fluctuations at temperatures in the range of 100–300 K. Similarly to superconductivity, in which the supercurrent arising from a condensate has no entropy, which leads the thermopower to drop to zero below the superconducting transition, a sliding CDW was expected to carry no heat. Early experiments on $NbSe_3$, however, showed that, when the CDW was depinned by an electric field, the thermopower recovered (like the resistivity) the value it would have had without the CDW transition. These results have been shown to be at variance with the sliding CDW model [163]. On the other hand, recent experiments performed on orthorhombic TaS_3 clearly show that the Peltier heat coefficient is zero at temperatures $T > 150$ K [293] which is consistent with conduction by a sliding ordered structure such as a CDW or an array of discommensurations. At lower temperature the ordered moving structure acquires phonon drag. It was also shown [293] that the heat current dragged is proportional to the excess current.

Finally, the CDW motion has also clearly been shown in Hall effect measurements

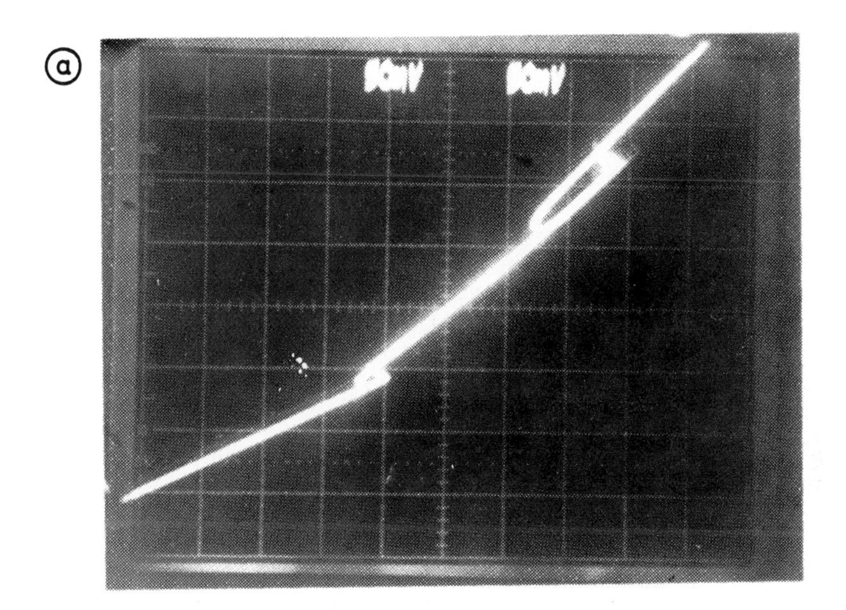

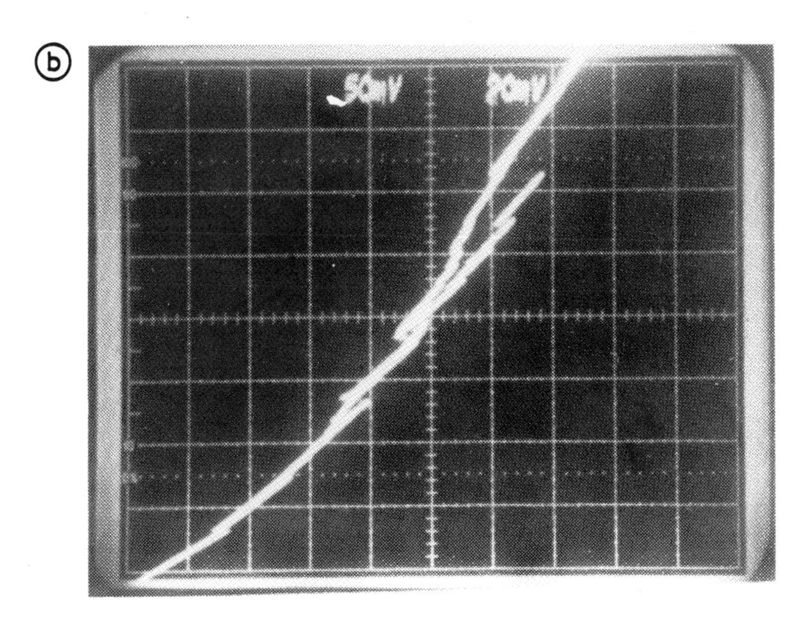

Fig. 38. (a) Oscilloscope traces of I *vs* V for $Fe_{0.03}NbSe_3$ recorded at $T = 43.5$ K at a drive frequency of 100 Hz. Two thresholds with hysteresis loop are detected. (b) Magnified oscilloscope trace of the upper hysteresis loop detected in (a) recorded at 10 Hz showing multiple transitions (after [217]).

performed in $NbSe_3$. The Hall voltage, V_H, as a function of the applied longitudinal current is not linear beyond the threshold current. When plotted as a function of the electric field, however, V_H is independent of E for fields much higher than the threshold

[145, 294]. These experiments allow us to conclude that the CDW slides along the chain axis in the direction where its wavelength is incommensurate with the lattice.

6. Theoretical Models

Many models try to account for the remarkable dynamical properties of the one-dimensional systems described in the preceding section. As has already been mentioned, J. Bardeen [229, 230] has ascribed the nonlinear properties of $NbSe_3$ to the Fröhlich conductivity resulting from the CDW motion when the latter is depinned by an electric field. In general, the phase of the CDW is dependent on time and coordinates: $\phi(r, t)$. One group of theories consider that the sample under investigation is broken up into domains with finite sizes in which ϕ is independent of the position [232, 295]. The equation of motion of such a rigid CDW has been treated classically [296, 238, 258] and with a quantum approach [297, 298]. The classical motion of the CDW taking into account deformations resulting from its interaction with impurities has also been studied [299–302]. Another group of theories connects the nonlinear properties to the motion of local phase defects such as discommensurations or solitons [303, 306]. These different theories are described at length in Part I of this work. Therefore, in this section, we only recall their fundamental basis.

6.1. CDW DOMAINS

Below the Peierls transition, the local charge density is modulated as expressed in (1-1):

$$\rho(r) = \rho_0 + \rho_1 [\cos Qr + \phi(r)]$$

Fukuyama and Lee [295] have shown that the interaction between the CDW and randomly distributed impurities involves two competitive energies [307]:

– An elastic energy, $\frac{1}{2}K\int(\partial\phi/\partial z)^2\,dz$, which indicates the ability of the CDW to be stretched or compressed by pinning centers.

– An impurity potential given by the Hamiltonian

$$-V\rho_1 \sum_i \cos[Qr_i + \phi(r_i) - \phi_i] \tag{6-1}$$

where the summation is over impurities at sites r_i, and ϕ_i is the phase that the impurity tends to accommodate. Lee and Rice [232] have also introduced an impurity strength parameter:

$$\epsilon = \frac{\rho_1 V}{\Delta(T)} \tag{6-2}$$

For non-isoelectronic impurites, $\rho_1 V$ is estimated to be large (a few tenths of eV) and, therefore, $\epsilon \gg 1$ whereas for weak pinning impurities, $\epsilon \ll 1$. In the latter case, the impurity potential is not strong enough to fix the phase at pinning centers but it was shown that the system breaks into anisotropic domains in which the phase varies smoothly. These domains have a length, L, along the chain axis (the so-called Fukuyama–Lee–Rice domain length) and dimensions ξ_x, ξ_y in the transverse directions, ξ_x, ξ_y and ξ_z being the coherence lengths for the CDW amplitude. ξ_z, the coherence length along the chain axis is equal to:

$$\xi_z \sim a_z \frac{E_F}{\Delta} \tag{6-3}$$

where E_F is the Fermi energy, Δ: the Peierls gap and a_z: the unit cell parameter along z. A random walk summation over impurities leads to a gain in impurity pinning energy as follows:

$$-V\rho_1 \left[\left(\frac{\xi_x \xi_y}{\xi_z} \right)^2 L^3 n_i \right]^{1/2} \tag{6-4}$$

The elastic energy to be paid in the volume of a domain is $\Delta \times L/\xi_z$. The distance, L, is deduced by minimizing the sum of the elastic and the impurity pinning energy which leads to:

$$L = \frac{1}{\epsilon^2} \frac{1}{n_i} \frac{1}{\xi_x \xi_y} \tag{6-5}$$

The threshold field is the field at which the electric energy supplied by the electric field overcomes the pinning energy per unit volume to dislodge the CDW over one wavelength $\lambda = 2\pi/Q$. The latter energy is calculated from (6-4) with the expression for L given in (6-5); consequently,

$$eE_c\lambda = \frac{\Delta^2}{E_F} (\xi_x \xi_y \xi_z n_i^2) \epsilon^4 \tag{6-6}$$

Therefore, for weak pinning impurities, E_c varies as the square of the impurity concentration and the domain length as the inverse of the impurity concentration. Explicit values of L and E_c from the theoretical expressions (6-5) and (6-6) are dubious because of the uncertainty of the unknown parameter, ϵ.

On the other hand, in the case of strong pinning caused either by large amplitude pinning potential or by dilute impurity concentration, the phase of the CDW is fixed at the impurity site as follows: $Qr_i + \phi(r_i) = -\pi$. In this case, the domain size along the chain is $1/n_i \times 1/\xi_x \xi_y$ and the threshold field given by:

$$eE_c\lambda = \frac{\Delta^2}{E_F} (\xi_x \xi_y \xi_z n_i) \tag{6-7}$$

For strong pinning impurities, E_c is proportional to the impurity concentration. The differentiation between strong and weak pinning impurities is derived from the value of ϵ. Near T_c, the Peierls gap, Δ goes to zero; therefore, a weak pinning situation at low temperatures can become a strong pinning one in the vicinity of T_P which may explain the increase of E_c in this temperature range for all the compounds studied in Section 5.

It has to be noted that no finite threshold in the thermodynamic limit can be obtained without elasticity in the CDW [195]. If the CDW is strictly rigid, the electric energy supplied by E is proportional to the volume, whereas the pinning energy, because of the random-walk summation, is proportional to the root of the volume. Therefore, if the volume becomes infinite, E_c decreases to zero. The CDW can be considered as a rigid unit only in a finite size domain. If the sample is made up of many independent domains, each one with a different E_c, it is possible to define a distribution of E_c with a mean value E_{cM} and a standard deviation, s, inversely proportional to the root of the number of domains. E_c thus becomes an extensive quantity, at least if the number of domains is large enough. It would be possible to establish size effects on the threshold value only if the volume of the sample consisted of a unique or a very small number of Fukuyama–Lee–Rice domains.

In the case of strong pinning, Fukuyama and Lee [295] have defined a frequency, ω, that they call pinning frequency, as follows:

$$\omega = C_0 n_i \xi_x \xi_y \tag{6-8}$$

where C_0 is the phason velocity defined in (2-4). This frequency corresponds to the standing wave with a wave-vector equal to the average spacing between impurities. Similarly, in the weak pinning case, the pinning frequency is the excitation of the phase mode with a wavelength of the order of the domain size, L.

6.2. Rigid Motion of the CDW

In the models of rigid CDW motion, although the CDW interacts with impurities, it is assumed that its wave-vector is completely uniform. Consequently, the CDW can be described with a unique dynamical variable, $\phi(t)$. But, as mentioned above, the rigidity of the CDW implies that this description is only valid in a finite volume. The results of experiments described in [257] were the first to lead to the assumption that the periodic voltage oscillations discovered by Fleming and Grimes [231] could be attributed to the CDW motion in the anharmonic pinning potential created by impurities. Then the so-called classical model was worked out by Grüner *et al.* [296] and further in [238, 258]. Quantification of the same model was established by Bardeen [297, 298].

6.2.1. *Overdamped Harmonic Oscillator Approximation*

Electrical forces, damping forces and pinning forces are acting on the CDW. Since the current carried by the CDW at the velocity v_s is $-nev_s$ (1-5), the electric force per unit volume is:

$$F_{el} = -neE$$

Damping forces may arise from various mechanisms: excitation of low-energy phasons by the motion, coupling of the lattice oscillations when the CDW is in motion with very low-energy phonons by anharmonic terms, mutual friction between CDW and free electron motion [232, 308] .... In any case, the damping forces are proportional to the velocity, v_s, and defined by a phenomenological viscosity coefficient, η, which is a function of temperature. Per unit volume,

$$F_{\text{damp}} = -\eta v_s$$

In [238], it was assumed that the main interaction between impurities and the CDW is via Friedel screening oscillations which have an oscillatory behaviour with $Q = 2k_F$, the same as for the CDW distortion. Nucleation of a CDW in the presence of impurity-induced Friedel oscillations has been studied by Tsuzuki and Sasaki [309]. If the two oscillations have the same phase, they will help each other and the energy will be a minimum. For the opposite situation, phase opposition, the energy will be a maximum. For a given position of the impurity r_i, the interaction energy is *periodic* and can be written as follows [238]:

$$W_{\text{interact}} = -V[\cos \phi(r_i) - \psi(r_i)] \tag{6-9}$$

where ϕ is the CDW phase at the impurity site and ψ the ideal value taking into account the relative charge, Z, of the impurity in the matrix. At T_P, CDW domains nucleate at places where, in the random distribution of $\psi(r_i)$, the interaction energy is minimum. At slightly lower temperatures, the domains grow, encompassing less favourable ψ. When two domains come in contact, if the phase difference is small, they give a single domain with a well-defined phase, but if the phase difference is large (of the order of π), a normal wall appears and subsists between the two adjacent independent domains. These domains are not exactly the same as those of Fukuyama, Lee and Rice [232, 295]. Nevertheless, the sample is supposed to consist of independent domains with a zero order parameter inside the domain walls. The latter assumption only requires conservation of the electric current and continuity of potential without any coupling between domains such as a Josephson potential. In each independent domain, the phase of the CDW in motion is:

$$\phi(r, t) = Qr + \phi_0(t)$$

and, at equilibrium

$$\phi = Qr + \phi_0 \tag{6-10}$$

The pinning forces in each domain can be expressed according to (6-9) and (6-10) as follows:

$$F_{\text{pinning}} = QV \sum_i \sin[\phi_0 + (Qr_i - \psi_i)]$$

where ϕ_0 is the phase, say, at the centre of the domain. The summation of the random forces over (i) impurities belonging to the domain gives a net pinning force with a *periodic* dependence in ϕ_0:

$$F_{\text{pinning}} = F_0 \sin \phi_0 \tag{6-11}$$

The equation of motion for the jth domain with a volume Ω_j will be:

$$\Omega_j(mn\gamma + \eta v_s + neE) - F_j \sin \phi_j = 0$$

The velocity is associated to the translation of the phase,

$$v_s = \frac{1}{Q} \frac{\mathrm{d}\phi_j}{\mathrm{d}t} \tag{6-12}$$

As the inertial term is much smaller than the dissipative one (at least for frequencies smaller than 10^{10} s^{-1}), the equation of motion is that of an overdamped oscillator:

$$\frac{\eta}{Q}\frac{d\phi_j}{dt} - \frac{F_j}{\Omega_j}\sin\phi_j = -neE$$

With slight changes in the coefficients, the equations of motion for the CDW phase in a unique domain are as follows:

$$\eta\frac{d\phi}{dt} - f\sin\phi = neE \tag{6-13a}$$

$$J = nev + \sigma_a E \tag{6-13b}$$

where σ_a is the Ohmic conductivity in the limit $E \to 0$. The equation of motion obtained by Grüner *et al.* [296] results from the phenomenological model proposed by Rice *et al.* for the CDW motion [19]. The equation of motion of the centre of mass, x, of the rigid CDW or of the phase $\phi = Qx$ when displaced from its equilibrium position is assumed to correspond to that of a harmonic damped oscillator (2-5). From the frequency dependence of the conductivity it is concluded that the inertial term is negligible. Therefore, from (2-5) and (2-7), the equation of motion is:

$$\Gamma\dot{x} + \omega_0^2 x = \frac{eE}{M^*} \tag{6-14}$$

Then, it is assumed that the pinning potential is periodic with the same periodicity as the CDW itself; (6-14) can be written as follows:

$$\Gamma\dot{x} + \frac{\omega_0^2}{Q}\sin Qx = \frac{eE}{M^*} \tag{6-15}$$

The nonlinear equation (2-5), taking into account thermal fluctuations, was studied by Wonnenberger and Gleisberg [310] and applied to the nonlinear dc conductivity in TTF–TCNQ (see Section 2.3 and Figure 3).

(i) Response to a dc Field – Single Domain Behaviour

The threshold field is obtained for $d\phi/dt = 0$: $neE_c = f$ according to (6-13a) or $M^*\omega_0^2/Q = eE_c$ according to (6-15). Therefore,

$$E_c = \frac{f}{ne} = \frac{M^*\omega_0^2}{Qe} \tag{6-16}$$

The threshold is the field beyond which the electric force overcomes the pinning force. Equation (6-13) can be transformed into dimensionless variables as follows: $\tau = \eta/f$, $e = E/E_c$, $J_c = \sigma E_c$ and β defined by:

$$\beta = \frac{\sigma_{E\to\infty} - \sigma_a}{\sigma_a} \tag{6-17}$$

One gets:

$$\tau \frac{d\phi}{dt} - \sin \phi = e \tag{6-18a}$$

$$J = \sigma_a \left[E + \beta E_c \tau \frac{d\phi}{dt} \right] \tag{6-18b}$$

The shape of the characteristic $E(J)$ depends on the impedance, Z_ω, of the external circuit at the frequency ω.

Behaviour at a given field $E > E_c$

In this case, Z_ω is $\ll R$, the resistance of the sample. Above E_c, the current is the superposition of a continuous current and a modulation. The continuous current is:

$$\bar{J} = \sigma_a [E + \beta(E^2 - E_c^2)^{1/2} \tag{6-19}$$

and the modulation has periodic components with frequencies which are multiple of the fundamental one, such as:

$$\omega = \frac{1}{\tau} \left[\left(\frac{E}{E_c} \right)^2 - 1 \right]^{1/2} = \frac{1}{\tau} (e^2 - 1)^{1/2} \tag{6-20}$$

More generally, it is found that,

$$\tau \frac{d\phi}{dt} = (e^2 - 1)^{1/2} \{ [1 + 2K \cos(\omega t + \pi/2 + \psi) + 2K^2 \cos[2(\omega t + \pi/2 + \psi)] +$$

$$+ \ldots 2K^n \cos[n(\omega t + \pi/2 + \psi)]] \tag{6-21}$$

where $K = e - (e^2 - 1)^{1/2}$ and ψ is an arbitrary phase factor. Equation (6-21) indicates that the amplitude of the fundamental frequency tends towards a finite value when e increases whereas other harmonics decrease rapidly when the value of e is approximately higher than 1.5.

Behaviour at a given current

In this case, $Z_\omega \gg R$, J is given and E is a function of time; accordlng to (6-18b),

$$E = \frac{J}{\sigma_a} - \beta E_c \tau \frac{d\phi}{dt}$$

(6-18a) can be written as follows:

$$\tau \frac{d\phi}{dt} = \left[\frac{J}{\sigma_a E_c} - \beta \tau \frac{d\phi}{dt} \right] + \sin \phi$$

$$\tau \frac{d\phi}{dt} (1 + \beta) = J/J_c + \sin \phi \tag{6-22}$$

The equation of motion is the same as that in the case in which E is imposed but with a different time constant. Therefore,

$$\bar{E} = \frac{J}{\sigma_a} - \frac{\beta E_c}{1+\beta}\left[\left(\frac{J}{J_c}\right)^2 - 1\right]^{1/2} \tag{6-23}$$

and the oscillations are multiple of the fundamental frequency:

$$\omega = \frac{1}{(1+\beta)\tau}\left[\left(\frac{J}{J_c}\right)^2 - 1\right]^{1/2} \tag{6-24}$$

The characteristics $J(E)$ and the normalized derivatives $(J_c/E_c)(\mathrm{d}E/\mathrm{d}J)$ in cases of both regulated current and regulated field are drawn in Figure 39. It should be noted that,

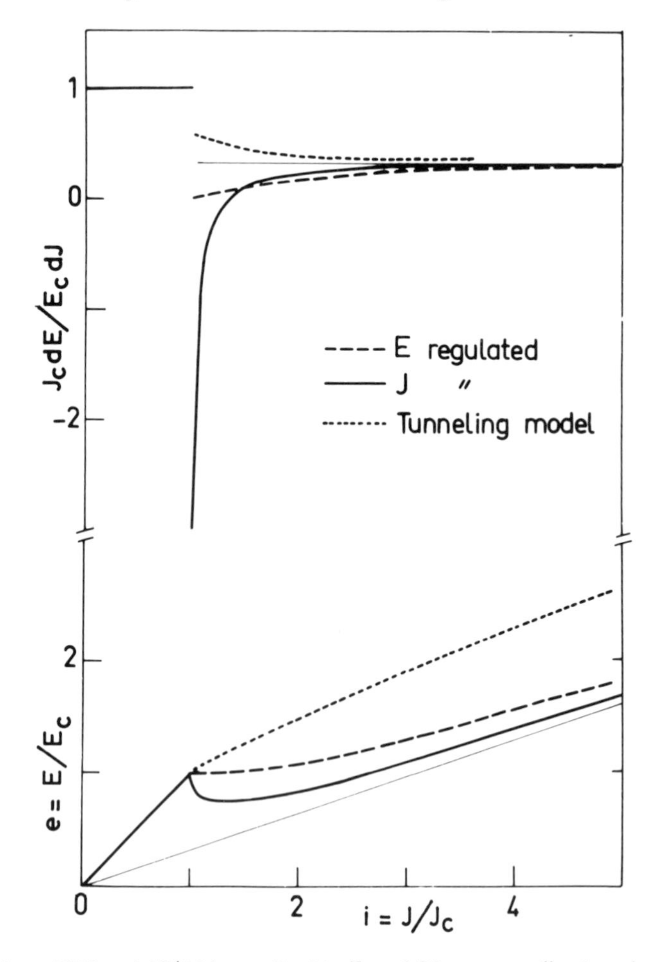

Fig. 39. Variation of $E(J)$ and $\mathrm{d}E/\mathrm{d}J$ (normalized to E_c and J_c) corresponding to a single rigid domain in the cases where the electric field [Equation (6-19)] or the current [Equation (6-23)] is regulated. Also shown is the variation of $E(J)$ and $\mathrm{d}E/\mathrm{d}J$ obtained by Bardeen [Equations (6-41) and (6-42)] in his quantum tunneling theory.

at the threshold, dE/dJ has an infinitely negative value when the current is regulated, whereas its value is zero when the electric field is regulated. For intermediate situations, the slope of dE/dJ at the threshold has the value, $-Z_\omega/R$. In the limit of an infinite electric field, the limit value of $J_c/E_c \times dE/dJ$ is $1/(1+\beta)$ for all the cases. Experiments are generally performed with a constant current source and, except perhaps for some insulating samples at low temperatures, the current through the sample is regulated. Although a sharp drop in dV/dI has been observed in some $NbSe_3$ samples in a narrow temperature range (see Figures 26 and 37), generally dV/dI does not show an infinite negative value at E_c which leads us to suppose that the model with a unique domain is unrealistic. Any sample is made up of many domains, each one with its own threshold field. It has been shown in [238] that the divergence at E_c is suppressed and that a well-defined threshold is still defined in the case of a small s/E_{cM} distribution (s: standard deviation and E_{cM}: mean value of E_c for a Gaussian distribution of threshold fields).

(ii) Response to a Small Amplitude ac Field

The driving force, $E = E_1 \exp(i\omega t)$, is applied with an amplitude, E_1, much smaller than the threshold field. The ac conductivity of a damped harmonic oscillator was given in (2-9). In the case of an overdamped oscillator, Re $\sigma(\omega)$ and Im $\sigma(\omega)$ are expressed as follows:

$$\mathrm{Re}\,\sigma(\omega) = \frac{ne^2}{M^*\Gamma}\left(1 + \frac{\omega_0^2}{\Gamma\omega}\right)^{-2} \tag{6-25a}$$

$$\mathrm{Im}\,\sigma(\omega) = \frac{ne^2}{M^*\Gamma}\,\frac{\omega_0^2}{\Gamma}\left(1 + \frac{\omega_0^2}{\Gamma\omega}\right)^{-2} \tag{6-25b}$$

The experimental variation of Re $\sigma(\omega)$ and Im $\sigma(\omega)$ as a function of ω for $NbSe_3$ and orthorhombic TaS_3 is largely in keeping with (6-25a) and (6-25b), as shown in Figure 30. The frequency, ω_0^2/Γ, at which Re $\sigma(\omega)$ has a value half the maximum extra-conductivity (Re σ at saturation $- \sigma_{dc}$) is called the crossover frequency. At this frequency, the pinning forces have the same magnitude as the viscous forces. Following Shapira and Neuringer [311] who have studied a similar problem in the case of superconductors, this crossover frequency can be called the *depinning* frequency.

From $\epsilon(\omega) = 4\pi\,\mathrm{Im}\,\sigma(\omega)/\omega$, the dielectric constant in the low frequency range is given by:

$$\epsilon_{\omega\to 0} = \frac{4\pi ne^2}{M^*\omega_0^2} \tag{6-26}$$

Combined with (6-16), one can derive the following relationship:

$$\epsilon_{\omega\to 0}E_c = \frac{4\pi ne}{Q} = 2ne\lambda_{\mathrm{CDW}} \tag{6-27}$$

Within this model, the product of the dielectric constant (in the limit of $\omega \to 0$) by the threshold is independent of the temperature, which corresponds fairly well to the experimental results [219, 255].

(iii) DC Response to a Large Amplitude ac Excitation

The overdamped oscillator approximation has been used by Grüner *et al.* [280] in order to estimate the critical amplitude of an ac driving exciting field which, at a given frequency, induces a dc current. The depinning of the CDW and, consequently, a dc conductivity appears, without dc bias, if the displacement of the CDW in the harmonic potential is larger than half its periodicity, i.e., half the CDW wavelength. The maximum displacement corresponds to:

$$x_0 = \left(e\frac{E}{M^*}\frac{1}{\omega_0^2}\right)\left[1+\left(\frac{\Gamma\omega}{\omega_0^2}\right)\right]^{-1/2}$$

With the expression (6-16) for the dc threshold, one gets:

$$E_c(\omega) = E_c\ (\omega = 0)\left[1+\left(\frac{\Gamma\omega}{A\omega_0^2}\right)^2\right] \tag{6-28}$$

where A is a function of the shape of the potential. Equation (6-28) shows that, for $\omega < \omega_0^2/\Gamma$, the ac amplitude for depinning the CDW is equivalent to the dc value, but when $\omega > \omega_0^2/\Gamma$, the depinning occurs only at a much larger amplitude.

(iv) AC + dc Coupling – rf Steps Induced in the dc Characteristics

Equation (6-18a) has been solved when a rf exciting field is superposed on a dc field such as:

$$\tau\frac{\mathrm{d}\phi}{\mathrm{d}t} = e + \sin\phi + e_1\cos\omega_1 t \tag{6-29}$$

with $e_1 = E_1/E_c \ll 1$. Let us call a the reduced dc electric field which, if applied alone, would exactly give ω_1 for the CDW motion frequency, and ϕ_1 the solution of:

$$\tau\frac{\mathrm{d}\phi_1}{\mathrm{d}t} = a + \sin\phi_1$$

Because e_1 is small, one can try the solution:

$$\phi = \phi_1 + \epsilon \tag{6-30}$$

It was shown in [258] that a solution can be found for ϵ, synchronized with ω_1 and a relative phase, α, given by:

$$\cos\alpha = -\frac{2a(e-a)}{e_1} \tag{6-31}$$

This solution exists only between two fields, e_m and e_M corresponding to $|\cos\alpha| < 1$. The variation of ω as a function of e is plotted in Figure 40(a). For a given rf amplitude, synchronisation occurs in the range between e_m and e_M. But, since ω is proportional to the velocity, it is also proportional to the dc current carried by the CDW. If small variations of e are made between e_m and e_M, the current carried by the CDW will not change,

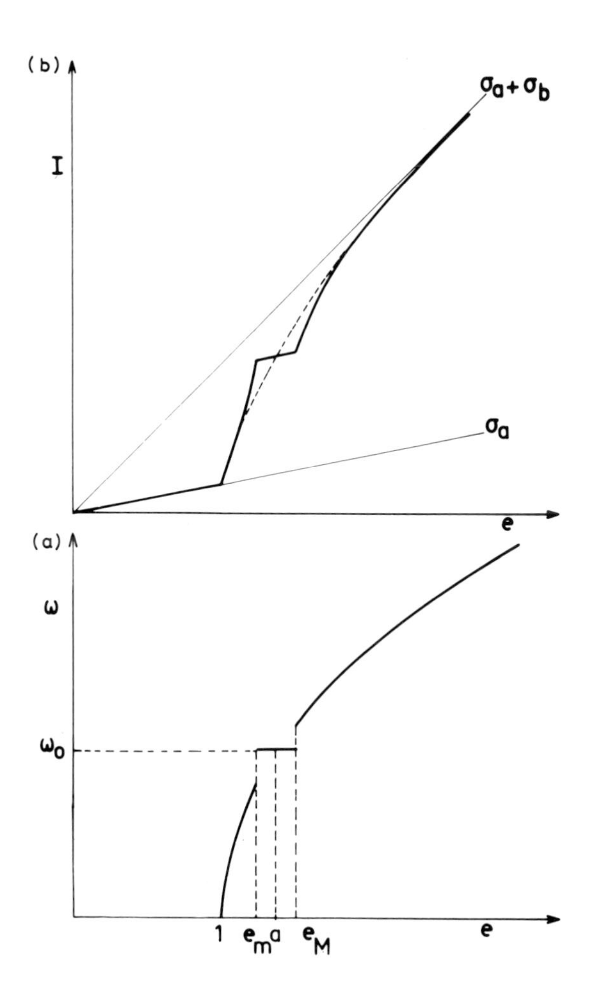

Fig. 40. (a) Variation of the ac frequency, ω, generated by the CDW motion as a function of the electric field normalized to E_c [Equation (6-20)]. If a is the electric field corresponding to the applied rf frequency, ω_0, synchronization takes place in the range between e_m and e_M defined in (6-31). (b) Schematic variation of the I–E (normalized to E_c) characteristics in the presence of radiation of frequency ω_0. Steps occur at $e = a$ with a width comprised between e_m and e_M; the slope of this step is σ_a, the Ohmic value for $e < 1$.

$\delta J_{\text{CDW}} = 0$. Therefore, the δI associated with the δV is only due to the Ohmic conduction (supplied by normal carriers); consequently, a peak appears in the $\mathrm{d}V/\mathrm{d}I$ characterististic equal to the Ohmic value below the threshold. Experiments shown in Figures 33 and 34 largely correspond to such a description. The center of the peak is independent of the rf amplitude but the width of the peak increases with the rf amplitude. If e_1 is increased, the linearization made in (6-30) is no longer justified and subharmonics and harmonics appear. The interference can also be detected by a step in the $I(V)$ characteristic, as shown in Figure 40(b); the slope of the step is equal to the Ohmic conductivity, σ_a.

This frequency locking results from nonlinear equations and it is quite model-independent. Such interferences have been largely studied in the context of Josephson junctions [273]. Russer [312] has calculated that the amplitude of the step height has a Bessel behaviour going through a series of maxima when the amplitude of the rf excitation is increased. Such a behaviour has also been observed with the nonlinear CDW characteristic [253].

(v) Harmonic Mixing

Equation (6-18a) including the inertia term has been solved by Wonnenberger [313] when two rf signals at ω_1 and ω_2 are applied. The driving electric field is:

$$e = e_0 + e_1 \cos(\omega_1 t + \phi) + e_2 \cos \omega_2 t$$

with e_1 and $e_2 \ll 1$. If the pinning potential has inversion symmetry, each ac field cannot individually provide any rectification. The rectification is only possible when the symmetry is broken by a dc bias. Harmonic mixing measures the dc compensation voltage across the sample when $\omega_2 = 2\omega_1$. The result for zero dc bias is as follows:

$$e^{\mathrm{dc}}_{\mathrm{H.M.}} = -\tfrac{1}{8} e_1^2 e_2 \cos(2\phi - \psi) F(\omega) \tag{6-32}$$

where expressions for the internal phase shift ψ and $F(\omega)$ are given by Equations (9) and (10) in [313]. Moreover, the internal phase shift is expected to occur for any dc bias below threshold.

When a dc field is applied above the threshold, $e_0 > 1$. Each ac field can individually lead to a rectification signal. But another dc signal comes from the harmonic mixing between both frequencies, ω_1 and ω_2. In the limit of high electric field ($e \gg 1$) and in the case of a constant voltage source, Wonnenberger has calculated that the change in the dc current due to the dc harmonic mixing voltage is [313]:

$$\frac{\Delta I_0}{I_0} = 9 \cos 2\phi \frac{e_1^2 e_2}{e e_0^2} F(\omega, e) \tag{6-33}$$

with

$$F(\omega, e) = \frac{1}{16} \left[\left(\frac{\Gamma\omega}{\omega_0^2} \right)^2 - e^2 \right]^{-1} \left[4 \left(\frac{\Gamma\omega}{\omega_0^2} \right)^2 - e^2 \right]^{-1}$$

Therefore, there is no internal phase shift in the nonlinear state, in contrast to the situation below threshold. Moreover, it can be seen from the expression for $F(\omega, e)$ that resonances are expected to occur at the frequency and twice the frequency of the voltage oscillations induced by the dc field.

(vi) Solid State Turbulence

The equation of motion (2-8) is that of a damped harmonic oscillator which has also been used for describing Josephson junctions or a driven pendulum. Solutions of (2-8) numerically calculated as a function of the amplitude, F, and the frequency (normalized to the natural resonant oscillation, ω_0) of the driving force have revealed quite a complicated behaviour. In the case of an anharmonic [314] and of a periodic potential

[315], Huberman *et al.* have found regions in the F, ω/ω_0 plane with periodic solutions which are running solutions for the phase, but they have also found regions with chaos preceded by sequences of period-doubling bifurcations. The latter mechanism is one of the routes to chaos. The broad band noise generated in Josephson junctions used as parametric amplifiers has been explained by the solid state turbulence [315]. The nature of the chaos has also been investigated in the forced pendulum case [316, 317]. Because the motion equation for CDW systems is isomorphic with the one for resistively-shunted Josephson junctions or forced pendulum (see Section 7) it is very likely that turbulence and chaos might be observed when a CDW system is submitted to a high amplitude and high frequency rf excitation. However, it was shown by McDonald and Plischke [317] that, in the frequency range in which the CDW response is overdamped, no chaotic behaviour is expected to appear.

6.2.2. *Quantum Model*

A complete description of this model is given by J. Bardeen in Chapter 3, Part I. The CDW with a mass M^* and an effective charge $e^* = \alpha e = e(m^*/M^*)$ (M^* is the Fröhlich mass defined in (2-3)) is submitted to a periodic potential which has the same variation, in $\sin\phi$, as for the classical model. But the dynamical variable, ϕ, is quantified and described by eigenfunctions, $\psi(\phi)$, such as:

$$\psi_{n,k}(\phi) = e^{ik\phi}u_n(\phi)$$

where u_n has the potential periodicity, i.e., 2π. $\psi(\phi)$ are solutions of the Schrödinger equation:

$$-\frac{\hbar^2}{2m}\frac{d^2\psi}{d\phi^2} + V(\phi) = \mathcal{E}\psi(\phi)$$

The pinning potential leads to bands with gaps:

$$\epsilon_g = \hbar\omega_p \tag{6-34}$$

where ω_p is the pinning frequency of the CDW. When a ac electric field is applied, similarly to transport properties in semiconductors, the eigenfunctions which characterize the CDW follow the field such as:

$$\hbar\frac{dk}{dt} = e^*E$$

At $k = +\pi$ a Bragg reflection occurs towards $k = -\pi$. Therefore, the phase in real space oscillates around an average value without carrying a current. However, Zener tunneling from a band to an excited band with the displacement in real space of a coherent portion of the CDW gives rise to a dc current [297, 298]. The tunneling probability is:

$$P = \exp-(\xi(E)/2\xi_0) = \exp-(E_0/E) \tag{6-35}$$

with the activation energy, E_0, given by the Zener expression:

$$E_0 = \frac{\pi\epsilon_g^2}{4\hbar e^* v_F} \tag{6-36}$$

ξ_0 is the CDW coherence length associated to the pinning gap such as:

$$\xi_0 = \frac{2\hbar v_F}{\pi \epsilon_g} \qquad (6\text{-}37)$$

Therefore, the tunneling distance, $\xi(E)$, is:

$$\xi(E) = \frac{\epsilon_g}{e^* E} \qquad (6\text{-}38)$$

In the preceding section, it was shown that the CDW motion is overdamped. Caldera and Leggett [318] have calculated that the tunneling probability is reduced in the presence of viscous dissipation. Their theory has been extended to the CDW tunneling model by Wonnenberger [319]. Whereas the theory of Caldera and Leggett is for $T = 0$, its application to the CDW problem for finite temperatures is thought to be valid because thermal fluctuations, at least for T not in the vicinity of the Peierls temperature transition, are suppressed by the Peierls gap. When dissipation is taken into account, Wonnenberger finds that the tunneling probability is decreased by a factor $[2(\alpha + 1)]^{1/2}$. Therefore, the correlation length, ξ_0, is reduced to a value, L_0 such as:

$$L_0 = 2\xi_0 [2(\alpha + 1)]^{1/2} = \frac{2\hbar v_F}{\pi \epsilon_g [2(\alpha + 1)]^{1/2}} \qquad (6\text{-}39)$$

In Section 2, the phason velocity was shown (2-4) to be $C_0 = v_F/(\alpha)^{1/2}$. As mentioned in Section 6-1, Fukuyama and Lee have defined the pinning frequency, ω_p, in the case of a weak pinning situation, as $\omega_p \sim C_0/L$, where L is the size of a Fukuyama–Lee–Rice domain. Therefore, it can be seen from (6-34), (6-37) and (6-39) that the new reduced coherence length, L_0, in the tunneling probability is equivalent to a coherent CDW domain length, L. If the CDW coherence extends to L, a tunneling distance larger than L is meaningless; so $\xi(E)$ has a cut-off at $\xi(E) = L$ which leads to a threshold field, E_c, defined from (6-38) such as:

$$e^* L E_c = \epsilon_g = \hbar \omega_p \qquad (6\text{-}40)$$

This threshold is the consequence of a finite CDW coherence. The tunneling distance for a perfect CDW coherence would be infinite.

(i) Response to a dc Field

A dc conductivity occurs when tunneling takes place. Bardeen has found that the $I(V)$ characteristics can be expressed as follows:

$$I(E) = \sigma E = \sigma_a E + \sigma_b P(E) E \qquad (6\text{-}41)$$

with:

$$P(E) = (1 - E_T/E) \exp-(E_0/E) \qquad (6\text{-}42)$$

where σ_a is the Ohmic conductivity when $E \to 0$, and $\sigma_a + \sigma_b$ is the conductivity limit for $E \to \infty$. The activation field, E_0, can be expressed in terms of the threshold field such as:

$$E_0 = k E_c \qquad (6\text{-}43)$$

The theoretical expression (6-41) for the conductivity, σ, is well followed for all the compounds studied in Section 5 as shown in Figure 25. Experimental data are fitted to Equation (6-41) with two adjustable parameters: the saturation value of the extra-conductivity, σ_b, and the parameter k. Typically $k = 1$ for the upper CDW transition [320] and $k = 2$ for the lower CDW transition [251] in $NbSe_3$, 5 for both forms of TaS_3 [221, 241], 9 for $(NbSe_4)_{10}I_3$ [99] and 5 for $(TaSe_4)_2I$ [97]. The value and the temperature dependence of k will be discussed in Section 8. For the upper CDW in $NbSe_3$, σ has been found to follow (6-41) up to 50 E_c for temperatures higher than approximately 100 K [320]. It can be noted that dE/dI derived from (6-41) has a small discontinuity at E_c as is shown in Figure 39. Similarly to the classical model, the discontinuity results from a single-domain behaviour which is suppressed if a distribution of domains is taken into account.

(ii) Response to an ac Field – Scaling

The CDW response to an ac voltage has also been treated quantically [297, 298]. The photon-assisted tunneling theory that Tucker [321] has developed for quasi-particle tunneling in a superconductor–insulator–superconductor junction has been applied to the CDW problem by Bardeen. In the presence of the voltage, $V = V_0 + V_1 \cos \omega t$, the response is:

$$I = \sum_{n=-\infty}^{+\infty} J_n^2(V_1/\alpha\omega) I(V_0 + n\alpha\omega) \tag{6-44}$$

where $I(V)$ is the nonlinear characteristic (6-41) and J_n the Bessel function of order n. For a small value of V_1 one gets:

$$I_1(\omega, V_0) = (V_1/2\alpha\omega)[I(V_0 + \alpha\omega) - I(V_0 - \alpha\omega)] \tag{6-45}$$

With zero dc bias, $V_0 = 0$, and,

$$I_1(\omega) = (V_1/\alpha\omega) I(\alpha\omega) \tag{6-46}$$

Therefore $\sigma(\omega)$ and $\sigma(E)$ can be scaled by the relation

$$V = \alpha\omega \tag{6-47}$$

The scaling parameter, α, is deduced from (6-40) as follows:

$$\alpha = \frac{\hbar}{e^*} \frac{l}{L} = \frac{\hbar}{e} \frac{M^*}{m^*} \frac{l}{L} \tag{6-48}$$

where l is the length of the sample and L the Lee–Rice domain length.

Figure 41(a) shows the variation of the nonlinear conductivity, Re $\sigma(\omega) - \sigma_a$ and $\sigma(E) - \sigma_{a'}$ (normalized to the Ohmic room temperature conductivity) of an orthorhombic TaS_3 sample at $T = 170$ K as a function of the frequency and of the dc voltage, respectively [255]. The scaling between $\sigma(\omega) - \sigma_a$ and $\sigma(E) - \sigma_a$ is only achieved far beyond the threshold. At low frequencies Re $\sigma(\omega) > \sigma(E)$. The insert of Figure 41(a) shows

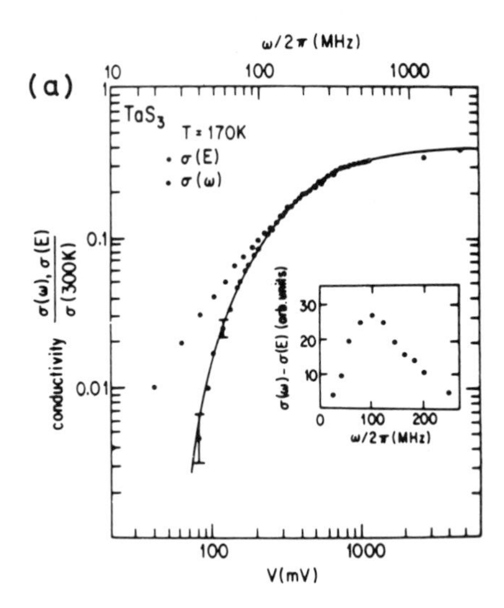

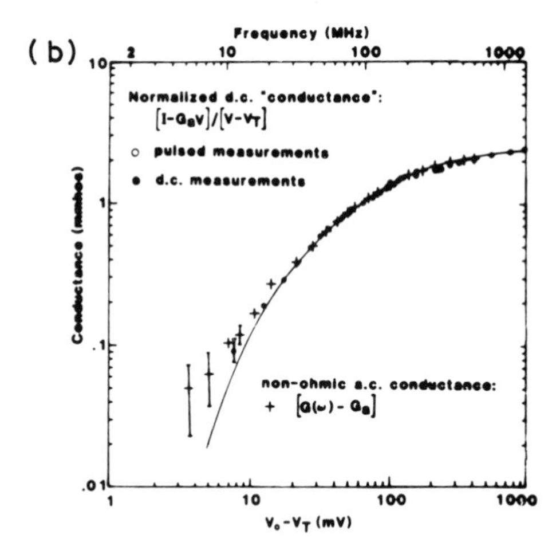

Fig. 41. (a) Variation of the non-Ohmic conductivity Re $\sigma(\omega) - \sigma_a$ and $\sigma(E) - \sigma_a$ normalized to room temperature conductivity as a function of frequency (upper scale) and electric field (lower scale), respectively, for an orthorhombic TaS_3 sample at $T = 170$ K. The insert shows the variation of Re $\sigma(\omega) - \sigma(E)$ as a function of frequency (after [255]). (b) Variation of the non-Ohmic ac conductance $G(\omega) - G_a$ and of the non-Ohmic dc conductance $I - G_a V/V - V_c$ as a function of frequency (upper scale) and the voltage $V - V_c$ (lower scale) for an orthorhombic TaS_3 sample at $T = 190$ K (after [287]).

the variation of Re $\sigma(\omega) - \sigma(E)$ as a function of frequency. It has been thought that an extra contribution to the frequency-dependent conductivity may take place in the vicinity of the threshold. A similar discrepancy in the scaling near E_c has also been found in $NbSe_3$ [251]. In terms of conductance, the extra conductance can be obtained from (6-41) such as:

$$G = \frac{I - G_a V_0}{V_0} \tag{6-49}$$

where $G_a = \sigma_a/l$ is the Ohmic conductance and V_0 is the applied voltage. Miller *et al.* [287] have shown that the scaling is improved if the ac conductance $G(\omega) - G_a$ is not compared to (6-49) but to

$$G' = \frac{I - G_a V_0}{V_0 - V_c} \tag{6-50}$$

Therefore, as far as the response to a small amplitude ac excitation is concerned, the effective dc voltage would not be the total dc bias, V_0, but the voltage $V_0' = V_0 - V_c$. Figure 41(b) shows the variation of G' and $G(\omega) - G_a$ for an orthorhombic TaS_3 sample at $T = 190$ K as a function of V_0' and the frequency, respectively [287]. Comparison between Figure 40(a) and Figure 40(b) clearly shows the improvement in scaling provided by (6-50). An excellent relationship between G' and $G(\omega) - G_a$ is found over two orders of magnitude in both field and frequency. The scaling parameter of the given sample, the data of which are shown in Figure 40(b), is found to be $\alpha = 0.7$ mV/MHz. This value has to be compared to $\hbar/e = 4 \times 10^{-6}$ mV/MHz for single-particle tunneling. It has to be noted that the determination of the tunneling distance, L, from the value of this parameter, α, (6-48) needs some assumption on the value of the Fröhlich mass, M^*.

(iii) Ac + dc Response

The rectification dc current resulting from the superimposition of an ac voltage $V_1 \cos \omega t$ on a dc bias V_0 has been calculated by Tucker [321] as follows:

$$\Delta I_{ac} = \frac{V_1^2}{(2\alpha\omega)^2} \left[I(V_0 + \alpha\omega) - 2I(V_0) + I(V_0 - \alpha\omega) \right] \tag{6-51}$$

(6-51) shows that a dc current appears, even if V_0 is lower than the threshold, provided that:

$$V_0 + \alpha\omega > V_c \tag{6-52}$$

This photon-assisted tunneling, essentially a quantum phenomenon, has not been observed in experiments on $NbSe_3$ and TaS_3 with large excursions of V_0 and ω [251, 255], but a rectification current has been obtained for any $V_0 > V_c$ [251, 287]. These experimental results can be understood if the nonlinear $I(V)$ characteristic is modified following the suggestions of Miller *et al.* [287]. Figure 42 shows a schema of the variation of the nonlinear current, or I_{CDW}, as a function of the experimentally measured applied voltage (curve a): I_{CDW} is zero up to the threshold $\pm V_c$. Miller *et al.* assume that, in the ac response, the part of the nonlinear characteristic $I(V)$ between $-V_c$ and $+V_c$ collapses to zero because fields less than $|V_c|$ are thought to be ineffective. It was shown above that this assumption has strongly improved the scaling between frequency-dependent

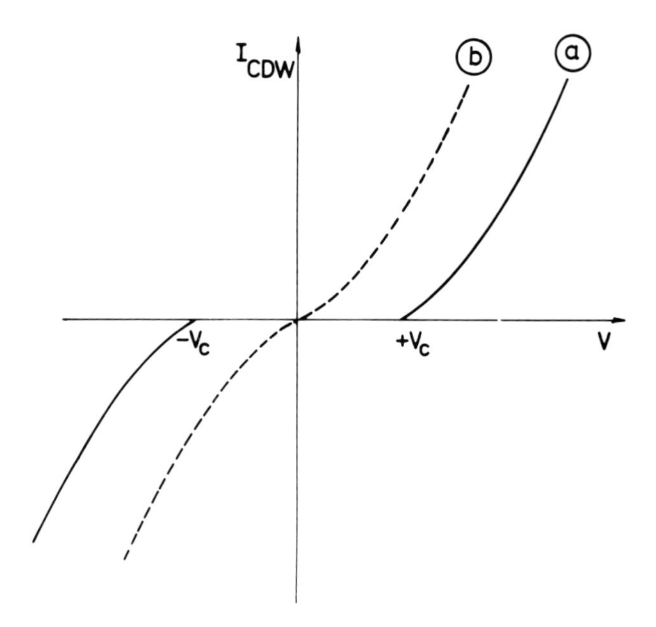

Fig. 42. Schematic variation of the nonlinear $I_{CDW}(V)$ characteristics. (a) In the dc measurements; V_c is the threshold field. (b) In the photon-assisted tunneling theory; in the ac response, all voltages below threshold are equivalent to zero bias.

and field-dependent conductivity. The modified characteristic $I'(V_0')$ (curve b in Figure 42) is such as:

$$\begin{aligned} &\text{for} \quad |V| < V_c \qquad V_0' = 0 \\ &\text{and} \quad |V| > V_c \qquad V_0' = \text{sign}(V_0)(|V_0| - V_c) \end{aligned} \tag{6-53}$$

The $I'(V')$ thus has inversion symmetry which cannot be broken by a dc bias as long as $|V| < V_c$. Therefore there is no possibility for any rectification as long as $V_0' = 0$.

The ac conductivity in the presence of a dc field can be derived from (6-44). With the same assumptions as in (6-53) it is expected that the ac conductivity is independent of the V_0 bias for any $|V_0| < V_c$. Recent bias-dependent ac conductivity measurements performed on orthorhombic TaS_3 verify these calculations [282]. Finally, the dc conductivity induced by a large amplitude ac field has also been calculated from (6-44) [322].

(iv) Harmonic Mixing

When a voltage $V_0 + V_1 \cos(\omega_1 t + \phi) + V_2 \cos \omega_2 t$ is applied to a sample the $I(V)$ characteristic of which is nonlinear, a quasi-dc current, $\Delta I \cos(\omega_0 t + 2\phi - \psi)$ is generated by harmonic mixing, ω_0 being $\omega_2 - 2\omega_1 \simeq 0$. In the case of $V_1, V_2 \ll V_c$ the tunneling model [287] leads to:

$$\Delta I = \frac{1}{16} V_1^2 V_2 \frac{[I(V_0 + 2\alpha\omega) - 2I(V_0 + \alpha\omega) + 2I(V_0 - \alpha\omega) - I(V_0 - 2\alpha\omega)]}{(\alpha\omega)^3} \tag{6-54}$$

If $I(V)$ is modified to $I'(V'_0)$ following (6-53), then a dc current is expected to be detected for $V < V_c$ resulting from a non-zero summation of $I(2\alpha\omega) - I(-2\alpha\omega)$. As has already been mentioned in Section 5.4.3, no quasi-dc current has been measured in $NbSe_3$ [279] and TaS_3 [287] for any $V < V_c$ when $\omega_0 \sim 1$ kHz and ω_1 varied between 1 and 250 MHz; a current at ω_0 appears only when ω_2 is not equal to $2\omega_1$ (a pseudo-harmonic mixing situation). Above V_c, a quasi-dc mixing signal has been measured with the absence of internal phase shift, ψ, when ω_1 is varied. In Section 5.4.3 it was mentioned that this phase shift, ψ, is related to the characteristic time constant associated with the nonlinear mechanism. A finite ψ may appear if time constants are of the order of the inverse angular frequency of the rf applied signal. The tunneling mechanism is assumed to be instantaneous and, consequently, the tunneling model predicts $\psi = 0$ at all applied frequencies. Therefore, the absence of an internal phase shift has been thought to lend experimental support to the tunneling model of depinning. Wonnenberger, however, has shown that the classical model also leads to $\psi = 0$ when the applied voltage $V > V_c$ [313].

6.2.3. *Josephson-Type Oscillations*

A microscopic approach of the pinning potential in CDW systems was proposed by Barnes and Zawadowski [322a] by analogy with the Josephson effect. Again, the CDW is thought to be rigid and the sample is made up of Fukuyama–Lee–Rice domains. The CDW is described as a superposition of two macroscopic quantum states characterized by $\pm Q = 2\pi/\lambda_{CDW}$. These states are degenerate when the CDW is at rest and they are formed by electron–hole pairs with total momentum $\pm Q$. In Josephson junctions, Cooper pairs undergo tunneling through a barrier and an oscillating behaviour, the ac Josephson effect, is obtained when a voltage is applied to the junction. In the CDW problem, two subsequent impurity scatterings transfer one electron–hole state with momentum $+Q$ into an electron–hole state with momentum $-Q$ when the degeneracy of the state is lifted, resulting from the motion of the CDW with a velocity, v_s. These two impurity scatterings take place, locally, in a volume ξ^3, where ξ is the amplitude coherence length of the CDW. This process leads to a pinning energy in $\cos 2\phi$. This result can also be obtained from a second order perturbation theory. Naturally, the first order term also exists and Barnes and Zawadowski find the same pinning energy as Lee and Rice (see Section 6.1). Consequently, the pinning energy consists of two terms: a 'Josephson' term in $\cos 2\phi$ and a 'classical' term in $\cos \phi$. The CDW equation of motion is treated classically as in Section 6.2.1. Therefore, the ac voltage produced by the CDW motion has two periodicities: one at the frequency, ν, (and harmonics) and the other at the frequency, 2ν, (and harmonics). If in some temperature range both processes have similar strength, one expects that even harmonics detected in Fourier noise analysis will have larger amplitude than odd harmonics.

6.2.4. *Coupling Between Domains*

The models described in the preceding sections concern the CDW equation of motion in a unique domain. The divergence at E_c in both classical and tunneling models is suppressed when a collection of independent domains with a distribution of threshold fields is considered. If the correlation between domains is neglected, the current at the electrodes is the sum of currents, each one reflecting its own frequency at the given electric field with arbitrary phases the result of which is equivalent to a noise with a spectrum centered at the frequency associated with the mean value of the probability distribution and a width reflecting its standard deviation. Well-defined discrete frequencies, however, are detected through the voltage leads. In order to explain these frequencies in a multidomain sample, interaction between domains has to be introduced. It was shown in [258] that self-synchronization between many domains may occur on their own rf field even in the absence of applied external ac current. However, the mechanism of coupling between domains which leads to a rigidity of the CDW in motion has not as yet been studied. The equation for the conservation of current at the domain walls is not known and electrostatic interaction between charges built at the domain walls have probably to be introduced.

6.3. Classical Motion of a Deformable CDW

Dynamics of the CDW have been studied in the frame of the Fukuyama–Lee–Rice model (see Section 6.1) in which the CDW is deformable to minimize its interaction with impurities. The forces which act on the CDW are: elastic restoring forces, pinning forces and electric forces. A microscopic calculation of this problem has been made by Klemm and Schrieffer [302].

The overdamped equation of motion has been studied by treating the impurity potential as a perturbation; this assumption is valid when the velocity of the CDW is large, i.e., when the local distorsions of the CDW are small. In this approximation, Sneddon *et al.* [299] have calculated the asymptotic behaviour of the conductivity. The deviation from the limit $E \to \infty$ follows the law:

$$j = \sigma_{E \to \infty} E - C\sqrt{E} \tag{6-55}$$

When an ac field is superposed on a large dc current, the response shows interferences in the $I(V)$ or dI/dV characteristics, which has been observed experimentally. An extension of the same model in the hydrodynamic approximation (low frequencies and long wavelengths) has been made by Sneddon [301]. He shows that, for an *incommensurate* CDW, there is no ac voltage generated by a dc excitation in the infinite volume limit for any order in perturbation. Furthermore, he treats the problem of charge accumulation caused by the CDW distortions and their screening by the remaining normal electrons. For compounds in which the conductivity strongly decreases below the Peierls transition, such as TaS_3 and $(MSe_4)_nI$, this screening is strongly reduced which leads to long-range Coulomb interaction of the CDW with itself. The coefficient C in (6-55) is thus expected to increase rapidly when T is reduced below T_P. Consequently, the conductivity will follow (6-55) over a relatively large range of E only in the vicinity of T_P. In the case in which pinning is provided by commensurability between the CDW and the lattice, Sneddon [301] finds that:

$$j = \sigma_{E \to \infty} E - (C'/E) \tag{6-56}$$

with C' being reduced when T decreases below T_P.

The same problem has also been solved in the vicinity of the threshold by Fisher [300] in a regime in which the CDW distortions are quite large. He introduces a time-dependent mean field thermodynamic potential $\Phi(\phi_j, t)$, a function of the local phase, ϕ_j, of the CDW at site, j, and of time. In the static configuration, $E < E_c$, the potential Φ is multivalued with many minima and maxima, which leads to metastable states. When $E > E_c$ there is a unique state with long range order and an average phase $\overline{\Phi} = vt$. Fisher considers the CDW depinning in the frame of critical phenomena. He finds that, above E_c, the velocity of the CDW follows the law:

$$v \sim (E - E_c)^{3/2} \tag{6-57}$$

Naturally, for $E \gg E_c$, v is proportional to E. The power law coefficient, 3/2, in (6-57) is the consequence of collective pinning. If the number of impurities decreases and becomes a finite but small number, tending eventually to one, the result for rigid CDW motion, $v \sim (E - E_c)^{1/2}$, is recovered.

Numerical methods have been used by Sokoloff [323], Pietronero and Strässler [324]. When the size of the system increases, it is found that the curvature of $v(E)$ has a definite tendency to become concave upwards, as expected in the thermodynamic limit, and that the singularity at the threshold is confined to a very narrow region. Figure 43

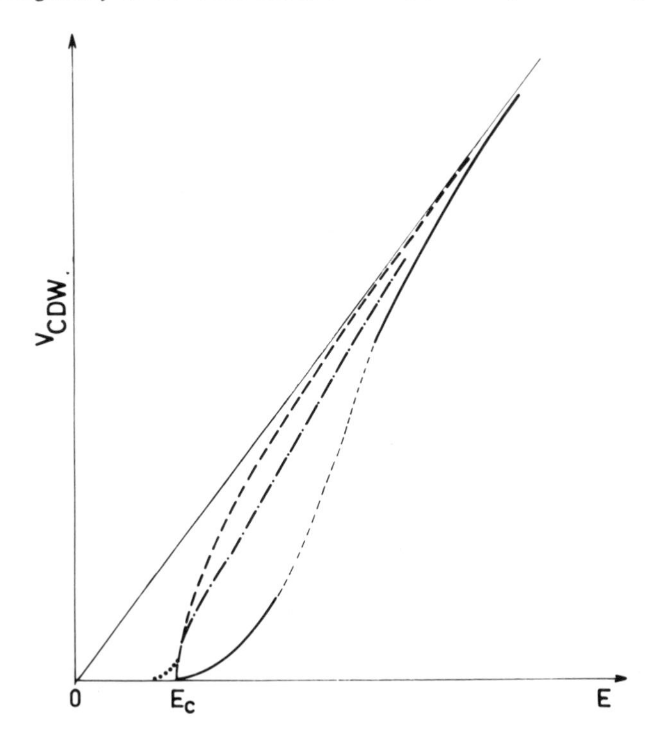

Fig. 43. Schematic variation of the CDW velocity as a function of electric field – – – for a classical rigid single domain motion (Equation 6-20) [238, 296], · · · · for self-synchronized independent domains [258], —— for a deformable motion of the CDW (Equations 6-56 and 6-57) [299, 300], – · – · – for a sample with a finite length [323, 324].

shows a schematic variation of $v(E)$ according to the different models mentioned above, namely, (a) the rigid single domain, (b) self-synchronized independent domains, (c) the collective pinning and (d) the finite size sample.

6.4. INHOMOGENEOUS $\phi(x, t)$ – LOCAL PHASE DEFECT MOTION

Instead of a coherent motion of the CDW phase, $\phi(t)$, independently of the position, another approach ascribes the nonlinear transport properties described in Section 5 to local variations of the phase, $\phi(x, t)$, in motion by the current. These phase defects – discommensurations or solitons created by the electric field – allow a general sliding of the phase without having to overcome the lattice-CDW potential. This mechanism is similar to plastic flow in crystals associated with the motion of dislocations, which give rise to a lengthening without breaking the internal atomic order.

6.4.1. *Discommensurations*

If the wavelength of the CDW distortion is in the vicinity of commensurability of order M, the ground state, instead of being homogeneous along the chain axis, may consist of commensurate regions separated by defects. At the defect site, the phase varies rapidly to insure the average phase at long distances: the phase is compressed or stretched depending on whether the incommensurability is above or below the commensurate value. The coordinate dependence of the phase has the shape of a staircase, each step of which gives a phase increment of $2\pi/M$ and has a charge, $\pm 2e/M$. These defects are called discommensurations or soliton lattice [92, 325]. These discommensurations are topological and constitute the ground state; they have, therefore, to be distinguished from thermally excited solitons in commensurate systems, some properties of which were mentioned in Section 2.3. The distance between discommensurations is a measure of the incommensurability. This distance is given by $l = (\pi/k_F - Md)^{-1}$, where d is the distance between atoms along the chain axis.

This discommensuration lattice – soliton in the case of a single chain, domain wall if many chains are coupled in the transverse directions – can be pinned by impurities. A threshold field is needed to overcome pinning forces. So, Bak [303] has proposed the explanation that the ac voltage in $NbSe_3$ may result from charge impulses at the electrodes provided by the discommensuration motion. This suggestion has subsequently been critized by Rice [326]. The motion of a discommensuration lattice has also been studied by Burkov *et al.* [305, 306]. Above the threshold electric field, new solitons are periodically created at the crystal boundary which then move into the bulk.

On the other hand, Weger and Horovitz [304] consider that, in a system without impurities, the soliton lattice is unpinned and consequently mobile. The local excess charge, thus, contributes to normal conductivity. They ascribe the threshold field to the depinning of the commensurate regions between discommensurations. Weger and Horovitz derive the equation of motion as follows:

$$\frac{1}{\omega_0^2}\frac{\partial^2\phi}{\partial t^2} + \frac{\Gamma}{\omega_0^2}\frac{\partial\phi}{\partial t} - \xi^2\frac{\partial^2\phi}{\partial x^2} + \sin\phi = \frac{E}{E_1} \tag{6-58}$$

Equation (6-58) is of the same type as (2-13) with a damping term added; ξ is the coherence length and E_{c_1} the threshold value for the strictly commensurate case when the density, n_k, of discommensurations is zero. For a finite density n_k, the threshold is given by:

$$E_c = E_{c_1}\left[1 + \left(2\pi n_k \xi \frac{\Gamma}{\omega_0}\right)^2\right]^{1/2} \tag{6-59}$$

According to (6-59), the threshold is expected to decrease when the system becomes commensurate ($n_k = 0$). An explanation of this result, which at first sight appears to be surprising, is given in [304]: the electric force needed to overcome the pinning force acts only on the volume in which the CDW is commensurate; therefore, at a given field, this force is lower in the incommensurate situation than in the commensurate one because of the volume occupied by the mobile discommensurations; consequently, depinning requires a higher field in the incommensurate case. The discommensuration lattice is also expected to yield to broaden the frequency dependence of Re $\sigma(\omega)$ and Im $\sigma(\omega)$ [327]. The commensurability pinning energy had been calculated by LRA [9]. For $M = 4$ this energy is equal to: $\Delta^4/E_F W^2$, with 2Δ: the Peierls gap; E_F: the Fermi energy; and W: the bandwidth. Horovitz and Krumhansl [328], however, have shown that commensurability 4 is particular; in this case, the pinning energy is reduced to the one of an $(M + 2)$ system.

6.4.2. *Creation of Excitations by an Electric Field*

It was mentioned in Section 2.3 that large amplitude excitations may be generated at low temperatures from the pinned CDW condensate. These excitations are called solitons, ϕ particles, kinks .... Their properties are reviewed by K. Maki in Chapter 4, Part I. Solution of Equation (2-13) are solitons with a charge, $\pm 2e/M$ (if the commensurability is of order M), a mass, M_ϕ, and a rest energy $M_\phi C_0^2$. These excitations are separated from the ground state by an energy gap, E_ϕ, such as:

$$E_\phi = 2(2/M)^2 N_0 C_0 \omega_0 \tag{6-60}$$

where $N_0 = n_s M^*/Q^2$ and n_s is the density of condensed electrons. These solitons can be thermally activated and can contribute to the electrical conductivity as in TTF–TCNQ or KCP at low temperatures (see Section 2.3). When an electric field is applied which tilts the periodic pinning potential, quantum tunneling between two minima of this potential may be possible. Such a tunneling creates a soliton–antisoliton pair. For a strictly one-dimensional case, Maki [329] has calculated the conductivity associated with this process. He finds that

$$\sigma_{\text{tun}} = 2\pi(2/M)^2 e^{*2}(C_0/E_\phi)\exp-(E_0/E) \tag{6-61}$$

with

$$E_0 = ME_\phi^2/2\hbar e^* C_0 \tag{6-62}$$

Compared to the thermally-activated conductivity, this tunneling conductivity is expected to operate only at low temperatures. Similarity between soliton–antisoliton pair creation and CDW tunneling has been noted by Bardeen in note 3 of [297b] and also in [319]. According to (6-62) and (2-4) for $M = 4$, (6-62) becomes as follows:

$$E_0 = (32/\pi^2)(M^*/m^*)^{1/2}\,[(\hbar\omega_0)^2/2\hbar e^* v_F] \qquad (6\text{-}63)$$

(6-63) is formally identical to (6-36), giving the activation energy calculated in the Bardeen theory, but modified by (6-39) to take account of damping. However, in the soliton pair creation process, ω_0 is the frequency for small amplitude oscillations in the potential which corresponds to the pinning frequency in the case of undamped motion, whereas in the Bardeen theory, $\hbar\omega_0$ is replaced by the pinning gap (6-40) where ω_p corresponds approximately to the pinning frequency in the damped situation, namely ω_0^2/Γ. Moreover, e^* in the soliton pair creation is the soliton charge, $e/2$, whereas in the tunneling theory $e^* = e(m^*/M^*)$.

Maki has not incorporated the effect of impurities in his model. Interaction between solitons and impurities, however, has been worked out by Larkin and Lee [330]. They have studied tunneling of solitons through an impurity site. But the rest energy of solitons on one chain is found to be much less than kT; therefore, a great number of thermally excited solitons should smear out the tunneling probability. Consequently, Maki [331] has considered that two- or three-dimensional domain walls or bubbles may be created by thermal activation. The sliding CDW is achieved not by the motion of the whole CDW but only by continuously creating and expanding domain walls. For a strictly two-dimensional CDW, Maki finds that the conductivity which appears by this process varies as $\exp(-E_0/E)$. This mechanism of carrying the CDW phase by bubbles is very similar to the generation of dislocations by the analog of Frank–Read sources, as proposed by Lee and Rice [232].

6.4.3. *Phase-Slippage at the Electrodes*

In the theories and experiments described in the preceding sections, boundary conditions have been neglected. But, as soon as the condensate is in motion, there is a boundary at which this condensate has to be transformed in normal electrons collected by the electrodes. The electrical contacts are thought to be equipotentials and, below them, the local electric field is lower than the threshold. Consequently, the boundary at which the CDW stops has been taken by Ong and Verma [260, 332] at some surface where the local electric field equals the threshold one. This boundary condition has recently been studied within a microscopy theory by Gor'kov [333]: at the boundary at which the CDW stops, the accumulation of charges requires elastic energy up to the point where, the CDW losing its stability, the order parameter decreases to zero within the following distance: $\xi \sim \hbar v_F/\Delta$; then the process can start again. However, Ong *et al.* [332] and Gor'kov [333] have independently shown that, instead of having a complete normal sheet, it needs much less energy if phase-slippage centers or vortices are created in the boundary wall (the gain in energy with regard to a complete normal sheet is $\pi\xi/e$ where e is the thickness of the sample). The phase-slippage center has a core of diameter ξ and the phase difference around it is $\pm 2\pi$. These vortices can be seen as edge dislocations in the CDW lattice. They move in a perpendicular direction to the CDW motion; when the CDW phase has advanced by 2π, a vortex collapses at the surface giving rise to a periodic voltage impulse at the electrode. A schematic representation of the phase slippage

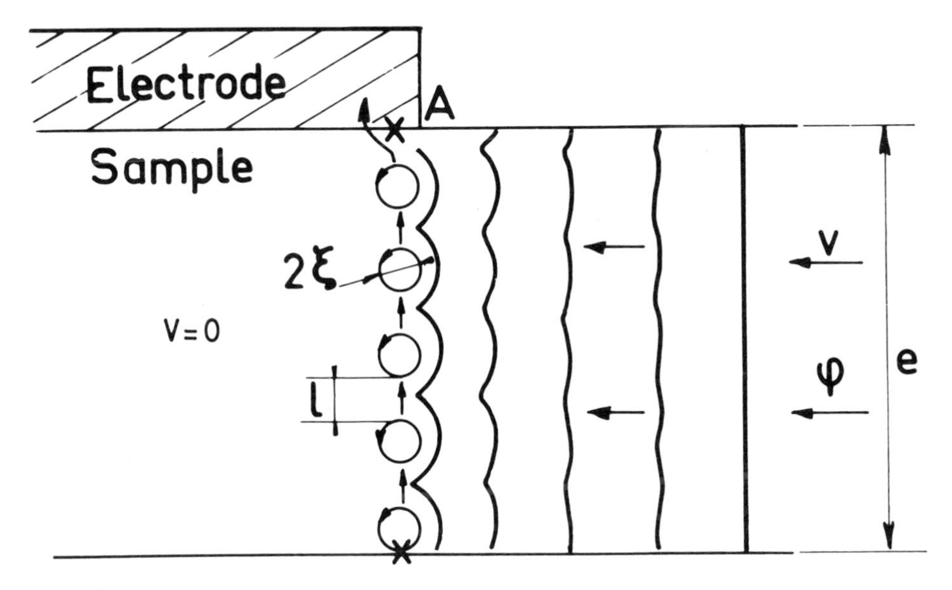

Fig. 44. Schematic representation of the interface at a voltage electrode at which the CDW stops to move. Each time the CDW advances by λ_{CDW} i.e., the CDW phase by 2π a phase-slip defect or vortex disappears at *A*.

sheet is drawn in Figure 44. The velocity of these vortices can be determined by current conservation. If the CDW velocity is v_{CDW} and its period λ_{CDW}, one has:

$$2\pi v_{CDW} = \lambda_{CDW}\Omega \tag{6-64}$$

For the vortices, the relationship between the velocity, v, and the distance, l, between vortices is as follows:

$$2\pi v = l\Omega \tag{6-65}$$

where Ω is the rate at which the phase advances by 2π. The phase-slip annihilation rate is the same as the frequency in the CDW motion. Gor'kov [333] has performed calculations for the vortex creation by using a time-dependent equation for the CDW amplitude similar to Ginsburg-Landau equations for superconductors. He has solved the problem when the CDW is gapless in the case of dirty samples. He shows that the location of the phase-slippage process takes place at a distance X_0 from the interface. X_0 has been numerically estimated to vary as $E^{-\alpha}$ with $\alpha \sim 0.28$ [334]. This intrinsic pinning effect due to contacts can be demonstrated by measurements of the increase of E_c when electrodes are nearer than the distance, X_0. Neglecting impurity bulk pinning, which is predominant in samples with a length $\gg X_0$, Batistic *et al.* [334] have calculated that in a sample with a finite length, L, the threshold, E_c, varies as $L^{-1.23}$. This increase is attributed to the interference between the two phase-slippage sheets which interact strongly when L is reduced below $2X_0$.

In the phase-slippage process, impurities play a minor role and voltage oscillations would be detected in perfectly pure samples. This model can be considered as a hybrid model: for long enough samples, with no interference between phase-slippage sheets at the electrodes, pinning is a bulk effect caused by impurities, which requires a threshold for the CDW motion; the extra conductivity, therefore, takes its origin from the Fröhlich mechanism. But the ac voltage is generated at the electrodes. This periodic voltage the source of which is the vortex creation and the vortex displacement requires well-defined CDW equiphase planes and, therefore, long range order and large transverse coherence.

6.5. Unpinned–pinned transition in incommensurate systems

The Peierls transition is one among other physical problems in which dynamical properties are governed by competition between two periodicities: the lattice periodicity and the CDW periodicity. One phenomenological model for such a study is the one-dimensional discrete Frenkel–Kontorova model which consists of atoms or balls connected by springs interacting with a sinusoidal potential. The potential energy of such an assembly is [335, 337]:

$$\phi\{u_i\} = \sum_i \left\{ \tfrac{1}{2} K(u_{i+1} - u_i - b)^2 + \tfrac{1}{2}\lambda[1 - \cos(\pi u_i/a)] \right\} \tag{6-66}$$

where K is the spring constant, u_i the position of the ith ball, b the unstretched spring length, λ the amplitude of the potential with a period $2a$. The mean distance, l, between balls:

$$l = \lim_{N - N' \to \infty} (u_N - u_{N'})/(N - N') \tag{6-67}$$

is determined by boundary conditions in order to fix the commensurability ratio, $W = l/2a$, which is the ratio between the two competing periods. The ground state of such a system is the solution of the equation: $\partial\phi/\partial u_i = 0$, i.e.

$$2u_i - u_{i+1} - u_{i-1} + (\lambda\pi/2a)\sin(\pi u_i/a) = 0 \tag{6-68}$$

The continuum limit of (6-68) is identical to the sine-Gordon model already described above. Solutions have principally been studied by Aubry [335, 336] also by Sacco and Sokoloff [338] and by Coppersmith and Fisher [337]. Without going into details outside the scope of this review, one can say that Aubry has numerically found that, if W is an irrational number, a transition between two incommensurate states takes place between an unpinned solution and a pinned solution by increasing the interaction strength between the two periodicities. In this case, the pinning is intrinsic and is the result of the lattice discreteness, the impurities playing no role. This transition has been called by Aubry, transition by 'breaking of analycity'. For weak values of the parameter λ, the two periodicities are unlocked and can move freely, one in relation to the other. A critical value, $\lambda = \lambda_c$, exists above which the CDW–PSD is locked to the lattice. This transition between an unpinned and a pinned phase has been connected to critical phenomena in dynamical systems [339]. Physical quantities such as the gap in the phonon spectrum, the coherence length of the ground state, the Peierls–Nabarro barrier (which is the smallest energy barrier to be overcome for continuously translating the chains on the periodic potential) and the depinning forces beyond which the chain can slide indefinitely have been shown to become critical when $\lambda \to \lambda_c$ from upper values [335].

The same transition has been found in other models as Peierls chains [340]. In this case the competition between periodicities is between the Fermi wave vector, $2k_F$, determined by the electron concentration per chain and the lattice periodicity. The strength of the interaction is the effective electron–phonon coupling, λ. If $\lambda < \lambda_c$, there is no impurity pinning and the phase of the CDW can slide freely (Fröhlich mode). But when $\lambda > \lambda_c$, although incommensurable, the CDW is pinned by the lattice and the electrons become exponentially localized. The system becomes insulating and can be represented as formed by equidistant localized polarons. The critical exponents are similar to those obtained in the Frenkel–Kontorova model. Physically, it can be said that, for $\lambda < \lambda_c$, the position of each atom continuously follows any infinitesimal change of the CDW phase. On the contrary, for $\lambda > \lambda_c$, for any infinitesimal change of the CDW phase the position of the atoms changes *discontinuously* after a jump beyond finite energy barriers.

These theories clearly show effects due to the lattice having been overlooked in the other theories. But, in order to account for CDW transport, λ should be approximately λ_c. The lattice properties are strongly temperature-dependent which may lead to a temperature dependence for the threshold field at low temperatures.

7. Analogy Between CDW Motion and Other Nonlinear Physical Systems

7.1. Analogy with Relaxation Oscillators

Weger *et al.* [261] have proposed to simulate the nonlinear transport properties in CDW systems by an electrical relaxation oscillator. An external driving voltage, V_{dc}, is applied to an electrical circuit formed with an RC element and a gas discharge tube. The breakdown voltage of this discharge tube is V_b and the cut-off voltage at which it ceases to conduct is V_c. When V_{dc} is increased from below V_c, no current flows until V_{dc} becomes higher than V_b. At this voltage, the relaxation oscillator discharges and V drops abruptly to V_c, then the charging of the capacitance starts again up to V_b and so on . . . thereby creating an oscillating voltage.

7.2. Analogy with Josephson Junction and Damped Pendulum

As has already been mentioned, similarity between CDW transport and some properties of Josephson junctions has been noted [272, 277, 278]. Rice *et al.* [9] have proposed the phenomenological equation (2-8) for the centre of mass of the CDW; (2-8) with (6-15) can be written as follows:

$$\frac{\mathrm{d}^2 x}{\mathrm{d}t^2} + \Gamma \frac{\mathrm{d}x}{\mathrm{d}t} + \frac{\omega_0^2}{Q} \sin Qx = \frac{eE}{M^*} \tag{7-1}$$

With the variable change: $\phi = Qx$, (7-1) becomes:

$$\frac{d^2\phi}{dt^2} + \frac{\Gamma}{\omega_0}\frac{d\phi}{dt} + \sin\phi = \frac{E}{E_c} \tag{7-2}$$

where the threshold, E_c, is given in (6-16) and the time is in units of ω_0^{-1}.

The equation of motion of a resistively-shunted Josephson junction (RSJ) is as follows [274]:

$$C\frac{dV}{dt} + \frac{V}{R} + I_J \sin\phi = I \tag{7-3}$$

with R, C the resistance and the capacitance, I_J the critical current of the junction, ϕ the phase difference across the junction, I the applied current. The time-dependence of ϕ is given by the Josephson relation:

$$\frac{d\phi}{dt} = \frac{2eV}{\hbar} \tag{7-4}$$

With a change of variable, (7-3) can be written as follows:

$$\frac{d^2\phi}{dt^2} + G\frac{d\phi}{dt} + \sin\phi = \frac{I}{I_J} \tag{7-5}$$

with $G = (\omega_J RC)^{-1}$ and the time in units of ω_J^{-1} ($\omega_J^2 = 2eI_J/\hbar G$).

Another similar nonlinear system is the forced pendulum. Its equation of motion is:

$$a\frac{d^2\theta}{dt^2} + b\frac{d\theta}{dt} + c\sin\theta = \Gamma(t) \tag{7-6}$$

with a: the inertia momentum; b: the damping constant; $c \sin\theta$: the restoring torque; and $\Gamma(t)$: the driving torque. With dimensionless variables (7-6) becomes:

$$\frac{d^2\theta}{dt^2} + \frac{1}{Q}\frac{d\theta}{dt} + \sin\theta = \gamma(\tau) \tag{7-7}$$

with $Q^2 = ac/b^2$ and the time in units of b/c. An electrical analog circuit has been proposed to simulate Equations (7-5) and (7-7) [326, 341].

Equations (7-2), (7-5) and (7-7) are formally equivalent: the applied current in Josephson junctions is equivalent to the electric field in CDW transport and to the torque in the driven pendulum. Consequently, it is not surprising to find the same phenomena in these different problems, especially interferences between the natural oscillation frequency and the frequency of the driving force. These interferences result from the similar nonlinear equation of motion.

The Josephson radiation frequency from (7-4) is:

$$\nu = \frac{1}{2\pi}\frac{2e}{\hbar}V \tag{7-8}$$

with V: the dc voltage bias. This frequency is 483.6 MHz/μV and its value is obtained only from universal constants. For CDW systems it can be deduced from (6-12) that the frequency radiated by the CDW motion is:

$$\nu = \frac{1}{ne\lambda} J_{\mathrm{CDW}} \tag{7-9}$$

with J_{CDW} the current density carried by the CDW, ne the total electron density condensed below the Peierls gap, and λ a periodicity the value of which will be discussed in Section 8.7. For the nonlinear CDW systems, ν is found to be between 25 and 50 kHz cm^2/A, but in contrast with (7-8) this frequency contains parameters which are dependent on the sample.

7.3. Analogy with Flux Flow Motion in Type II Superconductors

An analogy can also be made between CDW transport and flux flow in type II superconductors. For superconducting materials with $\kappa = \lambda/\xi$ larger than $\sqrt{2}$, where λ is the screening distance for supercurrent and ξ the BCS [8] coherence length, $\xi = \hbar v_F/\Delta$, the magnetic flux penetrates in the bulk above a critical field, H_{c_1}, in the form of a vortex-lattice. Each vortex has a normal core of diameter, ξ [20]. In a thin film with no demagnetization coefficient the vortex density is $N = B/\phi_0$ where B is the magnetic induction; if the vortices form a triangular lattice $N = 2/\sqrt{3}d^2$ with d the distance between vortices. The vortices are pinned by impurities and a critical current has to be applied to put into motion this vortex-lattice. The motion induces a flux-flow resistivity, ρ_f, found phenomenologically to be:

$$\frac{\rho_f}{\rho_n} = \frac{B}{H_{c_2}} \tag{7-10}$$

with ρ_n the resistivity in the normal state and H_{c_2} the critical field beyond which the superconductivity is suppressed. Already, the problem of the vortex pinning raises questions on pinning potential, equivalence between ac and dc fields, interference between ac and dc fields, commensurability between the vortex lattice and the potential . . . .

It was shown that the vortex lattice above H_{c_1} is a rigid unity because interactions between vortices are large in comparison to bending energy due to pinning centers. In a local region of the sample, the flux lattice finds an equilibrium position to minimize the local energy. In another part, another local arrangement to the pinning centers is found. The size of these crystallites is inversely proportional to the ratio between the pinning energy and the increase of the magnetic energy due to the lack of crystallographic perfection [343]. These crystallites are similar to the Fukuyama, Lee, Rice domains in the CDW case.

Gittleman and Rosenblum have derived a phenomenological equation of motion for the flux lattice in a periodic pinning potential when the superconductor is submitted to an ac field [344]. The pinning potential is assumed to be periodic such as:

$$V = A(1 - \cos 2\pi x/d) \tag{7-11}$$

A is deduced from the equality between the maximum Lorentz force on the flux lines and the pinning force:

$$\frac{2\pi A}{d} = \alpha_c \frac{\phi_0}{cH} \tag{7-12}$$

with ϕ_0 the flux quantum = $\hbar c/2e$ = 2×10^{-7} G cm^{-2}. The equation of motion of a flux tube of this rigid lattice is written as follows:

$$m\dot{x} + \eta\dot{x} + \frac{\alpha_c\phi_0}{cH} \sin\frac{2\pi x}{d} = \frac{J\phi_0}{c} \tag{7-13}$$

where the right term is the Magnus force provided by the applied current, m: the effective mass of the flux tube per unit length and η: the flow viscosity, $\phi_0 c^2/H_{c_2}\rho_n$. Stephen and Bardeen [345] have shown that this viscosity corresponds to the current dissipation in the normal vortex cores when the flux lattice is in motion.

For small displacements, (7-13) can be linearized, $(\alpha_c\phi_0/cH)\sin(2\pi x/d) \simeq kx$. Moreover, if $J = J_0\exp(i\omega t)$, the power absorbed by unit volume is:

$$P(\omega) = \frac{1}{2}\,\mathrm{Re}\left[\frac{J_0^* H}{c}\,\dot{x}_0\right] = \frac{J_0^2\phi_0 H\eta\omega^2}{2c^2\left[\omega^2\eta^2 + (\omega^2 m - k^2)^2\right]} \tag{7-14}$$

If $\omega^2\eta^2 \gg (\omega^2 m - k^2)^2$, the viscous forces dominate and $P = \frac{1}{2}J_0^2\rho_n H/H_{c_2}$ which is just the dc result. Therefore, at high frequencies the electromagnetic properties of a superconductor are those of the ideal mixed state and the pinning can be ignored [346]. The absorbed power (normalized to the infinite frequency limit) is plotted as a function of ω/ω_0 on a logarithmic scale with $\omega_0 = k/\eta$. At ω_0 the absorption reaches half its ideal dc value. This frequency has been called the *depinning frequency* by Shapira and Neuringer [311]. In the insert of Figure 45 it can be seen that $\omega_0/2\pi$ is typically

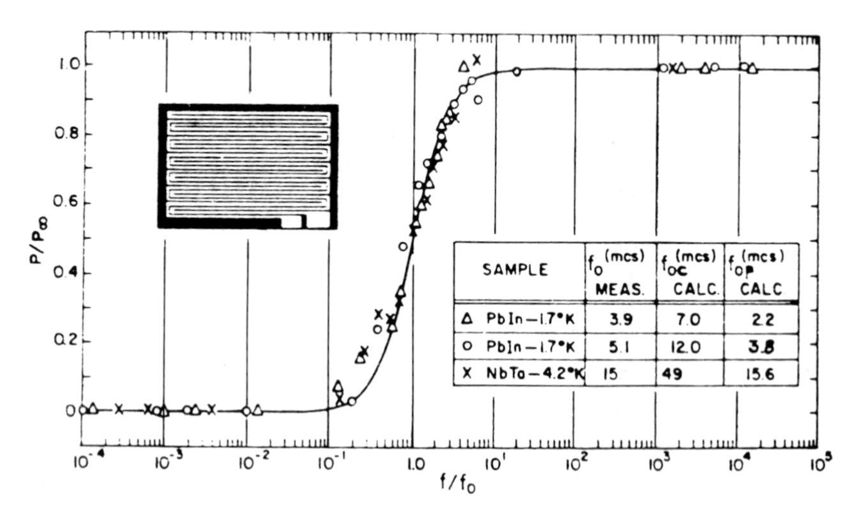

SAMPLE	f_0 (mcs) MEAS.	f_{0C} (mcs) CALC.	f_{0P} (mcs) CALC
Δ PbIn–1.7°K	3.9	7.0	2.2
O PbIn–1.7°K	5.1	12.0	3.8
X NbTa–4.2°K	15	49	15.6

Fig. 45. Power absorption in the mixed state of superconducting samples as a function of frequency [Equation (7-14)]. Pinning frequencies are listed in the table. The insert shows the non-inductive shape of the samples for rf measurements.

10–20 MHz. This frequency, ω_0, is equivalent to ω_0^2/Γ in the CDW problem (see Section 6.2.).

Steps in the dc current-voltage characteristics have been induced in the flux flow state by superimposed rf and dc fields [347]. These steps have been interpreted as quantum interference between the rf external field and the local supercurrent modulation generated by the moving vortex lattice. Interference occurs when the frequency, *mf*, associated with the vortex motion is a multiple of the rf field frequency, *f*, i.e.:

$$nf = m\frac{v}{d} \tag{7-15}$$

The velocity of the vortex is deduced from the relation

$$\mathbf{E} = -\frac{1}{c}\mathbf{v}\Lambda\mathbf{B} \tag{7-16}$$

Therefore, interferences are detected at fields, $E_{nn'}$, such as:

$$E_{nn'} = \frac{n'}{nc} f \left(\frac{2\phi_0 B}{\sqrt{3}}\right)^{1/2} \tag{7-17}$$

Fiory [347] has verified (7-17) with n' up to 8 and n up to 4. Schmid and Hauger [348] have suggested that the coupling between the local oscillations of the vortex lattice and the external rf field is produced by the pinning. Coherent velocity variations generated by the pinning potential modulate the velocity of the vortex lattice (this velocity should be uniform in the absence of pinning) and lead to a net ac supercurrent the existence of which is demonstrated by the ac + dc synchronization technique. With this model an ideal superconductor without pinning should not show interferences. Moreover, because impurities are randomly distributed, no direct ac radiation is expected [348] which has been experimentally confirmed [347].

It is, indeed, the experiment performed by Fiory [347] showing rf induced steps in the *V(I)* superconducting characteristics which inspired us to undertake similar experiments in the CDW transport problem [257]. Then, we interpreted the ac noise generated in the nonlinear state of $NbSe_3$ as resulting from the motion of the CDW in an anharmonic pinning potential, a motion which can be synchronised by an external rf field and lead to steps in the *V(I)* characteristic. This interpretation is exactly the one put forward by Schmid and Hauger [348] in the vortex-lattice motion case. Martinoli *et al.* [349, 350] have studied the vortex motion in superconducting thin films the thickness of which are periodically modulated. Thus the pinning is intentionally made periodic. By adjusting the magnetic induction, *B*, and therefore the lattice vortex parameter, matching or non-matching – i.e., commensurate or incommensurate conditions – between the wavelength of the pinning potential and the inter-vortex distance can be achieved. The critical current for the onset of the vortex motion is found to be higher in the commensurate situation [349]. When a rf field is superimposed, very sharp steps are induced in the *V(I)* curves [350]. Without an rf field, a direct electromagnetic radiation has been measured when the vortex lattice is in motion [351]. This radiation has only been detected in the commensurate situation and its amplitude sharply decreases when the commensurate condition is slightly lost [351]. The maximum amplitude of this rf radiation is several orders of magnitude smaller than the value expected from the phenomenological equation of motion. This result emphasizes the importance of uncorrelated vortex motion caused by random pinning (which exists even in these periodic structures) which reduces the radiated rf power. Interference experiments in the incommensurate and commensurate conditions show some differences which it is worthwhile to note: interferences occur at an electric field proportional to B in the commensurate case but proportional to $B^{1/2}$ in the incommensurate one. Furthermore the interference transitions are broadened in the latter situation because of dynamical excitations of strongly damped deformations in the vortex lattice. It has been shown that the width of the interference transition is proportional to the shear modulus of the vortex lattice [347] calculated from an elasticity theory of this lattice [352, 353]. However, the analogy between CDW motion and vortex lattice motion should not be unduly stressed, especially because of the essential role played by normal vortex cores for current dissipation.

8. Experiments Versus Theories

In this last section, we would like to analyse the experiments described in Section 5 clearly showing the non-linear properties of some CDW systems with regard to the different theories mentioned in the preceding section. Our aim is to try and point out the results which are unquestionable and to make a critical analysis of the controversial ones. In this kind of overall survey the judgment of the reviewer is necessarily somewhat subjective. Therefore some points may appear debatable and might be reinterpreted differently in the future. In the study of CDW transport, specialists use a few keywords. We would like to use a non-exhaustive list of these keywords as a basis for the following discussion. For each keyword we want to summarize what present theories and experiments allow us to establish.

8.1. Periodic Pinning Potential

In the class of models in which the CDW velocity is modulated by a potential, the origin of this potential is of some importance for the understanding of the depinning process. If the CDW is commensurate with the lattice, the pinning is provided by the lattice and the wavelength of the pinning potential is **b**, the distance between atoms along the chain axis. In the case in which the CDW consists of commensurate parts separated by discommensurations, the periodicity involved in the phase motion is the discommensuration lattice parameter (~180 Å in $NbSe_3$ (see Section 4.1.4)). The problem is much more complicated in the case in which the pinning is caused by randomly distributed impurities. For a unique domain and an undeformable phase, it is assumed that the interaction energy between the CDW and impurities is identical when the CDW is displaced by $\lambda_{CDW} = 2\pi/Q$. Therefore, the pinning periodicity is λ_{CDW}. It has been mentioned in Section 6.2.3 that the microscopic mechanism proposed by Barnes and Zawadowski [322a] leads to a pinning periodicity $\lambda_{CDW}/2$. The main problem, however, is the summation of pinning forces over a collection of domains. Sneddon [301] has shown that, in the thermodynamic limit, there is no ac voltage generated in the sample. Then, the ac voltage is thought to result from size effects. It has also to be noted that the amplitude of the ac voltage generated in the CDW motion is much larger than that measured in a similar problem, namely the vortex motion, although the vortex-lattice periodicity is commensurate with the pinning periodicity (see Section 7.3). Therefore, one is led to conclude that the ac voltage may be generated by a source other than the one resulting from the CDW motion in the pinning potential created by impurities. It has been proposed independently by Ong *et al.* [332] and Gork'ov [333] that the ac voltage may be the consequence of boundary conditions: the accumulation of the phase at the electrodes is evacuated by vortices or phase-slippage defects. Annihilation of one vortex occurs when the CDW phase increases by 2π, i.e., when the CDW advances by λ_{CDW}. The periodicity involved in this process, again, is λ_{CDW}. For the following discussion, the pinning potential periodicity is called λ_P; depending on the models, λ_P may have the value of λ_{CDW}, $\lambda_{CDW}/2$, $\lambda_{CDW}/4 = b$ (for a commensurability of 4) or $4b^2\ (\lambda_{CDW} - 4b)^{-1}$ for a discommensurability lattice.

Another mechanism may lead to a periodic interaction between the CDW and impurities. Whereas Fukuyama and Lee [295] have assumed that impurities are static and the CDW deformable, the opposite process can be considered. The charge modulation of the CDW may exert a force on impurities which are shifted in places which minimize this interaction. Impurity migration has been clearly shown to occur below the displacive transition in thiourea [354] but at a temperature approximately of 200 K. Such a mechanism is plausible for $(MSe_4)_nI$ compounds the Peierls temperature of which is in the vicinity of room temperature. But it is difficult to believe, because it is an activated process, that this impurity migration can still take place below 50 K in $NbSe_3$. Numerical calculations have been performed in both situations: a distortion of the CDW to accomodate a rigid lattice of impurities [324] or a distortion of the lattice to accomodate a rigid CDW [355].

8.2. CDW DOMAINS

The CDW is not coherent in phase on the whole sample. The domain structure of the CDW condensate has been revealed in dark field electron microscope imaging (see Section 4.1.3). The domain contours do not look like domain walls in magnetic substances such as ferromagnet. Typical size of a domain is a length of 1–2 μ and a width of 200 Å for $NbSe_3$ [134] and a length of 0.3 μ for orthorhombic TaS_3 [135].

When the pinning is provided by strong pinning impurities the domain length is the average distance between impurities. In the case of weak pinning, Lee and Rice [232] have calculated the domain length, L, the expression of which is given in (6-5); depending on the parameter ϵ, L is currently estimated to have a value between 1 and 100 μm.

It has been mentioned that the energy per electron associated with either the dc electric field or the rf frequency is much smaller than kT. Therefore, to be stable against thermal fluctuations, a domain must contain a large number of electrons. From the phenomenological equation of motion (2-8) and the pinning potential (2-11) with $M = 1$, the maximum pinning potential amplitude can be obtained from the comparison of the quadratic term in the potential with the small amplitude oscillation frequency, ω_0, as follows:

$$V_{\max} = \tfrac{1}{2} M^* \omega_0^2 \left(\frac{\lambda_{\mathrm{CDW}}}{2\pi} \right)^2 \tag{8-1}$$

Bardeen *et al.* [265] have estimated that the barrier amplitude is the same as in (8-1) but with ω_0, the free oscillation frequency in the underdamped harmonic oscillator model, replaced by ω_P defined by the scaling relation (6-40). Then, a domain containing N_c electrons is assumed to be stable if:

$$\tfrac{1}{2} N_c M^* \left(\frac{\lambda_{\mathrm{CDW}}}{2\pi} \right)^2 \omega_P^2 > \tfrac{1}{2} kT \tag{8-2}$$

With $M^* = 10^3 m^*$, $\omega_P = 2\pi \times 3$ MHz, $T = 40$ K, Bardeen *et al.* have calculated N_c to be 4×10^{10} for a given $NbSe_3$ sample, which corresponds to a domain the volume of which is 2×10^{-12} cm^3, or to a domain with a length of 100 μm and a cross-section of 0.4 μm $\times$ 0.4 μm. A much smaller domain size is found if M^* is estimated to be 100 or $20m^*$ (see the following section) and if ω_P in (8-2) is the pinning frequency, ω_0^2/Γ, which typically is $2\pi \times 70$ MHz. Another estimation of the domain volume has been made by Mozurkewich and Grüner [356] in their study of the volume dependence of voltage oscillations. They found a value of approximately 0.2 μm^3, one order of magnitude less than in [265]. Finally it was mentioned in Section 6.2.2 that the scaling parameter α in the CDW tunneling theory is several orders of magnitude the value $\hbar/e$ corresponding to a single particle. This enhancement factor can be considered as revealing the number of electrons (for this given sample, 10^5 [287]) which coherently tunnel. Experiments are currently performed on samples the dimensions of which are typically a length of 1 mm and a cross-section of 1–5 μm $\times$ 1–5 μm. Each sample, therefore, consists at least of 10^4–10^6 domains.

8.3. Long Range Phase Coherence

In a restricted temperature range, when a current pulse is applied on some $NbSe_3$ samples, the voltage oscillation is directly detected on an oscilloscope [265, 266]. Near the threshold, the amplitude of this oscillation may be as large as the average voltage. The sample is thought to act as a single domain. According to (6-18b) and (6-21) it can be seen that, at the threshold, the voltage oscillation generated in a single domain can reach twice the average voltage value. The sharp drop observed in the $V(I)$ characteristics of some $NbSe_3$ samples (see Figure 37) is reminiscent of the one which is calculated in the response of a single domain when the applied current is regulated (Equation (6-23) and Figure 39). It has also been shown that the slope of the rf-induced steps in the dc $V(I)$ characteristics is nearly that corresponding to the Ohmic low-field conductivity (or the peaks in dV/dI approximately reach the Ohmic value (see Figure 33)) which indicates that the whole sample is coherently synchronized by the rf field.

These experiments can be interpreted by assuming that the CDW phase is coherent over the entire volume of the sample. More probably, it can be said that the domains are strongly coupled when the CDW is in motion. This coupling mechanism is not as yet understood but impurities can help to couple domains as it has been shown in the case of $Fe_{0.03}NbSe_3$. Long range phase correlation has been clearly demonstrated in the experiments performed by Ong *et al.* [332]. They have studied the effect of a thermal gradient on the noise spectrum. One end of a $NbSe_3$ sample was heated with $\Delta T \sim 6$ K with regard to the fixed temperature of the other end. With $\Delta T = 0$ a fundamental frequency and harmonics are detected in the Fourier transformed voltage. When $\Delta T \neq 0$ the fundamental frequency keeps the same value without any broadening. Therefore, it seems that the CDW slides at a uniform velocity even in the presence of field or temperature inhomogeneities. Moreover, a new fundamental frequency appears which moves when ΔT is increased. Ong *et al.* ascribe this frequency to the hot end of the sample. Then, they interpret these two fundamental frequencies detected in the presence of a thermal gradient as the evidence that the ac voltage is generated at the electrodes. Each electrode acts as a localized pinning center and the phase-slip rate of the phase is only determined by the local temperature of the electrode. But the assumption that the CDW may slide at two distinct velocities implies a discontinuity in the electric field in some place along the sample in order to conserve the current. The pinning, however, does not need to be localized at the electrodes only. The periodic generation and recombination of the CDW condensate in normal electrons may take place at any domain wall. Electrostatic coupling between charges built at these domain boundaries may lead to a strong coupling between domains and induces a dynamical phase correlation on a distance corresponding to the length of the sample.

8.4. Inertia – Damping

The frequency dependence of the conductivity of $NbSe_3$ and TaS_3 (see Figure 30) shows that the CDW response to a small amplitude rf field is overdamped. Relevant frequencies in the CDW motion can be estimated. One first needs a value for the Fröhlich mass, M^*. Its theoretical expression calculated by LRA [9] was given in (2-3). However, M^* can be derived as follows: when the CDW is in motion, the electronic kinetic energy at the frequency, ω, is $m^*\omega^2\lambda^2$ and the kinetic energy associated with the lattice is $M\omega^2(\Delta b)^2$ where M is the ionic mass, λ: the CDW wavelength and Δb the maximum ionic displacement. The Fröhlich mass corresponding to the electron–phonon mode can be estimated as follows:

$$\frac{M^*}{m^*} = \frac{M(\Delta b)^2 + \lambda^2 m^*}{\lambda^2 m^*} = 1 + \frac{M}{m^*}\left(\frac{\Delta b}{\lambda}\right)^2 \tag{8-3}$$

For $NbSe_3$ $m^* = 0.3m$ [80] and $\lambda \sim 4b$, therefore $M^*/m^* = 1 + 3 \times 10^4(\Delta b/b)^2$. The displacement Δb is unknown but with $\Delta b/b = 1\%$, $M^* = 5m^*$ and $\Delta b/b = 5\%$, which is a large distortion, $M^* \sim 75m^*$. These numbers have to be compared to those recently determined for TTF–TCNQ: $M^* \sim 60m \sim 20m^*$ [23] and $M^* \sim 24m \sim 6m^*$ [24]. The free oscillation frequency, ω_0, in the harmonic oscillator model can be deduced from (6-16); with $M^* \sim 10m$ and $E_c \sim 0.1$ V/cm, one gets $\omega_0 \sim 10$ GHz. The damping frequency, Γ, can be obtained from the conductivity value when the pinning is negligible as expressed in (2-10). In the limit of $E \to \infty$, $\sigma_b = ne^2/\Gamma M^*$. In the case of $NbSe_3$ with $\sigma_b \sim 1000$ mhos (σ at room temperature is 4000 mhos), $M^* \sim 10m$ and $n = 2 \times 10^{21}$ electrons/cm^3 (see Section 4.2.1) $\Gamma \sim 500$ GHz. Consequently, $\omega_0^2/\Gamma \sim 200$ MHz. These estimates of ω_0, Γ, M^* are in keeping with the experimental values obtained from the frequency dependence of $\sigma(\omega)$ in the linear regime (see Section 5.2).

It was shown in Figure 23 that the maximum conductivity in $NbSe_3$, $\sigma_a + \sigma_b$, in the limit of $E \to \infty$ or $\omega \gg \omega_0^2/\Gamma$ is the same as if the CDW had not existed. This result has been explained by Bardeen [297c], Gork'ov and Dolgov [357], and Danino and Weger [358]. In the case of orthorhombic TaS_3 samples with the lowest threshold fields, however, the conductivity at saturation, $\sigma_a + \sigma_b$, reaches only approximately half the conductivity value at room temperature. The Fröhlich conductivity is very sensitive to impurities. It was mentioned in Section 5.1.2 that σ_b is strongly reduced in doped or irradiated samples which corresponds to an increase of the damping frequency, Γ.

Tessema and Ong [291], however, have shown that the $I_{CDW} - V$ curves measured at frequencies between 0.1 and 5 MHz have a strong inductive behaviour. Therefore, in the nonlinear state the CDW response is dominated by an inertia term the presence of which is not revealed in the linear $\sigma(\omega)$ conductivity measurements. Tessema and Ong attribute this inertia term to metastable states involved in the CDW depinning. It has to be noted, however, that the $I - V$ phase shift has been measured capacitively rather than inductively in the case of $K_{0.30}MoO_3$ [290]. In order to explain the large amplitude voltage oscillations in the vicinity of the threshold Bardeen *et al.* [265] have also introduced inertia effects in the CDW motion. They assume that the total energy, kinetic plus potential, of the CDW is constant and, thus, that the CDW slides without dissipation over peaks and valleys of the pinning potential. With this assumption they find that, at high electric field, the amplitude of the current density oscillations, ΔJ_{CDW}, is inversely proportional to the current density carried by the CDW, J_{CDW}, as follows:

$$\Delta J_{CDW} \times J_{CDW} = -\frac{V(\phi)}{M^* n^2 e^2} \tag{8-4}$$

where $V(\phi)$ is the pinning potential. Equation (8-4) predicts that the current oscillation amplitude is larger if the pinning potential is stronger. This result is a direct consequence of the absence of the damping term which seems unrealistic for impure samples, as mentioned above. Although experimental results reported in [265] are in keeping with (8-4), further experiments [356] show that, for $E > 2E_c$, ΔJ_{CDW} is constant when J_{CDW} increases.

8.5. SIZE EFFECTS

8.5.1. *Threshold Field as a Function of the Distance Between Electrodes*

It has been mentioned in Section 6.1 that the threshold field, E_c, is an extensive quantity only if the number of domains in the given sample is large. Size effects on E_c are expected to occur when the sample consists of one or a few Fukuyama–Lee–Rice domains. Consequently, measurements E_c as a function of the distance, l, between electrodes were thought to lead to an experimental determination of the domain size. E_c has been shown to increase when l is smaller than approximately 100 μm [189, 262, 253, 359]. The voltage across the sample if found to follow the relation:

$$V = V_0 + E_c l \tag{8-5}$$

where V_0 is typically 1 mV [189]. This potential, V_0, can be interpreted in the frame of the vortex model of Ong *et al.* [332] and Gor'kov [333] (see Section 6.4.3). The power dissipated at the electrode to create a normal sheet with a thickness $\xi = \hbar v_F/\Delta$ is $P = E_S A v$ where E_S is the condensation energy per area unit, A the cross-section of the sample, v the rate at which the CDW moves. E_S can be calculated from the condensation energy per volume unit: $E_{\text{cond}} = 1/2\Delta^2/E_F \log 2E_F/\Delta$ [307] such that $E_S = E_{\text{cond}} n\xi$ where n is the electron concentration per unit volume. At this power corresponds a voltage V_0 defined as: $P = V_0 J_{\text{CDW}} A$, therefore V_0 is [189]:

$$V_0 = E_S \frac{v}{J_{\text{CDW}}} \tag{8-5}$$

In the case of $NbSe_3$, with $J_{\text{CDW}}/v = 25$ A/MHz cm² (see Section 8.7), $\Delta = 700$ K (see Table II), and $E_F \sim 300$ K [80] one finds $V_0 = 120$ mV, which is too large to account for the experimental value of 1 mV. However, if the boundary does not consist of a complete normal sheet but of vortices with a diameter ξ, the dissipated power has to be reduced by $\sim\pi\xi/e$ where e is the thickness of the sample. With this assumption, for the given sample in [189] the voltage is calculated to be 0.4 mV per electrode or 0.8 mV for the sample in keeping with the experimental value. Therefore, the intrinsic pinning at the electrodes may account for the increase of E_c in small sized samples. It has, however, to be noted that the anisotropy in electrical conductivity can lead to erroneous results when the inter-electrode distance has a value comparable to the thickness of the sample.

8.5.2. *Volume Dependence of the Voltage Oscillations*

The amplitude of the voltage oscillations has been studied as a function of the volume of the sample. The initiation of this study was the theoretical result found by Sneddon [301] indicating the absence of voltage oscillations in the thermodynamic limit and the experimental results showing an ac voltage in a $(TaSe_4)_2I$ sample with a cross-section of 12 μ^2 [97] whereas no ac voltage was detected in another $(TaSe_4)_2I$ with a cross-section approximately two orders of magnitude larger [98]. Moreover, if the ac voltage has its origin at the electrodes, its amplitude is expected to be independent of the length of the sample. Two contradictory experimental results have been obtained on $NbSe_3$ at the same temperature by Mozurkewich and Grüner [356] and Ong *et al* [332]. The former authors have found that the oscillation amplitude, ΔV_1, measured with a Fourier spectrum analyser increases as the root of the length and decreases as the root of the cross-section, A, of the sample. From ΔV_1, they define an oscillating current density, $\Delta J_1 = \Delta V_1/RA$ where R, proportional to l, is the nonlinear dc resistance for the applied field. Consequently, $\Delta J_1 \propto \Omega^{-1/2}$ and the current oscillation is vanishing in the infinite volume limit. In order to interpret these amplitude measurements, the instrumental bandpass of the spectrum analyser has to be considered: Mozurkewich and Grüner have set the analyser band-width to 100 kHz, a value larger than the width of the frequency peak response, whereas Ong *et al.* have used a resolution of 1 kHz and integrated the area under the fundamental peak of the spectrum. The latter authors have found that the integrated ac voltage is independent of the length of the sample with a variation in the length by a factor 60. The conclusions derived from one of these two results have obviously to be corrected. Finally the broad band noise power has been found proportional to the volume of the sample and, therefore, to be a bulk property [239].

8.6. Classical Description versus Quantum Description

In early papers on the nonlinear properties in $NbSe_3$ [227, 78] the extra conductivity was attributed to Zener tunneling and shown to fit the expression

$$\sigma = \sigma_a + \sigma_b \exp-(E_0/E) \tag{8-6}$$

with E_0 being an activation energy. Fleming and Grimes [231] have proposed a modified expression in which the threshold, E_c, is introduced as follows:

$$\sigma = \sigma_a + \sigma_b \exp-[E_0/(E - E_c)] \tag{8-7}$$

Brill *et al.* have shown that (8-7) fitted their experimental results in $NbSe_3$ for $T < 30$ K [212]. The following empirical functional form:

$$\sigma = \sigma_a + \sigma_b(1 - E_c/E) \exp-[E_0/(E - E_c)] \tag{8-8}$$

has also been shown by Fleming [247] to fit the experimental data for $NbSe_3$ at T = 51 K until $E = 6E_c$ and at T = 125 K until $E = 3E_c$. The variation of σ for both forms of TaS_3, $(TaSe_4)_2I$ and $(NbSe_4)_{10}I_3$ at a given temperature has been fitted in Figure 25 with the theoretical expressions (6-41) and (6-42) found by Bardeen in his tunneling model. The parameters used in these fits are $k = E_0/E_c = 5.7$ and $\sigma_b/\sigma_a = 73.5$ for monoclinic TaS_3, $k = 4.3$ and $\sigma_b/\sigma_a = 41.5$ for orthorhombic TaS_3, $k = 5.6$ and $\sigma_b/\sigma_a = 5.1$ for $(TaSe_4)_2I$, $k = 9.7$ and $\sigma_b/\sigma_a = 31.3$ for $(NbSe_4)_{10}I_3$. σ has been shown to follow (6-41) in $NbSe_3$ for temperatures between 100 K and T_2 until $E = 50E_c$ with $k = 1$ [320]. Because of self-heating problems, measurements of the conductivity are currently performed up to an electric field which corresponds to $\sigma(E)/\sigma_a \sim 0.5\sigma_b/\sigma_a$. Thus, σ_b is a fitting parameter. In this restricted electric field excursion, it is always found a value of the parameter, k, which gives a good fit between experimental data and the Bardeen expression (6-41). However, the parameter $k = E_0/E_c$ is sample-dependent and temperature-dependent. It becomes larger when E_c increases sharply at low temperatures; thus, k is approximately constant and equal to 5 in orthorhombic TaS_3 between 210 K and 130 K in the temperature range in which E_c is only slightly temperature-dependent but its value becomes 7 at T = 80 K when E_c has greatly increased. In the same way, k increases continuously when T is reduced below the Peierls transition temperature of $(NbSe_4)_{10}I_3$ and $(TaSe_4)_2I$, compounds in which it has been shown that E_c varies exponentially with T (see Section 5.1.1); for instance for $(NbSe_4)_{10}I_3$ $k = 5$ at T = 230 K and $k = 9.7$ at T = 160 K.

The increase of the conductivity in the nonlinear state has also been calculated in the frame of classical theories. Portis [360] has suggested that the exponential dependence of the non-ohmic dc conductivity may be accounted for by a Poisson distribution of pinning strength for strong pinning centers. Functional dependence of the conductivity has been derived from an elastic theory of the CDW in two limits:

– near the threshold by Fisher [300] in a theory of critical phenomena, $\sigma \sim (E - E_c)^{3/2}/E$ according to (6-57) and $J = nev$,

– in the limit of very high electric fields ($E/E_c \gg 1$) by Sneddon *et al.* [299] in a perturbation theory,

$$\sigma = \sigma_b - CE^{-1/2} \tag{8-9}$$

Sneddon *et al.* have shown that the same conductivity data (for $NbSe_3$ at $T = 51$ K) that Fleming had fitted with (8-8) [247] can also be fitted with (8-9); the latter fit, however, is made in an electric field range in which their theory seems to be not applicable because the condition $E/E_c \gg 1$ is not obeyed.

The variation of the fundamental frequency in the CDW motion near the threshold as a function of $(E - E_c)$ for an orthorhombic TaS_3 sample is fitted in Figure 46 at

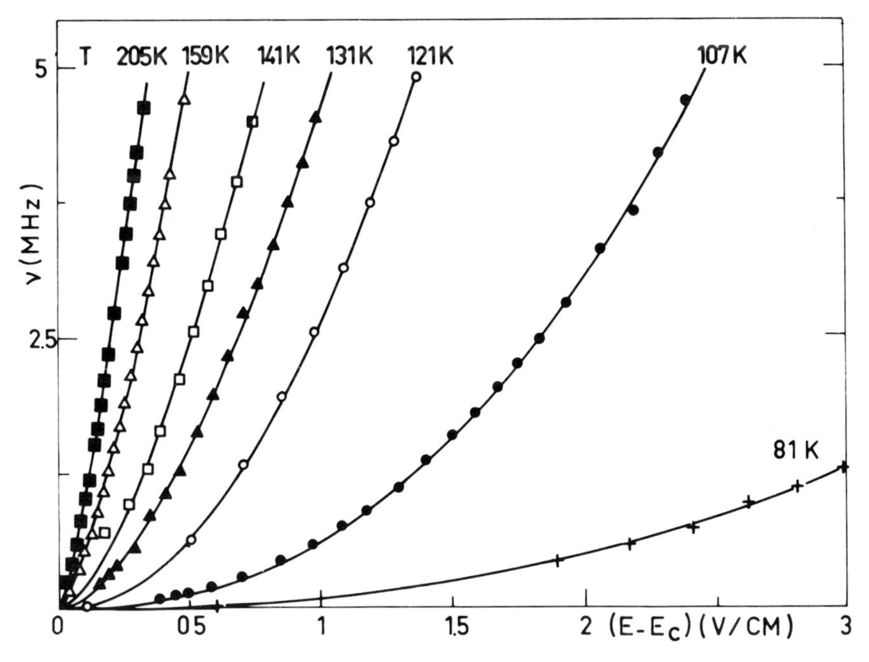

Fig. 46. Variation of the fundamental noise frequency, ν, measured in the Fourier-transformed voltage as a function of $(E - E_c)$ at different temperatures for an orthorhombic TaS_3 sample. The lines are fits to the expression $(E - E_c)^\gamma$. At $T = 205$ K, $\gamma = 1.33$; $T = 159$ K, $\gamma = 1.45$; $T = 141$ K, $\gamma = 1.44$; $T = 131$ K, $\gamma = 1.63$; $T = 121$ K, $\gamma = 2$; $T = 107$ K, $\gamma = 2.3$ and $T = 81$ K, $\gamma = 2.3$.

several temperatures with the law $(E - E_c)^\gamma$. The exponant, γ, is chosen as a fit parameter. It is found that γ corresponds approximately to the value 1.5 proposed in (6-57). γ is temperature-dependent and increases in the same temperature range in which k has also been found to increase. For instance, $\gamma \sim 1.5 \pm 0.2$ in orthorhombic TaS_3 for temperatures between 210 K and 130 K but its value is 2.2 ± 0.2 at $T = 81$ K. For the upper transition in $NbSe_3$ γ is found to be 1.2 ± 0.1.

One can conclude that, in the vicinity of the threshold, the conductivity can be fitted by several functional forms and, consequently, these data alone cannot allow to distinguish between the different theories. The expression of the conductivity given by the tunneling model is valid for the total range of variation in electric field and the agreement with experimental results is quantitatively good. The confirmation of the tunneling model, however, has to be searched for in the joint ac + dc experiments described in Sections 5.4.5 and 6.2.2 [279, 282, 287] the results of which can be compared with the theoretical predictions.

8.7. ELECTRONIC CHARGES CARRIED BY THE CDW IN MOTION

The current carried by the CDW in motion has been derived in (1-5). The CDW velocity can be written as the product of the fundamental frequency of the ac voltage generated in the sample and the pinning periodicity, λ_p, as follows:

$$J_{\mathrm{CDW}}/\nu = ne\lambda_p \tag{8-10}$$

An estimation of λ_p depending on the type of pinning has been given in Section 8.1. The total electron concentration, ne, in the bands affected by the CDW distortion has been evaluated for each compound in Section 4.2.1. In a general way, ne can be expressed as:

$$ne = p\,\frac{2b}{\lambda_{\mathrm{CDW}}}\,\frac{1}{v_{\text{unit cell}}} \tag{8-11}$$

where p is the number of bands affected by the CDW and $v_{\text{unit cell}}$ is the volume of the unit cell. Equation (8-11) is simply derived by considering that the Fermi level is at $Q = 2k_F = 2\pi/\lambda_{\mathrm{CDW}}$ and that, in the absence of the Peierls transition, there are two electrons in the band filled up to $b^* = 2\pi/b$. Thus,

$$\frac{J_{\mathrm{CDW}}}{\nu} = \frac{2ep}{\mathscr{A}}\,\frac{\lambda_{\text{pinning}}}{\lambda_{\mathrm{CDW}}} \tag{8-12}$$

with $\mathscr{A}$ the cross-section of the unit cell. J_{CDW} is directly obtained from the nonlinear $V(I)$ characteristics with the assumption that

$$J_{\mathrm{CDW}} = J(1 - R/R_n) \tag{8-13}$$

with J: the applied current density; R: the resistance for this J value; and R_n: the Ohmic resistance in the linear state below the threshold field. The cross-section of the samples taken for the calculation of J_{CDW} is deduced from the value of the absolute resistivity measured on samples from the same batch. ν is the fundamental frequency which appears the first in the Fourier-transformed voltage at the threshold.

The linear relationship (8-12) between J_{CDW} and ν is, remarkably, followed at temperatures well below the Peierls transition temperature for all the compounds exhibiting nonlinear transport properties, as is shown in Figure 47. The experimental values of J_{CDW}/ν are listed in Table III with, for each compound, the value of p, the Peierls transition temperature and the section of the unit cell.

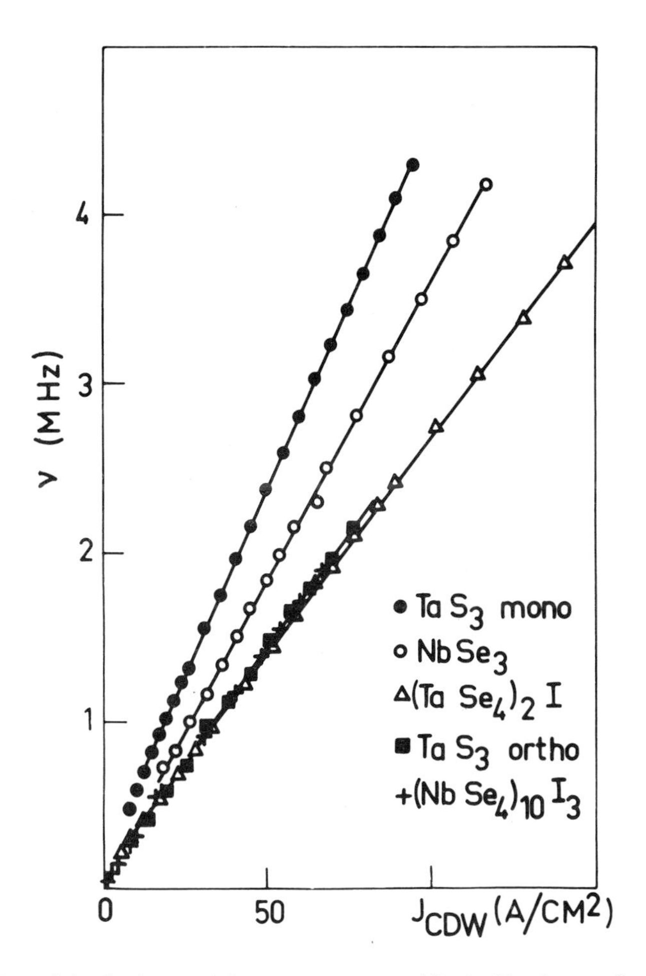

Fig. 47. Variation of the fundamental frequency ν measured in the Fourier-transformed voltage as a function of the current carried by the CDW at a low temperature compared to the Peierls temperature transition for the transition metal tri- and tetrachalcogenides exhibiting CDW transport (measurements at $T = 110$ K for monoclinic TaS_3, $T = 45$ K for $NbSe_3$, $T = 81$ K for orthorhombic TaS_3, $T = 150$ K for $(TaSe_4)_2I$ and $T = 192$ K for $(NbSe_4)_{10}I_3$.

It is found that J_{CDW}/ν has the same value for each independent CDW in $NbSe_3$ or in monoclinic TaS_3. For orthorhombic TaS_3, J_{CDW}/ν below $T_0' = 130$ K is twice the value measured between T_0' and T_0. It was shown in Sections 4.1.1 and 4.1.2 that T_0' is the temperature at which the CDW wavelength along the chains becomes locked to the commensurate value of four lattice distances. The increase of J_{CDW}/ν at T_0' has been explained by a locking between strongly coupled CDWs on two types of chains [221]. From the J_{CDW}/ν values and without adjustable parameters it is found that $\lambda_{pinning}/\lambda_{CDW}$ is nearly 0.5 for all the compounds. This $\lambda_{CDW}/2$ periodicity is that predicted by Barnes and Zawadowski [322a] in their two impurity scattering process. But, as mentioned in Section 6.2.3, the single impurity scattering leads to a λ_{CDW}

TABLE III

Parameters and experimental data for the determination of $\lambda_{pinning}/\lambda_{CDW}$ following (8-12). P is the estimated number of bands affected by each CDW. In orthorhombic TaS_3 for $T < T_0' = 130$ K the CDW wavelength is locked to the commensurate value of four lattice distances along the chain axis.

	Peierls transition temperature (K)	Section of the unit cell (Å^2)	J/ν (A MHz^{-1} cm^{-2})		P	$\lambda_{pinning}/\lambda_{CDW}$
$NbSe_3$	145 59	147.48	25 25	[238]	2 2	0.575 0.575
Orthorhombic TaS_3	210 (130)	558.43	38 20	[117, 120]	12 6	0.55 0.58
Monoclinic TaS_3	240 160	133.33	20 20	[120, 222]	2 2	0.42 0.42
$(TaSe_4)_2I$	262.5	90.84	38	[97]	2	0.54
$(NbSe_4)_{10}I_3$	285	88.51	36.5	[99]	2	0.51

periodicity and both scatterings may be effective together. If the fundamental frequency, however, is associated with the periodicity $\lambda_{CDW}/2$, the periodicity λ_{CDW} would be detected as a subharmonic at $\nu/2$. This subharmonic is rarely visible, which leads us to conclude that the process suggested by Barnes and Zawadowski may be the predominant one.

In the vicinity of the Peierls transition temperature, J_{CDW}/ν has been shown to decrease to zero at T_2 for $NbSe_3$ [265] and at T_0 for orthorhombic TaS_3 [264]. This temperature dependence is attributed to the temperature variation of the electron density, $n(T)$, condensed in the CDW state. The variation of J_{CDW}/ν with temperature is similar to the temperature-dependent CDW order parameter, $\Delta(T)$, evaluated from thermopower measurements or X-rays studies. The measurements reported in [221], however, show that J_{CDW} does not vary linearly with ν in the vicinity of the Peierls transition which may indicate the failure of (8-13) for calculating J_{CDW}. Equation (8-13) is based on the assumption that normal carriers and the CDW condensate are independent. This model with two independent fluids is probably not appropriate near T_P. Experimentally it is found that $\nu(J_{CDW})$ shows a curvature for small J_{CDW} and that for current density in the order of 100 A cm^{-2} the slope between J_{CDW} and ν is the same as that measured at lower temperatures for any J_{CDW}. Experimental data of $\nu(J_{CDW})$ are drawn in Figure 48 for a monoclinic TaS_3 sample at several temperatures below T_2. These measurements indicate that, depending on the CDW velocity, there are different electron concentrations carried along with the CDW. As discussed by Rice *et al.* [361], Weger and Horovitz [304] Boriack and Overhauser [308], one has to distinguish between ρ_c, the fraction of condensed electrons, and $\rho_{eff} > \rho_c$ which includes the normal electrons carried along with the CDW. At low temperatures with regard to the Peierls temperature $\rho_{eff} = \rho_c = \rho$. In the vicinity of the Peierls transition, the charge carried by the CDW is $\rho_c(T)$ and this value decreases to zero at the transition, but at high velocity or for current

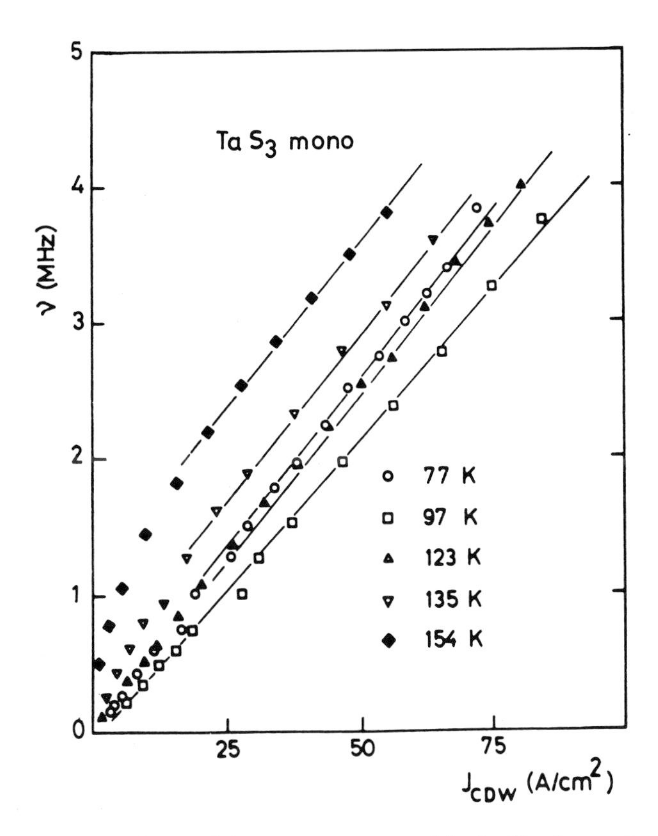

Fig. 48. Variation of the fundamental frequency, ν, measured in the Fourier spectrum as a function of the current carried by the CDW at different temperatures below the onset of the lower CDW (T_2 = 160 K) of a monoclinic TaS_3 sample.

density higher than 50 A cm^{-2}, the CDW drags the quasi-particles and the charge density is $\rho_{eff} \sim \rho$.

9. Conclusions

Some one-dimensional transition metal tri- and tetrachalcogendies exhibit nonlinear transport properties at any temperature below the Peierls transition temperature. The non-Ohmic electrical current carries no heat, which rules out any mechanism based on charge excitations, such as hot electrons, but which is consistent with a conduction by a condensate, such as a CDW or a lattice of discommensurations [293]. Although the ac voltage detected in the nonlinear state has been associated with shot noise generated at the electrodes of the sample by the impact of charged periodic solitons [303], it has to be noted that any coherent voltage may be suppressed by irregularities of the electrodes with regard to the discommensuration lattice. It has been shown that the charges put in motion in the nonlinear state correspond to the electronic concentration (or half the electronic concentration) in the bands affected by the CDW. One can therefore conclude that the extra dc conductivity, which appears beyond a threshold field, is attributed to the CDW motion. This charge transport mechanism, probably a quantum mechanism, results from the coherent motion of a macroscopic number of electrons at temperatures in the vicinity of room temperature.

Several problems, however, remain unsolved. Although current theories permit an estimation of the threshold field amplitude, none of them predicts the temperature dependence of E_c, especially the large increase of E_c at low temperatures. Thermal activation in the CDW dynamics has certainly to be considered. It has been shown that any sample is made up of many domains. Experiments under a thermal gradient have revealed that the CDW velocity is conserved along the whole sample. But the coupling between domains has as yet not been studied appropriately. Electrostatic coupling between charges built at each domain boundary may lead to a strong coupling between domains. Most of the theories developed so far are phenomenological and they are largely derived from the analysis of the experimental data. A microscopic approach has been suggested, based on a Josephson-type theory. A theory built on fundamental principles is clearly needed. Finally, it has been shown that the velocity, v, of the CDW is the product of the fundamental frequency, ν, measured in the Fourier transformed voltage, by the pinning periodicity, λ_P. A direct measurement of v would lead directly to the determination of λ_P without any possible controversial assumptions about the electronic concentration and it would give the true mechanism involved in the CDW motion. Several methods may be considered to measure the CDW velocity: motional narrowing in the NMR line or variation of T_1 with the applied electric field [362], Doppler shift of the neutron wavelength in neutron scattering measurements [363] or oscillations in the tunneling current between two Peierls conductors separated by an insulating barrier [364]. The measurement of the CDW velocity may well represent one of the main challenges facing researchers in the months or years ahead.

Acknowledgments

I wish to thank J. Rouxel and his research team from Nantes (France) for their relentless collaboration during this work and P. Gressier, L. Guemas, F. A. Lévy, A. Meerschaut, for providing us with the samples. I am deeply grateful to my colleagues R. Ayrolles, A. Briggs, P. Haen, M. Nunez-Regueiro, M. Papoular, M. Renard, J. Richard, C. Roucau, M. C. Saint-Lager, H. Salva and Z. Z. Wang without whose help this research would not have been possible. I especially wish to thank M. Renard for many an enlightening discussion. I have profited from discussions and correspondence with S. Aubry, J. Bardeen, S. Barišic, L. P. Gor'kov, B. Horovitz, P. Lederer, K. Maki, P. Martinoli, N. P. Ong, L. Pietronero and W. Wonnenberger. My acknowledgments are due to J. W. Brill, R. M. Fleming, J. C. Gill, G. Grüner, B. Horovitz, R. A. Klemm, N. P. Ong, L. Sneddon, J. R. Tucker, M. H. Whangbo and W. Wonnenberger for sending me their manuscripts before publication. I am grateful to Otto Samson and D. Devillers for their technical assistance.

I would like to thank the authors, publishers and copyright holders of the journals mentioned below for permission to reproduce the figures mentioned in their publications.

Fig. 2: *J. Physique Lett.* **40** L385 (1979), A. Andrieux, H. J. Schultz, D. Jerome, K. Bechgaard and the Société Française de Physique.
Fig. 3: *Phys. Rev. B* **16** 688 (1977), M. J. Cohen and A. J. Heeger and the American Physical Society.
Fig. 10a: *Inorg. Chemistry* **23**, 1305 (1984), M. H. Whangbo and P. Gressier.
Fig. 10b: *J. Phys. C* **15** 3069 (1982), D. W. Bullett and the Institute of Physics.
Fig. 11: *J. Phys.* **43** 1489 (1982), F. Devreux and the Société Française de Physique.
Fig. 19: *Solid State Commun.* **39** 553 (1981), P. M. Chaikin, W. W. Fuller, R. Lacoe, J. F. Kwak, R. L. Greene, J. C. Eckert, N. P. Ong and Pergamon Press Ltd.
Fig. 30a: *Phys. Rev. B* **29** 755 (1984), A. Zettl, G. Grüner, and the American Physical Society.
Fig. 30b: *Phys. Rev. B* **26** 5773 (1982), A. Zettl, C. M. Jackson, G. Grüner and the American Physical
and 41a Society.
Fig. 35: *Phys. Rev. B.* **28** 6996 (1983), R. M. Fleming, L. F. Schneemeyer and the American Physical Society.
Fig. 38: *Phys. Rev. B* **28** 6659 (1983), M. P. Everson, R. V. Coleman and the American Physical Society.
Fig. 41b: *Phys. Rev. Letters* **51** 1592 (1983), J. H. Miller, J. Richard, J. R. Tucker, J. Bardeen and the American Physical Society.
Fig. 45: *Phys. Rev. Lett.* **16** 734 (1966), J. I. Gittleman, B. Rosenblum and the American Physical Society.

Finally, I wish to dedicate this chapter to the memory of my father, who departed this life while I was writing this manuscript.

References

1. J. Solyom, *Adv. in Physics* **28**, 201 (1979).
2. B. Horovitz, H. Gutfreund and M. Weger, *Phys. Rev. B* **8**, 3174 (1975).
3. R. E. Peierls, *Ann. Phys.* **4**, 121 (1930) and *Quantum Theory of Solids*, Oxford University Press (1955), p. 108.
4. R. Comes and G. Shirane in *Highly Conducting One-Dimensional Solids*, ed. J. T. Devreese, R. P. Evrard and V. E. Van Doren, Plenum Press, London (1979), p. 17.
5. For a review, J. A. Wilson, F. J. Di Salvo and S. Mahajan, *Adv. in Physics* **24**, 117 (1975). See also *Physics and Chemistry of Materials with Layered Compounds*, D. Reidel Publishing Company, Dordrecht, Holland, Vols. 1 to 6.
6. H. Fröhlich, *Proc. R. Soc. A.* **223**, 296 (1954).
7. D. Allender, J. W. Bray and J. Bardeen, *Phys. Rev. B* **9**, 119 (1974).
8. J. Bardeen, L. N. Cooper and J. R. Schrieffer, *Phys. Rev.* **108**, 1175 (1957).
9. P. Lee, T. M. Rice and P. W. Anderson, *Solid State Commun.* **14**, 703 (1974).
10. W. Kohn, *Phys. Rev. Lett.* **2**, 293 (1959).
11. A. W. Overhauser, *Phys. Rev. B* **3**, 3173 (1971).
12. For a review see V. Dvořák, *Modern Trends in the Theory of Condensed Matter*, Lecture Notes in Physics, edited by A. Pekalski and J. Przystawa, Springer-Verlag, Berlin, **115**, 447 (1980).
13. H. Cailleau, F. Moussa, C. M. E. Zeyen and J. Bouillot, *Solid State Commun.* **33**, 407 (1980).
14. L. Bernard, R. Currat, P. Delamoye, C. M. E. Zehen, S. Hubert and R. de Kouchkovsky, *J. Phys. C* **16**, 433 (1983).
15. D. Kuse and H. R. Zeller, *Phys. Rev. Letters* **27**, 1060 (1971).
16. B. Renker, H. Rietschel, L. Pintschovius, W. Gläser, P. Brüesch, D. Kuse and M. J. Rice, *Phys. Rev. Letters* **30**, 1144 (1973).
17. B. Renker and R. Comes, in *Low-Dimensional Cooperative Phenomena*, ed. H. J. Keller, Plenum Press (New York) (1975), p. 235.
18. A. J. Heeger in *Highly Conducting One-Dimensional Compounds*, ed. J. T. Devreese, R. P. Evrard and V. E. Van Doren, Plenum Press (New York) (1979), p. 69.

19. M. J. Rice, S. Strässler and W. R. Schneider in *One-Dimensional Conductors*, edited by H. G. Schuster, Springer-Verlag, Berlin, **34**, 282 (1975).
20. P. G. De Gennes in *Superconductivity of Metals and Alloys*, W. A. Benjamin, Inc., New York (1966).
21. P. Brüesch, S. Strässler and H. R. Zeller, *Phys. Rev. B* **12**, 219 (1975).
22. D. B. Tanner, C. S. Jacobsen, A. F. Garito and A. J. Heeger, *Phys. Rev. B* **13**, 3381 (1976).
23. D. B. Tanner, K. D. Cummings and C. S. Jacobsen, *Phys. Rev. Lett.* **47**, 597 (1981).
24. J. E. Eldridge and F. E. Bates, *Phys. Rev. B* **28**, 6972 (1983).
25. J. Bardeen, *Solid State Commun.* **13**, 357 (1973).
26. R. H. Friend, M. Miljak and D. Jerome, *Phys. Rev. Lett.* **40**, 1048 (1978).
27. A. Andrieux, H. J. Schulz, D. Jerome and K. Bechgaard, *Phys. Rev. Lett.* **43**, 227 (1979). Also *J. Physique-Lettres* **40**, L385 (1979).
28. J. Friedel and D. Jerome, *Contemp. Phys.* **23**, 583 (1982).
29. H. J. Schulz, *Solid State Commun.* **34**, 455 (1980).
30. J. F. Thomas and D. Jerome, *Solid State Commun.* **36**, 813 (1980).
31. M. J. Rice, A. R. Bishop, J. A. Krumhansl and S. E. Trullinger, *Phys. Rev. Lett.* **36**, 432 (1976).
32. M. J. Cohen and A. J. Heeger, *Phys. Rev. B* **16**, 688 (1977).
33. J. F. Thomas, *Phys. Lett.* **85A**, 457 (1981).
34. J. F. Thomas, *Solid State Commun.* **42**, 567 (1982).
35. M. Mehring, O. Kaneart, M. Mali and D. Brinkmann, *Solid State Commun.* **33**, 225 (1980).
36. A. Meerschaut, P. Gressier and J. Rouxel in *Quasi One-Dimensional Structures*, ed. J. Rouxel, D. Reidel Publishing Company, Dordrecht, (1985) followcoming.
37. W. Kröniert and K. Plieth, *Z. Anorg. Allg. Chem.* **336**, 207 (1965).
38. F. Jellinek, R. A. Pollack and M. W. Shafer, *Mater. Res. Bull.* **9**, 845 (1974).
39. J. Rijnsdorp and F. Jellinek, *J. Solid State Chem.* **25**, 325 (1978).
40. T. Cornelissens, G. Tendeloo, J. Landuyt and S. Amelincks, *Phys. Stat. Sol. a* **48**, K5 (1978).
41. F. W. Boswell and A. Prodan, *Physica* **99B**, 361 (1980).
42. C. Roucau, *J. Physique* **44**, C3–1725 (1983).
43. A. Zettl, C. M. Jackson, A. Janossy, G. Grüner, A. Jacobsen and A. H. Thompson, *Solid State Commun.* **43**, 345 (1982).
44. M. Izumi, T. Nakayama, K. Uchinoküra, T. Iwazumi, T. Seino, and F. Matsuura, *Proc. of the Int. Symposium on Nonlinear Transport in Quasi-One-Dimensional Inorganic Conductors*, Sapporo (Japan) October 1983, p. 301.
45. E. Bjerkelund, J. H. Fermor and A. Kjeskshus, *Acta Chem. Scand.* **20**, 1836 (1966).
46. A. Meerschaut and J. Rouxel, *J. Less Comm. Metals* **39**, 197 (1975).
47. J. L. Hodeau, M. Marezio, C. Roucau, R. Ayrolles, A. Meerschaut, J. Rouxel and P. Monceau, *J. Phys. C* **11**, 4117 (1978).
48. A. Meerschaut, L. Guemas and J. Rouxel, *J. Solid State Chem.* **36**, 118 (1981).
49. E. Bjerkelund and A. Kjekshus, *Z. Anorg. Allg. Chem. B* **328**, 235 (1964).
50. A. Zwick and M. A. Renucci, *Phys. Stat. Sol. b* **96**, 757 (1979); A. Zwick, M. A. Renucci and A. Kjekshus, *J. Phys. C* **13**, 5603 (1980); A. Zwick, G. Landa, M. A. Renucci, R. Carles and A. Kjekshus, *Phys. Rev. B* **26**, 5694 (1982).
51. A. Grisel, F. Lévy and T. J. Wieting, *Physica* **99B**, 365 (1980); T. J. Wieting, A. Grisel and F. Lévy, *Physica* **105B**, 366 (1981).
52. T. J. Wieting, A. Grisel and F. Lévy, *Mol. Cryst. Liq. Cryst.* **81**, 117 (1982).
53. K. Yamaya and Y. Abe, *J. Phys. Soc. Japan* **51**, 3512 (1982).
54. K. Yamaya and G. Oomi, *J. Phys. Soc. Japan* **52**, 1886 (1983).
55. S. J. Hillenius, R. V. Coleman, R. M. Fleming and R. J. Cava, *Phys. Rev. B* **23**, 1567 (1981).
56. R. J. Cava, V. L. Himes, A. D. Mighell and R. S. Roth, *Phys. Rev. B* **24**, 3634 (1981).
57. A. Meerschaut, P. Gressier, L. Guemas, and J. Rouxel, *Mat. Res. Bull.* **16**, 1035 (1981).
58. R. J. Cava, F. J. Di Salvo, M. Eibschutz and J. V. Waszczak, *Phys. Rev. B* **27**, 7412 (1983).
59. A. Ben Salem, A. Meerschaut, L. Guemas and J. Rouxel, *Mat. Res. Bull.* **17**, 1071 (1982).
60. A. Meerschaut, A. Ben Salem, L. Guemas, J. Rouxel, P. Monceau and H. Salva, *J. Physique* **44**, C3–1681 (1983).

61. P. Gressier, A. Meerschaut, L. Guemas, J. Rouxel and P. Monceau, *J. Solid State Chem.* **51**, 141 (1984).
62. L. Guemas, P. Gressier, A. Meerschaut, D. Lauër and D. Grandjean, *Rev. Chim. Miner.* **18**, 91 (1981).
63. A. Meerschaut, P. Palvadeau and J. Rouxel, *J. Solid State Chem.* **20**, 21 (1977).
64. P. Gressier, L. Guemas and A. Meerschaut, *Acta Cryst. B* **38**, 2877 (1982).
65. A. Meerschaut, P. Gressier, L. Guemas, and J. Rouxel, *J. Solid State Chem.* **51**, 307 (1984).
66. J. Graham and A. D. Wadsley, *Acta Cryst.* **20**, 93 (1966).
67. R. Allman, L. Baumann, A. Kutoglu, H. Räsch and E. Hellner, *Naturwiss.* **51**, 263 (1964).
68. K. Selte and A. Kjekshus, *Acta Chem. Scand.* **18**, 690 (1964).
69. F. W. Boswell, A. Prodan and J. K. Brandon, *J. Phys. C* **16**, 1067 (1983).
70. S. Fusureth, L. Brattas and A. Kjekshus, *Acta Chem. Scand.* **27**, 2367 (1973).
71. A. F. J. Ruysnik, F. Kadijk, A. J. Wagner and F. Jellinek, *Acta Cryst. B* **24**, 1614 (1968).
72. F. Lévy and H. Berger, *J. Crystal Growth* **61**, 61 (1983).
73. F. K. McTaggart, *Austral. J. Chem.* **11**, 471 (1959).
74. S. Takahashi, T. Sambongi and S. Okada, *J. Physique* **44**, C3–1733 (1983).
75. W. Schairer and M. Shafer, *Phys. Stat. Sol. a* **17**, 181 (1973).
76. P. Haen, P. Monceau, B. Tissier, G. Waysand, A. Meerschaut, P. Molinie and J. Rouxel, *Proceedings Fourteenth International Conference on Low Temperature Physics* (Otaniami, Finland 1975), ed. M. Krusius and M. Vuorio, North-Holland Vol. 5 (1975), p. 445.
77. J. Chaussy, P. Haen, J. C. Lasjaunias, P. Monceau, G. Waysand, A. Waintal, A. Meerschaut, P. Molinie and J. Rouxel, *Solid State Commun.* **20**, 759 (1976).
78. N. P. Ong and P. Monceau, *Phys. Rev. B* **16**, 3443 (1977).
79. N. P. Ong and J. W. Brill, *Phys. Rev. B* **18**, 5265 (1978).
80. H. P. Geserich, G. Scheiber, F. Lévy and P. Monceau, *Solid State Commun.* **49**, 335 (1984).
81. K. Yamaya, T. H. Geballe, J. V. Acrivos and J. Code, *Physica* **105B**, 444 (1981).
82. T. Sambongi, M. Yamamoto, K. Tsutsumi, Y. Shiozaki, K. Yamaya and Y. Abe, *J. Phys. Soc. Japan* **42**, 1421 (1977).
83. S. Etemad, E. M. Engler, T. D. Schultz, T. Penney and B. A. Scott, *Phys. Rev. B* **17**, 513 (1978).
84. P. A. Lee, T. M. Rice and P. W. Anderson, *Phys. Rev. Lett.* **31**, 462 (1973).
85. C. M. Varma and A. L. Simons, *Phys. Rev. Lett.* **51**, 138 (1983).
86. A. Meerschaut, J. Rouxel, P. Haen, P. Monceau and M. Nuñez-Regueiro, *J. Physique-Lettres* **40**, L157 (1979).
87. C. Roucau, R. Ayroles, P. Monceau, L. Guemas, A. Meerschaut and J. Rouxel, *Phys. Stat. Sol. a* **62**, 483 (1980).
88. T. Sambongi, K. Tsutsumi, Y. Shiozaki, M. Yamamoto, K. Yamaya and Y. Abe, *Solid State Commun.* **22**, 729 (1977).
89. M. Ido, K. Tsutsumi, T. Sambongi and N. Mori, *Solid State Commun.* **29**, 399 (1979).
90. J. C. Tsang, C. Hermann and M. W. Shafer, *Phys. Rev. Lett.* **40**, 1528 (1978).
91. A. W. Higgs and J. C. Gill, *Solid State Commun.* **47**, 737 (1983).
92. W. L. McMillan, *Phys. Rev. B* **14**, 1496 (1976).
93. R. Allgeyer, B. H. Suits and F. C. Brown, *Solid State Commun.* **43**, 207 (1982).
94. W. Biberacher and H. Schwenk, *Solid State Commun.* **33**, 385 (1980).
95. E. Amberger, K. Polborn, P. Grimm, M. Dietrich and B. Obst, *Solid State Commun.* **26**, 943 (1978).
96. C. Roucau, R. Ayroles, P. Gressier and A. Meerschaut, *J. Phys. C* **17**, 2993 (1984).
97. Z. Z. Wang, M. C. Saint-Lager, P. Monceau, M. Renard, P. Gressier, A. Meerschaut, L. Guemas and J. Rouxel, *Solid State Commun.* **46**, 325 (1983).
98. M. Maki, M. Kaiser, A. Zettl and G. Grüner, *Solid State Commun.* **46**, 497 (1983).
99. Z. Z. Wang, P. Monceau, M. Renard, P. Gressier, L. Guemas and J. Rouxel, *Solid State Commun.* **47**, 439 (1983).
100. H. Fujishita, M. Sato and S. Hoshino, *Solid State Commun.* **49**, 313 (1984).
101. M. Izumi, T. Iwazumi, T. Seino, *Solid State Commun.* **49**, 423 (1984).
102. A. Ben Salem, A. Meerschaut, H. Salva, Z. Z. Wang and T. Sambongi, *J. Physique* **45**, 771 (1984).

103. S. Okada, T. Sambongi and M. Ido, *J. Phys. Soc. Japan Letters* **49**, 839 (1980).
104. M. Izumi, K. Uchinokura and E. Matsuura, *Solid State Commun.* **37**, 641 (1981).
105. F. J. Di Salvo, R. M. Fleming and J. V. Waszczak, *Phys. Rev. B* **24**, 2935 (1981).
106. J. L. Genicon, personal communication.
107. W. Fogle and J. H. Perlstein, *Phys. Rev. B* **6**, 1402 (1972).
108. R. Brusetti, B. K. Chakraverty, J. Devenyi, J. Dumas, J. Marcus and C. Schlenker, *Recent Developments in Condensed Matter Physics*, ed. J. T. Devreese, L. F. Lemmens, V. E. Van Doren and J. Van Royen, Plenum Publishing Corporation, Vol. 2 (1981), p. 181.
109. G. Travaglini, P. Wachter, J. Marcus and C. Schlenker, *Solid State Commun.* **37**, 599 (1981).
110. K. Tsutsumi, T. Takagaki, M. Yamamoto, Y. Shiozaki, M. Ido, T. Sambongi, K. Yamaya and Y. Abe, *Phys. Rev. Lett.* **39**, 1675 (1977).
111. R. M. Fleming, D. E. Moncton and D. B. McWhan, *Phys. Rev. B* **18**, 5560 (1978).
112. J. P. Pouget, R. Moret, A. Meerschaut, L. Guemas and J. Rouxel, *J. Physique* **44**, C3–1729 (1983).
113. R. M. Fleming, C. H. Chen and D. E. Moncton, *J. Physique* **44**, C3–1651 (1983).
114. V. Emery and D. Mukamel, *J. Phys. C* **12**, L677 (1979).
115. R. Bruinsma and S. E. Trullinger, *Phys. Rev. B* **22**, 4543 (1980).
116. M. Papoular, *Phys. Rev. B* **25**, 7856 (1982).
117. Z. Z. Wang, H. Salva, P. Monceau, M. Renard, C. Roucau, R. Ayrolles, F. Lévy, L. Guemas and A. Meerschaut, *J. Physique-Lettres* **44**, L–311 (1983).
118. G. Van Tendeloo, J. Van Landuyt, S. Amelinckx, *Phys. Stat. Sol. a* **43**, K137 (1977).
119. K. Tsutsumi, T. Sambongi, S. Kagoshima and T. Ishiguro, *J. Phys. Soc. Japan* **44**, 1735 (1978).
120. P. Monceau, H. Salva and Z. Z. Wang, *J. Physique* **44**, C3–1639 (1983).
121. B. Horovitz, personal communication.
122. F. W. Boswell and A. Prodan, *Physica* **99B**, 361 (1980).
123. C. Roucau, Z. Z. Wang, H. Salva, P. Monceau, to be published.
124. R. Moret, J. P. Pouget, A. Meerschaut and L. Guemas, *J. Physique-Lettres* **44**, L93 (1983).
125. H. Fujishita, M. Sato and S. Hoshino, *Solid State Commun.* **49**, 313 (1984).
126. F. W. Boswell, A. Prodan and J. K. Brandon, *J. Phys. C* **16**, 1067 (1983).
127. J. Mahy, J. Van Landuyt and S. Amelinckx, *Phys. Stat. Sol. a* **77**, K1 (1983).
128. J. P. Pouget, S. Kagoshima, C. Schlenker, and J. Marcus, *J. Physique-Lettres* **44**, L113 (1983).
129. R. M. Fleming and L. F. Schneemeyer, *Phys. Rev. B* **28**, 6996 (1983).
130. S. Okada, T. Sambongi, M. Ido, Y. Tazuche, R. Aoki and O. Fujita, *J. Phys. Soc. Japan* **51**, 460 (1982).
131. A. Fournel and J. P. Sorbier, personal communication; see also A. Fournel, C. More, G. Roger, J. P. Sorbier and C. Blang, *J. Physique* **44**, C3–879.
132. W. A. Challener and P. L. Richards, *Solid State Commun.* **52**, 117 (1984).
133. K. K. Fung and J. W. Steeds, *Phys. Rev. Lett.* **45**, 1696 (1980).
134. J. W. Steeds, K. K. Fung and S. McKernan, *J. Physique* **44**, C3–1623 (1983).
135. C. H. Chen and R. M. Fleming, *Solid State Commun.* **48**, 777 (1983).
136. K. K. Fung, S. McKernan, J. W. Steeds and J. A. Wilson, *J. Phys. C* **14**, 5417 (1981).
137. C. H. Chen, J. M. Gibson and R. M. Fleming, *Phys. Rev. B* **26**, 184 (1982).
138. C. Haas, *Solid State Commun.* **26**, 709 (1978).
139. J. A. Wilson, *Phys. Rev. B* **19**, 6456 (1979).
140. J. A. Wilson, *J. Phys. F* **12**, 2469 (1982).
141. N. P. Ong and P. Monceau, *Solid State Commun.* **26**, 487 (1978).
142. D. W. Bullett, *J. Phys. C* **12**, 277 (1979).
143. D. W. Bullett, *J. Phys. C* **15**, 3069 (1982).
144. R. Hoffmann, S. Shaik, J. C. Scott, M. H. Whangbo and M. J. Foshee, *J. Solid State Chem.* **34**, 263 (1980).
145. G. X. Tessema and N. P. Ong, *Phys. Rev. B* **23**, 5607 (1981).
146. A. Meerschaut, *J. Physique* **44**, C3–1615 (1983).

147. N. P. Ong, G. X. Tessema, G. Verma, J. C. Eckert, J. Savage, and S. K. Khanna, *Mol. Cryst. Liq. Cryst.* **81**, 41 (1982).
148. R. M. Fleming, J. A. Polo and R. V. Coleman, *Phys. Rev. B* **17**, 1634 (1978).
149. M. H. Whangbo, R. J. Cava, F. J. Di Salvo and R. M. Fleming, *Solid State Commun.* **43**, 277 (1982).
150. M. H. Whangbo and P. Gressier, *Inorganic Chemistry* **23**, 1305 (1984).
151. N. Shima, *J. Phys. Soc. Japan* **51**, 11 (1982), **52**, 578 (1983).
152. P. Monceau, *Solid State Commun.* **24**, 331 (1977).
153. P. Monceau and A. Briggs, *J. Phys. C* **11**, L465 (1978).
154. S. Wada, M. Sasakura, R. Aoki and O. Fujita, *J. Phys. F* **14**, 1515 (1984).
155. P. Gressier, M. H. Whangbo, A. Meerschaut and J. Rouxel, *Inorganic Chemistry* **23**, 1222 (1984).
156. M. H. Whangbo and P. Gressier, *Inorganic Chemistry* **23**, 1228 (1984).
157. D. W. Bullett, *Solid State Commun.* **42**, 691 (1982).
158. M. H. Whangbo, F. J. Di Salvo and R. M. Fleming, *Phys. Rev. B* **26**, 687 (1982).
159. F. Devreux, *J. Physique* **43**, 1489 (1982).
160. P. M. Chaikin in *The Physics and Chemistry of Low-Dimensional Solids*, ed. L. Alcácer, D. Reidel, Publishing Company, Dordrecht, Holland (1980), p. 53.
161. T. D. Schultz and R. A. Craven in *Highly Conducting One-Dimensional Solids*, ed. J. T. Devreese, R. P. Evrard and V. F. Van Doren, Plenum Press, London (1979), p. 147.
162. T. Tagaki, M. Ido and T. Sambongi, *J. Phys. Soc. Japan Letters* **45**, 2039 (1978).
163. R. H. Dee, P. M. Chaikin and N. P. Ong, *Phys. Rev. Lett.* **42**, 1234 (1979).
164. P. M. Chaikin, W. W. Fuller, R. Lacoe, J. F. Kwak, R. L. Greene, J. C. Eckert and N. P. Ong, *Solid State Commun.* **39**, 553 (1981).
165. R. Allgeyer, B. H. Suits and F. C. Brown, *Solid State Commun.* **43**, 207 (1982).
166. D. C. Johnston, J. P. Stokes, P. L. Hsieh and G. Grüner, *J. Physique* **44**, C3–1749 (1983). (The sign of the thermopower is recognized to be positive in [293]).
167. H. Mukta and A. Meerschaut, *Mol. Cryst. Liq. Cryst.* **81**, 125 (1982).
168. B. Fisher, *Solid State Commun.* **46**, 227 (1983).
169. B. Fisher, *Solid State Commun.* **48**, 437 (1983).
170. N. P. Ong, *Phys. Rev. B* **18**, 5272 (1978).
171. J. D. Kulick and J. C. Scott, *Solid State Commun.* **32**, 217 (1979).
172. F. J. Di Salvo, J. V. Waszczak and K. Yamaha, *J. Phys. Chem. Solid.* **41**, 1311 (1980).
173. J. W. Brill, C. P. Tzou, G. Verma and N. P. Ong, *Solid State Commun.* **39**, 233 (1981).
174. M. Nunez-Regueiro, C. Ayache and M. Locatelli, *Physica* **108B**, 1035 (1981).
175 D. Djurek and S. Tomic, *Phys. Letters* **85A**, 155 (1981).
176 S. Tomic, K. Biljakovic, D. Djurek, J. R. Cooper, P. Monceau and A. Meerschaut, *Solid State Commun.* **38**, 109 (1981).
177. S. Tomic, *Solid State Commun.* **40**, 321 (1981).
178. J. C. Lasjaunias and P. Monceau, *Solid State Commun.* **41**, 911 (1982).
179. J. W. Brill and N. P. Ong, *Solid State Commun.* **25**, 1075 (1978); J. W. Brill, *Mol. Cryst. Liq. Cryst.* **81**, 107 (1982).
180. J. W. Brill, *Solid State Commun.* **41**, 925 (1982).
181. M. Barmatz, L. R. Testardi and F. J. Di Salvo, *Phys. Rev. B* **12**, 4367 (1975).
182. R. S. Lear, M. J. Stove, E. P. Stillwell and J. W. Brill, *Phys. Rev. B* **29**, 5656 (1984).
183. A. Briggs, P. Monceau, M. Nunez-Regueiro, J. Peyrard, M. Ribault and J. Richard, *J. Phys. C* **13**, 2117 (1980).
184. M. Ido, K. Tsutsumi, T. Sambongi, and N. Môri, *Solid State Commun.* **29**, 399 (1979).
185. P. M. Horn and D. Guiddoti, *Phys. Rev. B* **16**, 491 (1977).
186. J. Richard, H. Salva, M. C. Saint-Lager and P. Monceau, *J. Physique* **44**, C3–1685 (1983).
187. R. A. Craven and S. F. Meyer, *Phys. Rev. B* **16**, 4583 (1977).
188. N. P. Ong, *Phys. Rev. B* **17**, 3243 (1978).
189. M. C. Saint-Lager, Thesis 3° cycle, University of Grenoble, 1983, unpublished.

190. P. Monceau, J. Peyrard, J. Richard and P. Molinie, *Phys. Rev. Lett.* **39**, 160 (1977).
191. A. Briggs, P. Monceau, M. Nunez-Regueiro, M. Ribault and J. Richard, *J. Physique* **42**, 1453 (1981).
192. W. W. Fuller, P. Chaikin and N. P. Ong, *Solid State Commun.* **30**, 689 (1979), and **39**, 547 (1981).
193. K. Kawabata, M. Ido and T. Sambongi, *J. Physique* **44**, C3–1647 (1983).
194. W. W. Fuller, P. M. Chaikin and N. P. Ong, *Phys. Rev. B* **24**, 1333 (1981).
195. K. Nishida, T. Sambongi and M. Ido, *J. Phys. Soc. Japan* **48**, 331 (1980).
196. D. Jerôme, C. Berthier, P. Molinie and J. Rouxel, *J. Physique* **C4**, 125 (1977).
197. J. Friedel, *J. Physique-Lettres* **36**, L279 (1975).
198. G. Bilbro and W. L. McMillan, *Phys. Rev. B* **14**, 1887 (1976).
199. K. Yamaya, T. H. Geballe, J. F. Kwak and R. L. Greene, *Solid State Commun.* **31**, 623 (1979).
200. P. Haen, F. Lapierre, P. Monceau, M. Nuñez-Regueiro and J. Richard, *Solid State Commun.* **26**, 725 (1978).
201. P. Haen, J. M. Mignot, P. Monceau and M. Nuñez-Regueiro, *J. Physique* **C6**, 703 (1978).
202. Y. Tajima and K. Yamaya, *J. Phys. Soc. Japan* **53** (1984).
203. R. A. Buhrman, C. M. Bastuscheck, J. C. Scott and J. D. Kulick, in *Inhomogeneous Superconductors*, ed. D. W. Gubser, T. L. Francavilla, S. A. Wolf and J. R. Leibowitz, New York, American Institute of Physics (1980); C. M. Bastuscheck, R. A. Buhrman, J. D. Kulick and J. C. Scott, *Solid State Commun.* **36**, 983 (1980).
204. K. Kawabata and M. Ido, *Solid State Commun.* **44**, 1539 (1982).
205. M. Yamamoto, *J. Phys. Soc. Japan* **45**, 431 (1978).
206. J. A. Wollam, R. B. Somoano and P. O'Connor, *Phys. Rev. Lett.* **32**, 712 (1974).
207. L. A. Turkevich and R. A. Klemm, *Phys. Rev. B* **19**, 2520 (1979).
208. W. E. Lawrence and S. Doniach, *Proceedings of the 12th International Conference on Low Temperature Physics*, ed. Eizo Kanda (Academic Press of Japan, Kyoto) 1971, p. 361.
209. L. J. Azevedo, W. G. Clark, G. Deutscher, R. L. Greene, G. B. Street and L. J. Suter, *Solid State Commun.* **19**, 197 (1976).
210. S. T. Ruggiero, T. W. Barbee and M. R. Beasley, *Phys. Rev. Lett.* **45**, 1299 (1980).
211. L. N. Bulaevskii and M. V. Sadovskii, *Fiz. Tverd. Tela* **16**, 1159 (1974); *Sov. Phys. Solid State* **16**, 743 (1974).
212. J. W. Brill, N. P. Ong, J. C. Eckert, J. W. Savage, S. K. Khanna and R. B. Somoano, *Phys. Rev. B* **23**, 1517 (1981).
213. M. Nuñez-Regueiro (1980) Thesis 3° cycle, unpublished, University of Grenoble.
214. P. Monceau, *Physica* **109** and **110B**, 1890 (1982).
215. P. L. Hsieh, F. De Czito, A. Janossy and G. Grüner, *J. Physique* **44**, C3–1753 (1983).
216. K. Yamaya and Y. Abe, *Mol. Cryst. Liq. Cryst.* **81**, 133 (1982).
217. M. P. Everson and R. V. Coleman, *Phys. Rev. B* **28**, 6659 (1983).
218a. H. Gruber and H. Sassik, *J. Physique* **44**, C3–1717 (1983).
218b. P. Monceau, J. Richard and R. Lagnier, *J. Phys. C* **14**, 2995 (1981).
219. W. W. Fuller, G. Grüner, P. M. Chaikin and N. P. Ong, *Phys. Rev. B* **23**, 6259 (1981).
220. G. Mihaly, N. Housseau, H. Mukta, L. Zuppiroli, J. Pelissier, P. Gressier, A. Meerschaut and J. Rouxel, *J. Physique-Lettres* **42**, L263 (1981).
221. H. Salva, Z. Z. Wang, P. Monceau, J. Richard and M. Renard, *Phil. Mag.* **49**, 385 (1984).
222. H. Mukta, S. Bouffard, G. Mihaly and L. Mihaly, *J. Phys. Letters* **45**, L–113 (1984).
223. F. J. Di Salvo, J. A. Wilson and J. V. Waszczak, *Phys. Rev. Lett.* **36**, 885 (1976).
224. F. J. Di Salvo and J. E. Graebner, *Solid State Commun.* **23**, 825 (1977).
225. D. Kuse and H. R. Zeller, *Phys. Rev. Lett.* **27**, 1060 (1971); R. Denton and B. Mühlschlegel, *Solid State Commun.* **11**, 1637 (1972); J. Bernasconi, D. Kuse, M. J. Rice and H. R. Zeller, *J. Phys. C* **5**, L127 (1972).
226. S. J. Hillenius and R. V. Coleman, *Phys. Rev. B* **25**, 2191 (1982).
227. P. Monceau, N. P. Ong, A. M. Portis, A. Meerschaut, and J. Rouxel, *Phys. Rev. Lett.* **37**, 602 (1976).
228. J. Richard and P. Monceau, *Solid State Commun.* **33**, 635 (1980).

229. J. Bardeen in *Quasi One-Dimensional Conductors* I, Lecture Notes in Physics, ed. S. Barisic, A. Bjelis, J. R. Cooper and B. Leontic, Springer-Verlag (Berlin) **95**, 3 (1979).
230. J. Bardeen in *Highly Conducting One-Dimensional Solids*, ed. J. T. Devreese, R. P. Evrard and V. E. Van Doren, Plenum Press London (1979), p. 373.
231. R. M. Fleming and C. C. Grimes, *Phys. Rev. Lett.* **42**, 1423 (1979).
232. P. A. Lee and T. M. Rice, *Phys. Rev. B* **19**, 3970 (1979).
233. R. M. Fleming in *Physics in One-Dimensional*, Solid State Sciences **23**, ed. J. Bernasconi and T. Schneider, Springer-Verlag (Berlin) (1981), p. 253.
234. G. Grüner, *Comments in Solid State Physics* **10**, 173 (1983).
235. G. Grüner, *Physica* **8D**, 1 (1983).
236. N. P. Ong, *Can. J. Phys.* **60**, 757 (1981).
237. P. Monceau in *Statics and Dynamics of Nonlinear Systems*, Solid State Sciences **47**, ed. G. Benedek, H. Biltz and R. Zeyher, Springer-Verlag (Berlin) (1983), p. 144.
238. P. Monceau, J. Richard and M. Renard, *Phys. Rev. B* **25**, 931 (1982).
239. J. Richard, P. Monceau, M. Papoular and M. Renard, *J. Phys. C* **15**, 7157 (1982).
240. A. Zettl and G. Grüner, *Phys. Rev. B* **26**, 2298 (1982).
241. A. H. Thompson, A. Zettl and G. Grüner, *Phys. Rev. Lett.* **47**, 64 (1981).
242. S. K. Zhilinskii, M. E. Itkis, F. Ya. Nad, I. Yu Kalnova and V. B. Presbrazhenskii, *Zh. Eksp. Teor. Fiz.* **85**, 362 (1983), *J.E.T.P. Soviet Physics* **58**, 211 (1983); S.K. Zhilinskii, M.E. Itkis and F. Ya Nad, *Phys. Status Solid* **a81**, 367 (1984).
243. T. Takoshima, M. Ido, K. Tsutsumi, T. Sambongi, S. Honma, K. Yamaya and Y. Abe, *Solid State Commun.* **35**, 911 (1980).
244. G. Mihaly and L. Mihaly, *Solid State Commun.* **48**, 449 (1983).
245. J. C. Gill, *J. Phys. F* **10**, L–81 (1980).
246. J. C. Gill, *Solid State Commun.* **37**, 459 (1981).
247. R. M. Fleming, *Phys. Rev. B* **22**, 5606 (1980).
248. J. Richard, unpublished.
249. H. Salva, J. Richard, Z. Z. Wang, P. Monceau, to be published.
250. G. Grüner, L. C. Tippie, J. Sanny, W. G. Clark and N. P. Ong, *Phys. Rev. Lett.* **45**, 935 (1980).
251. G. Grüner, A. Zettl, W. G. Clark and J. Bardeen, *Phys. Rev. B* **24**, 7247 (1981).
252. S. W. Longcor and A. M. Portis, *Bull. Am. Phys. Soc.* **25**, 340 (1980); and S. W. Longcor, Ph.D dissertation, 1980, University of California at Berkeley, unpublished.
253. A. Zettl and G. Grüner, *Phys. Rev. B* **29**, 755 (1984).
254. C. M. Jackson, A. Zettl, G. Grüner and A. H. Thompson, *Solid State Commun.* **39**, 531 (1981).
255. A. Zettl, C. M. Jackson and G. Grüner, *Phys. Rev. B* **26**, 5773 (1982).
256. G. Verma, N. P. Ong, S. K. Khanna, J. C. Eckert and J. W. Savage, *Phys. Rev. B* **28**, 910 (1983).
257. P. Monceau, J. Richard and M. Renard, *Phys. Rev. Lett.* **45**, 43 (1980).
258. J. Richard, P. Monceau, and M. Renard, *Phys. Rev. B* **25**, 948 (1982).
259. N. P. Ong and C. M. Gould, *Solid State Commun.* **37**, 25 (1981).
260. N. P. Ong and G. Verma, *Phys. Rev. B* **27**, 4495 (1983).
261. M. Weger, G. Grüner and W. G. Clark, *Solid State Commun.* **44**, 1179 (1982).
262. J. C. Gill, *Solid State Commun.* **44**, 1041 (1982).
263. G. Grüner, A. Zettl, W. G. Clark and A. H. Thompson, *Phys. Rev. B* **23**, 6813 (1981).
264. A. Zettl and G. Grüner, *Phys. Rev. B* **28**, 2091 (1983).
265. J. Bardeen, E. Ben Jacob, A. Zettl and G. Grüner, *Phys. Rev. Lett.* **49**, 493 (1982).
266. R. M. Fleming, *Solid State Commun.* **43**, 167 (1983).
267. J. C. Gill and A. W. Higgs, *Solid State Commun.* **48**, 709 (1983).
268. J. Dumas, C. Schlenker, J. Marcus and R. Buder, *Phys. Rev. Lett.* **50**, 757 (1983).
269. A. Zettl and G. Grüner, *Solid State Commun.* **46**, 29 (1983).
270. A. Maeda, M. Naito and S. Tanaka, *Solid State Commun.* **47**, 1001 (1983).
271. P. Monceau, J. Richard and M. Renard, *Solid State Commun.* **41**, 609 (1982).
272. A. Zettl and G. Grüner, *Solid State Commun.* **46**, 501 (1983).
273. P. E. Lindelof, *Reports on Progress in Physics* **44**, 949 (1981); and L. Solymar, *Superconducting Tunneling and Applications*, Chapman and Hall, London (1972).

274. D. E. McCumber, *J. Appl. Phys.* **39**, 3113 (1968).
275. W. C. Stewart, *Appl. Phys. Lett.* **12**, 277 (1968).
276. P. E. Gregers-Hansen, M. T. Levinsen and G. Fog Pedersen, *J. Low Temp. Phys.* **7**, 99 (1972).
277. M. Papoular, *Phys. Lett.* **76A**, 430 (1980).
278. S. N. Artemenko and A. F. Volkov, *Zh. Eksp. Teor. Fiz.* **81**, 1872 (1981); *Sov. Phys. J.E.T.P.* **54**, 992 (1981). See also S. N. Artemenko and A. F. Volkov, *Zh. Eksp. Teor. Fiz.* **80**, 2018 (1981), *Sov. Phys. J.E.T.P. Lett.* **33**, 147 (1981).
279. J. H. Miller, J. Richard, R. E. Thorne, W. G. Lyons, J. R. Tucker and J. Bardeen, *Phys. Rev. B* **29**, 2328 (1984).
280. G. Grüner, W. G. Clark and A. M. Portis, *Phys. Rev. B* **26**, 3641 (1981).
281. The increase of the threshold field at low frequencies below 100 kHz, shown in Figure 5 of [271], has not been measured again in further experiments.
282. R. E. Thorne, W. G. Lyons, J. H. Miller, J. Richard, and J. R. Tucker, *Solid State Commun.* **50**, 833 (1984).
283. D. Djurek, M. Prester and S. Tomic, *Solid State Commun.* **42**, 807 (1982).
284. W. Schneider and K. Seeger, *Appl. Phys. Lett.* **8**, 133 (1966).
285. W. Mayr, H. Kahlert, K. Seeger and P. Monceau, *Synthetic Metals* **4**, 91 (1981).
286. K. Seeger, W. Mayr and A. Philipp, *Solid State Commun.* **43**, 113 (1982). See also W. Mayr, A. Philipp and K. Seeger, *J. Physique* **44**, C3–1693 (1983).
287. J. H. Miller, J. Richard, J. R. Tucker and J. Bardeen, *Phys. Rev. Lett.* **51**, 1592 (1983).
288. J. Richard, R. E. Thorne, W. G. Lyons, J. H. Miller, Jr and J. R. Tucker, *Solid State Commun.* **52**, 183 (1984).
289. J. C. Gill, *Solid State Commun.* **39**, 1203 (1981).
290. R. M. Fleming and L. F. Schneemeyer, *Phys. Rev. B* **28**, 6996 (1983).
291. G. X. Tessema and N. P. Ong, *Phys. Rev. B* **27**, 1417 (1983).
292. B. Joos and D. Murray, *Phys. Rev. B* **29**, 1094 (1984).
293. J. P. Stokes, A. N. Bloch, A. Janossy and G. Grüner, *Phys. Rev. Lett.* **52**, 372 (1984).
294. K. Kawabata, M. Ido and T. Sambongi, *J. Phys. Soc. Japan* **50**, 1992 (1981).
295. H. Fukuyama and P. A. Lee, *Phys. Rev. B* **17**, 535 (1978).
296. G. Grüner, A. Zawadowski and P. M. Chaikin, *Phys. Rev. Lett.* **46**, 511 (1981).
297. J. Bardeen (a) *Phys. Rev. Lett.* **42**, 1498 (1979); (b) *Phys. Rev. Lett.* **45** 1978 (1980); (c) *Mol. Cryst. Liq. Cryst.* **81**, 1 (1982).
298. J. Bardeen, Lecture Notes for International School of Physics 'Enrico Fermi' (1983).
299. L. Sneddon, M. C. Cross and D. S. Fisher, *Phys. Rev. Lett.* **49**, 292 (1982).
300. D. S. Fisher, *Phys. Rev. Lett.* **50**, 1486 (1983).
301. L. Sneddon, *Phys. Rev. B* **29**, 719 (1984).
302. R. A. Klemm and J. R. Schrieffer, *Phys. Rev. Lett.* **51**, 47 (1983); **52**, 482 (1984). See also D. S. Fisher, S. N. Coppersmith and M. C. Cross, *Phys. Rev. Lett.* **52**, 481 (1984).
303. P. Bak, *Phys. Rev. Lett.* **48**, 692 (1982).
304. M. Weger and B. Horovitz, *Solid State Commun.* **43**, 583 (1982).
305. S. E. Burkov, V. L. Pokrovsky and G. V. Uimim, *Solid State Commun.* **40**, 363 (1981).
306. S. E. Burkov and V. L. Pokrovsky, *Sold State Commun.* **46**, 609 (1983).
307. For a review, A. J. Berlinsky, *Rep. Progr. Phys.* **42**, 1244 (1979).
308. M. L. Boriack and A. W. Overhauser, *Phys. Rev. B* **15**, 2847 (1977), and *B* **16**, 5206 (1977).
309. T. Tsuzuki and K. Sasaki, *Solid State Commun.* **33**, 1063 (1980) and **34**, 219 (1980) and *Prog. Theor. Phys.* **65**, 19 (1981).
310. W. Wonnenberger and F. Gleisberg, *Solid State Commun.* **23**, 665 (1977). W. Wonnenberger, *Lecture Notes in Physics*, ed. S. Barisic, A. Bjelis, J. R. Cooper and B. Leontic **95**, 311 (1979).
311. Y. Shapira and L. J. Neuringer, *Phys. Rev.* **154**, 375 (1967).
312. P. Russer, *J. Appl. Phys.* **43**, 2008 (1972).
313. W. Wonnenberger, *Z. Phys. B* **53**, 167 (1983).
314. B. A. Huberman and J. P. Crutchfield, *Phys. Rev. Lett.* **43**, 1743 (1979).
315. B. A. Huberman, J. P. Crutchfield and N. H. Packard, *Appl. Phys. Lett.* **37**, 750 (1980).

316. D. D'Humières, M. R. Beasley, B. A. Huberman and A. Libchaber, *Phys. Rev. A* **26**, 3483 (1982).
317. A. H. MacDonald and M. Plischke, *Phys. Rev. B* **27**, 201 (1983).
318. A. O. Caldeira and A. J. Leggett, *Phys. Rev. Lett.* **46**, 211 (198) and *Ann. Phys.* **149**, 374 (1983).
319. W. Wonnenberger, *Z. Phys. B* **50**, 2332 (1983).
320. M. Oda and M. Ido, *Solid State Commun.* **44**, 1535 (1982).
321. J. R. Tucker, *I.E.E.E. Quantum Electron.* **15**, 1234 (1979).
322. J. R. Tucker, J. H. Miller, K. Seeger and J. Bardeen, *Phys. Rev. B* **25**, 2979 (1982).
322a. S. E. Barnes and A. Zawadowski, *Phys. Rev. Lett.* **51**, 1003 (1983).
323. J. B. Sokoloff, *Phys. Rev. B* **23**, 1992 (1981).
324. L. Pietronero and S. Strässler, *Phys. Rev. B* **28**, 5863 (1983).
325. P. Bak, *Rep. Prog. Phys.* **45**, 587 (1982).
326. M. Rice, *Phys. Rev. Lett.* **48**, 1640 (1982).
327. B. Horovitz and S. E. Trullinger, *Solid State Commun.* **49**, 195 (1984).
328. B. Horovitz and J. A. Krumhansl, *Phys. Rev. Lett.* **B29**, 2109 (1984) and **B29**, 7022 (1984).
329. K. Maki, *Phys. Rev. Lett.* **39**, 46 (1977).
330. A. I. Larkin and P. A. Lee, *Phys. Rev. B* **17**, 1596 (1978).
331. K. Maki, *Phys. Lett.* **70A**, 449 (1979).
332. N. P. Ong, G. Verma and K. Maki, *Phys. Rev. Lett.* **52**, 663 (1984).
333. L. P. Gor'kov, *Pis'ma Zh. Eksp. Teor. Fiz.* **38**, 76 (1983), *J.E.T.P. Letters* **38**, 87 (1983).
334. I. Batistic, A. Bjelis and L. P. Gor'kov, *J. Physique* (1984), in press.
335. M. Peyrard and S. Aubry, *J. Phys.* **C16**, 1593 (1983) and references therein.
336. S. Aubry, in *Solitons and Condensed Matter*, ed. A. Bishop and T. Schneider, Solid State Sciences, Vol. 8, Springer-Verlag, Berlin (1978), p. 264.
337. S. N. Coppersmith and D. S. Fisher, *Phys. Rev. B* **28**, 2566 (1983).
338. J. E. Sacco and J. B. Sokoloff, *Phys. Rev. B* **18**, 6549 (1978).
339. S. J. Shenker and L. Kadanoff, *J. Stat. Phys.* **27**, 631 (1982).
340. P. Y. Le Däeron and S. Aubry, *J. Phys. C* **16**, 4827 (1983).
341. C. K. Bak and N. F. Pedersen, *Appl. Phys. Lett.* **22**, 149 (1973), and P. Y. Le Däeron, Thesis, Orsay University (1983).
342. Y. B. Kim, C. F. Hempstead and A. R. Strnad, *Phys. Rev. A* **139**, 1163 (1965).
343. J. I. Gittleman and B. Rosenblum, *J. Appl. Phys.* **39**, 2617 (1968).
344. J. I. Gittleman and B. Rosenblum, *Phys. Rev. Lett.* **16**, 734 (1966).
345. J. Bardeen and M. J. Stephen, *Phys. Rev. A* **140**, 1197 (1965).
346. J. le G. Gilchrist and P. Monceau, *J. Phys. C* **3**, 1399 (1970). See also: G. Fisher, R. D. McConnell, P. Monceau and K. Maki, *Phys. Rev. B* **1**, 2134 (1970) and Y. Brunet, P. Monceau and G. Waysand, *Phys. Rev. B* **10**, 1927 (1974).
347. A. T. Fiory, *Phys. Rev. Lett.* **27**, 501 (1971), *Phys. Rev. B* **8**, 5039 (1973), *Phys. Rev. B* **7**, 1881 (1973).
348. A. Schmid and W. Hauger, *J. Low Temp. Phys.* **11**, 667 (1973).
349. O. Daldini, P. Martinoli, J. L. Olsen and G. Berner, *Phys. Rev. Lett.* **32**, 218 (1974).
350. P. Martinoli, O. Daldini, C. Leemann and E. Stocker, *Solid State Commun.* **17**, 205 (1975).
351. P. Martinoli, O. Daldini, C. Leemann and B. Van Den Brandt, *Phys. Rev. Lett.* **36**, 382 (1976). For reviews see P. Martinoli, *Phys. Rev. B* **17**, 1175 (1978) and P. Martinoli, J. L. Olsen and J. R. Clem, *J. Less Common Metals* **62**, 315 (1978).
352. K. Yamafuji and F. Irie, *Phys. Lett. A* **25**, 387 (1967).
353. R. Labusch, *Phys. Rev.* **170**, 470 (1968).
354. J. P. Jamet and P. Lederer, *J. Physique-Lèttres* **44**, L257 (1983).
355. J. B. Sokoloff and B. Horovitz, *J. Physique* **44**, C3–1667 (1983).
356. G. Mozurkewich and G. Grüner, *Phys. Rev. Lett.* **51**, 2206 (1983).
357. L. P. Gor'kov and E. N. Dolgov, *J. Low Temp. Physics* **42**, 101 (1981).
358. M. Danino and M. Weger, *J. Phys. Chem. Solids* **44**, 691 (1983).

359. G. Mihaly, Gy Hutiray and L. Milaly, *Phys. Rev. B* **28**, 4896 (1983).
360. A. M. Portis, *Mol. Cryst. Liq. Cryst.* **81**, 59 (1982).
361. T. M. Rice, P. A. Lee and M. C. Cross, *Phys. Rev. B* **20**, 1345 (1976).
362. C. Berthier, personal communication.
363. F. Mezei, personal communication.
364. S. N. Artemenko and A. F. Volkov, *Pis'ma Zh. Eksp. Teor. Fiz.* **37**, 310 (1983), *Sov. Phys. J.E.T.P. Lett.* **37**, 368 (1983).

LATTICE DYNAMICAL STUDY OF TRANSITION METAL TRICHALCOGENIDES

F. A. LÉVY and A. GRISEL†

Institut de physique appliquée,
Ecole Polytechnique Fédérale, CH–1015 Lausanne (Switzerland).

1. Lattice Dynamics of Pseudo-One-Dimensional Crystals

The interest in crystals with low-dimensional structures is stimulated by the particular physical properties related to the anisotropy of the structure. From the structural anisotropy new phenomena can directly result such as lattice instabilities or charge redistributions. In at least several cases, significant differences appear with respect to the most investigated cubic or isotropic materials. Nevertheless, there are properties, like optical properties, for which the influence of the large bonding anisotropy is not enhanced with respect to other non-cubic crystals.

The dynamical properties of the crystals with layer or chain-like structures straightforwardly reflect the low-dimensional character of the lattice. Therefore, an investigation of the vibration spectra gives us an additional way to understand the crystal structure. Moreover, analysis of the lattice modes leads towards a striking picture of the bonding and of the relative strength of the valence forces in the crystal.

The most accurate representation of the nature of the bonding and of the bond strengths results from a repeated comparison between the calculated modes based on assumptions for the valence forces in the known crystal structure and the assigned frequencies of the vibration modes measured experimentally by Raman scattering or by infrared spectroscopic investigations.

In the case of the crystals with layered structures, lattice dynamical investigations have led to the distinction between internal modes characteristic of the displacements within the layers, and lower frequency rigid layer modes. In crystals with chain-like structure, the low frequency rigid chain modes are also distinct from the internal modes, even if the chain–chain interactions play an appreciable role. For the transition metal trichalcogenides, the structural anisotropy in the plane perpendicular to the chain axis suggests that the layered character is reflected by rigid layer modes too.

Infrared spectroscopy and Raman scattering are the most frequently used experimental techniques in order to investigate the lattice vibration modes at the center of the Brillouin zone $|\mathbf{k}| = 0$ or Γ. The restriction to the zone center inherently results from experiments involving the interaction of the lattice modes with photons of comparatively small wave vectors. Neutron scattering experiments allow us to extend the investigation for phonons with non-zero wave vectors. In the most favorable cases, the phonon dispersion law can be represented across the whole Brillouin zone.

† Present address: Fondation Suisse Pour la Recherche en Microtechnique, CH–2000 Neuchâtel (Switzerland).

P. Monceau (ed.), Electronic Properties of Inorganic Quasi-One-Dimensional Materials, II, 269–308.

The success of the experimental investigation depends on the availability of suitable samples. Tiny, well-oriented single crystals are sufficient for Raman scattering experiments. Infrared investigations need samples with one large surface. For neutron scattering, bulk crystals of the order of a cubic centimeter are preferable. In the case of crystals with chain-like structures, these requirements constitute the main drawback to the complete investigation of the vibration modes. Usually, single crystals occur in the shape of needles, more rarely in the shape of thin films or flakes.

Lattice dynamical studies of crystals with pseudo-one-dimensional structures are rather scarce, principally because of the necessity of having adequate samples. The first investigation of $(SN)_x$ crystals [1, 2] by Raman scattering and by infrared reflectivity was interpreted on the basis of a simplified three-dimensional model and, actually, the anisotropy of the $(SN)_x$ crystal structure was not emphasized by the investigation. The appropriate valence force model [3] was fitted in order to reproduce the strongest A_g Raman line at 660 cm^{-1} measured with incident and scattered light, both polarized parallel to the polymer chain axis [1]. However, the model appeared to be unable to give a good fit to the other features in the spectra.

This example suggests how difficult it is to find a suitable valence force model on the basis of the experimental investigation of one crystal only. It is more successful to develop a valence force model for a family of compounds. Then it can be checked by mass and force constant scaling. The assignments of the modes and the prediction of their frequencies can be optimized, always trying to keep the number of independent parameters (assumed force constants) as small as possible.

In the case of KCP salts, the experimental studies by infrared spectroscopy and by Raman scattering have been mainly carried out in relation to the investigation of the structural instabilities associated with the Peierls phase transitions occuring in these crystals of pronounced one-dimensional character [4]. Beside the modes connected with the charge-density-wave transition, the stretching modes of the H_2O and of the CN^- molecules have been measured at rather high vibration frequencies. They are affected by the increase in interchain correlations as the temperature decreases. The phonon dispersion relations were measured by inelastic neutrons scattering measurements [5] which made manifest the Kohn anomaly at the phase transition.

For the TTF compounds, an experimental investigation of the Raman spectra has been carried out [6]. The observed vibration modes have been interpreted in terms of lattice modes at low energy (around and below 100 cm^{-1}) [7] and in terms of intramolecular vibrations with strong characteristic Raman peaks in the 1500 cm^{-1} region, associated with the doubly-bonded carbon atoms in the TTF molecules [8]. The symmetry properties could be assigned to the most characteristic modes only, and the results have been mainly discussed with respect to the problem of charge transfer, closely related to the electronic properties.

The lattice dynamics in the transition metal trichalcogenides have been investigated by separate research groups [9–12]. Over and above the other quasi-one-dimensional compounds, these crystals present the advantage of constituting a homogeneous family. Various degrees of low dimensionality are represented in materials with different chemical compositions and with significant related variations in the crystal structure. The empirical valence force model can then be checked by the precise technique of mass scaling.

Moreover, the trichalcogenides comprise compounds with various bonding contributions and with a wide range of electronic properties, from superconductors like $TaSe_3$ [13] to wide-gap semiconductors like ZrS_3 [14]. Intermediate cases with lattice instabilities [15, 16] and charge-density-wave phases are typical of the materials with structures of lower dimension [17]. This diversity is reflected in the bonding in the crystals and consequently in the lattice dynamical properties.

Following this, the lattice dynamical properties have been mostly investigated in the transition metal trichalcogenides among the crystals with quasi-one-dimensional structures. A valence force model has been proposed which accounts for most of the vibrational frequencies in the semiconducting compounds of the $ZrSe_3$ type [18]. It constitues a reliable starting point for the assignment of the symmetries of the experimentally determined modes.

2. Transition Metal Trichalcogenides

The Group IVB transition metal trichalcogenides MX_3 all crystallize in the chain-like monoclinic structure of the $ZrSe_3$ type [19–23]. The crystallographic data have been reviewed in several papers, for example by Hulliger [23] and by Meerschaut and Rouxel [24]. The basic structural units are prismatic columns of MX_6 wedge-shaped trigonal prisms, linked together to build layers which themselves are bound by much weaker forces. Each metal atom lies at the center of a trigonal prism of six anions (Figure 1).

Each trigonal prismatic chain is shifted with respect to the two neighbour chains by half the lattice parameter along the chain axis. In each **(ab)** plane perpendicular to the **c** chain axis, anions belonging to one chain are linked to cations of neighbouring chains (Figure 2). Therefore, each metal atom is closely linked with the six anions of the prism within the chain; in addition it is bound to one anion of each two neighbour chains. The length of bond with the next chain is less than 10% longer than the internal anion–cation bond within the chain (e.g., for $ZrSe_3$: 2.889 with respect to 2.708 Å). In the compounds $HfSe_3$ and $HfTe_3$, the bond length is even shorter between the chains than within the chains. Therefore, the coordination of the metal atom with its two neighbour anions in the equatorial **(ab)** plane is responsible for the strong anisotropy of the interchain bonding and for the layered character of the structure. It plays a significant role for the interpretation of the dynamical vibrations spectra.

Two variants of the $ZrSe_3$ type of structure are distinguished [21]. Each arrangement is a kind of mirror image in the **(ab)** plane, of each other. Both types belong to the monoclinic space group C_{2h}^2 ($P2_1/m$). In the A-type structure, the base of the trigonal prism is an isoceles triangle but in the B-type structure it is scalene [25]. In both types, the short anion–anion bond within the prism nearly corresponds to the bond length in the diatomic molecule.

The primitive unit cell contains eight atoms distributed on two chains (Figure 2). The first setting for the assignment of the crystal axis in the monoclinic structure is choosen so that the **c**-axis (z-coordinate) is the C_2 symmetry axis, parallel with the direction of the chains. In agreement with the predicted electronic balance taking into account the double strong quasi-di-anionic bond within each trigonal prism (one anion–anion diatomic bond per MX_3 chemical formula), the Group IVB compounds are all

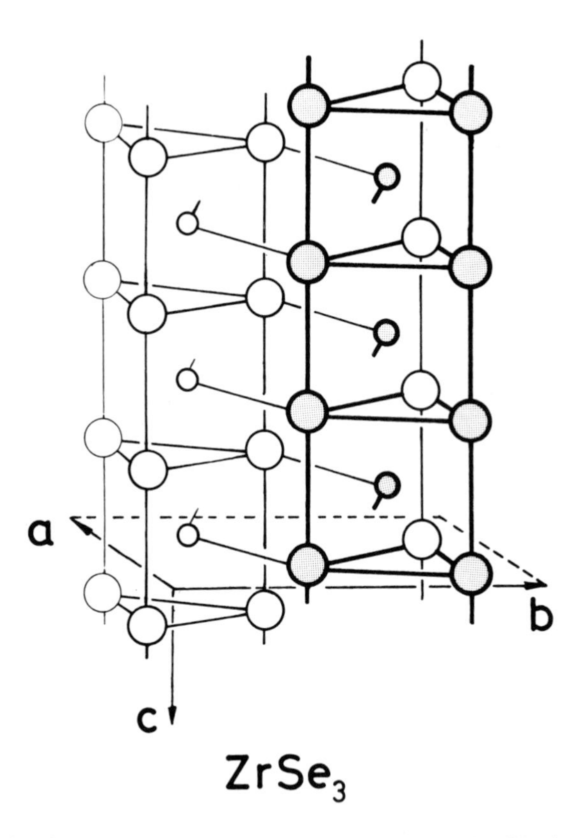

Fig. 1. $ZrSe_3$: Wedge-shape trigonal prismatic units along the c axis. (Metal atoms: small circles; Anions: large circles).

semiconductors. The structural data necessary for the discussion of the experimental results of the vibrational spectra and for the establishment of the valence force model are summarized in Tables I and II. Mixed phases have been synthesized in the systems ZrS_3Se_{3-x} and HfS_xSe_{3-x} while only limited solubility has been found in the Te compounds [20–27].

In the Group VB transition metal trichalcogenides, the effect of the additional metal electron is governed by the behaviour of the strong, nearly diatomic molecular bonds within the trigonal prismatic units, as well as by the interchain coupling emphasized in the $NbSe_3$ structure. In metallic $TaSe_3$, two different chains are stabilized so that the prismatic unit cell contains four chemical formula units. A schema of the structure projected on the equatorial (**ab**) plane is shown in Figure 3a. The relevant crystal lattice data are given in Table I. $TaSe_3$ is stable down to very low temperatures and becomes superconducting below 2.1 K [3].

The $NbSe_3$ type of structure is characterized by three different chains. The interaction is increased by a stronger bonding between anions belonging to the neighbouring chains. The layers become undulated (Figure 3b) and the different one-dimensional structural

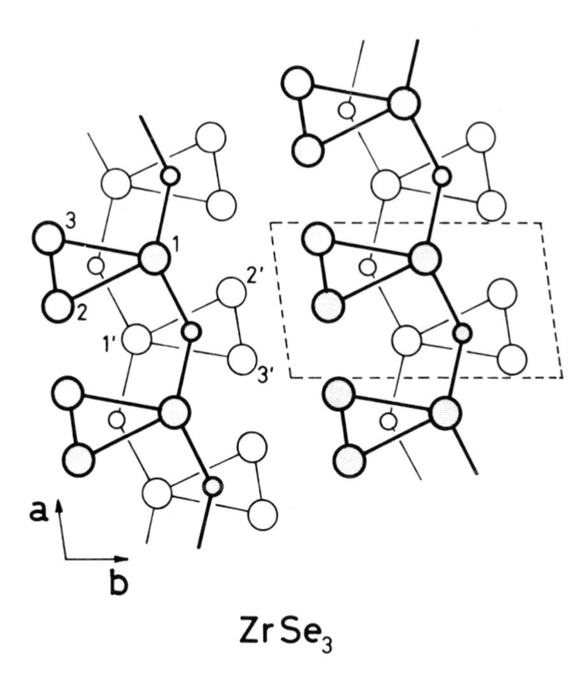

Fig. 2. $ZrSe_3$: Projection on the equatorial (**ab**) plane of the monoclinic structure of the type *A*. (small circles: metal atoms; large circles: anions). Two chemical formula units are in the primitive unit cell.

elements are responsible for the lattice instabilities and for the two charge-density-wave phase transitions at 142 and at 58 K [28].

The structure of NbS_3 is similar to that of $ZrSe_3$ [29–31], with a doubling of the cell dimension along the direction of the chain. The doubling is associated with the pairing of the Nb atoms along the prismatic chain direction, so that the compound is semiconducting at room temperature and below.

The case of TaS_3 appears to be more complex and several phases have been found [15, 16]. The orthorhombic form was synthesized first [32], and it undergoes a Peierls transition at 210 K [33]. The exact structure is not yet very well known. Monoclinic TaS_3 is isotypic with $NbSe_3$ [34] and undergoes two transitions: one at 240 K and the other at 160 K [15, 44].

In the $TaS_3/TaSe_3$ mixed systems several crystals have been synthesized, in particular $Ta(S_{0.15}Se_{0.85})_3$ [35] which becomes, at low temperature, a semiconductor with a periodic lattice distortion.

The two transition metal pentatellurides $HfTe_5$ and $ZrTe_5$ crystallize in a structure closely related to that of the trichalcogenides [36]. However, the bonding between the chains is achieved through an anion zig-zag chain running along the axis of the prismatic columns and in between (Figure 4). In these crystals, puzzling electrical anomalies have been observed at low temperature without, however, evidence of the usually associated structural instabilities [37].

TABLE I

Crystallographic data for the transition metal trichalcogenides and pentachalcogenides (the **c** axis is chosen along the pseudo-one-dimensional chain axis and corresponds to the crystallographic first setting).

MX_3	Symmetry	Space group	Lattice parameters						Ref.	First setting
			a	**b**	**c**	α	β	γ		
TiS_3	monoclinic	C_{2h}^2	4.973	8.714	3.433			97.74	[40]	B
ZrS_3	monoclinic	C_{2h}^2	5.1243	8.980	3.6244			97.28	[20]	A
$ZrSe_3$	monoclinic	C_{2h}^2	5.4109	8.444	3.7488			97.48	[20]	$P2_1/m$ A
$ZrTe_3$	monoclinic	C_{2h}^2	5.8939	10.100	3.9259			97.82	[20]	A
HfS_3	monoclinic	C_{2h}^2	5.09	8.97	3.59			97.38	[41]	B
$HfSe_3$	monoclinic	C_{2h}^2	5.388	9.428	3.7216			97.78	[20]	B
$HfTe_3$	monoclinic	C_{2h}^2	5.879	10.056	3.9022			97.98	[20]	
NbS_3	triclinic	C_i^1	4.963	9.144	6.730			97.17	[30]	$P\bar{1}$
NbS_3	monoclinic	–	9.68	14.83	3.37			109.9	[42]	
$NbSe_3$	monoclinic	C_{2h}^2	10.009	15.629	3.481			109.47	[43]	$P2_1/m$
TaS_3	orthorhombic	D_2^5	36.804	15.177	3.340				[44]	$C222_1$
TaS_3	monoclinic	C_{2h}^2	9.515	14.912	3.341			109.99	[44]	
$TaSe_3$	monoclinic	C_{2h}^2	10.042	9.829	3.495			106.26	[45]	
$TaSe_3$	monoclinic	–	10.02	15.65	3.48			109.6	[42]	$>2 GPa$
$NbTe_4$	tetragonal	C_{4v}^5/D_{4h}^2	6.449		6.836				[46]	
$TaTe_4$	tetragonal	C_{4v}^5/D_{4h}^2	6.514		6.810				[46]	
$ZrTe_5$	orthorhombic	D_{2h}^{17}	13.727	14.502	3.988				[47]	$Cmcm$
$HfTe_5$	orthorhombic		13.730	14.492	3.974				[47]	

TABLE II

Characteristic distances (in Å) in the transition metal compounds, useful for the description of the lattice dynamical properties (cf. Figures 1–4). [20, 39]

(a) MX_3 compounds of the $ZrSe_3$ structural type [20, 39]

	A			B		
Compound	ZrS_3	$ZrSe_3$	HfS_3	TiS_3	$HfSe_3$	$ZrTe_3$
Distances within the chains						
M_I-2X_1	2.601	2.707	2.588	2.496	2.752	3.030
M_I-2X_2	2.602	2.725	2.612	2.668	2.936	3.162
M_I-2X_3	2.605	2.737	2.590	2.358	2.586	2.771
X_2-X_3	2.089	2.351	2.102	2.038	2.333	2.761
X_1-X_2	3.568	3.740	3.581	3.758	4.149	4.601
X_1-X_3	3.577	3.773	3.575	3.285	3.611	3.959
Distances between the layers						
$M_1-X_{1'}$	2.724	2.889	2.697	2.416	2.624	2.829
X_2-X_3	3.035	3.061	2.991	2.921	3.057	3.134
Distances between the slabs						
$X_2-2X_{2'}$	3.660	3.886	3.601	3.508	3.734	3.899

(b) $TaSe_3$ [45]

Chain I		Chain II	
Ta_I-2Se_1	2.643	$Ta_{II}-2Se_5$	2.642
Ta_I-2Se_4	2.637	$Ta_{II}-2Se_3$	2.642
Ta_I-2Se_2	2.651	$Ta_{II}-2Se_6$	2.644
Ta_I-Se_6	2.823	$Ta_{II}-Se_2$	2.718
Ta_I-Se_1	2.779	$Ta_{II}-Se_6$	2.715
Se_1-Se_4	2.896	Se_3-Se_5	2.576

Distances between the slabs

Se_4-Se_5	2.653

Distances between the layers

Se_4-2Se_5	3.642

(c) TaS_3 (monoclinic) [44]

Chain I		Chain II		Chain III	
Ta_I-2S_1	2.495	$Ta_{II}-2S_4$	2.462	$Ta_{III}-2S_7$	2.512
Ta_I-2S_2	2.527	$Ta_{II}-2S_5$	2.529	$Ta_{III}-2S_8$	2.481
Ta_I-2S_3	2.509	$Ta_{II}-2S_6$	2.522	$Ta_{III}-2S_9$	2.520
Ta_I-S_1	2.564	$Ta_{II}-S_1$	2.747	$Ta_{III}-S_5$	2.566
Ta_I-S_6	2.566	$Ta_{II}-S_8$	2.870	$Ta_{III}-S_8$	2.589
S_2-S_3	2.105	S_4-S_5	2.835	S_7-S_9	2.068

Table II (*continued*)

S–S distances between the chains	
S_2-S_4	2.796
S_6-S_9	2.920

S–S distances between the slabs	
S_2-S_4	3.694
S_3-S_7	3.259
S_7-S_7	3.499

(d) $NbSe_3$ [48]

Chain I		Chain II		Chain III	
Nb_I-2Se_1	2.647	$Nb_{II}-2Se_4$	2.629	$Nb_{III}-2Se_7$	2.652
Nb_I-2Se_2	2.670	$Nb_{II}-2Se_5$	2.668	$Nb_{III}-2Se_8$	2.630
Nb_I-2Se_3	2.636	$Nb_{II}-2Se_6$	2.645	$Nb_{III}-2Se_9$	2.657
Nb_I-2Se_2	2.726	$Nb_{II}-Se_1$	2.857	$Nb_{III}-Se_5$	2.738
Nb_I-Se_6	2.729	$Nb_{II}-Se_8$	2.949	$Nb_{III}-Se_8$	2.750
Se_2-Se_3	2.485	Se_4-Se_5	2.909	Se_7-Se_9	2.374

Se–Se distances between the chains	
Se_2-Se_4	2.733
Se_6-Se_9	2.929

Se–Se distances between the slabs	
Se_2-Se_4	3.481
Se_3-Se_7	3.289
Se_7-Se_7	3.481

(e) $HfTe_5$ [47]

$Hf-2Te_1$	2.945	Te_2-Te_2	2.763
$Hf-4Te_2$	2.944	Te_3-Te_3	2.908
$Hf-2Te_3$	2.960		

Single crystals of the transition metal trichalcogenides have been generally obtained by vapour transport or by chemical transport reactions [38]. The compounds in the Group IVB chalcogenides crystallize mostly in the shape of flakes suitable for optical investigations. In contrast, the Group VB compounds grow in the shape of ribbons or of needles. The external shape of the crystals brings about restrictive conditions for the investigation of vibrational properties. The z axis lies in the plane of the flake or of the ribbon which are parallel with the (**ac**) planes of the structural layers.

Infrared reflectivity or transmission measurements are only possible with rather large samples, so that crystals of the Group IV chalcogenides only have been measured.

The geometry adopted for Raman scattering provides the incident-radiation beam inclined at about 30° with respect to the sample z axis. The scattered light is observed

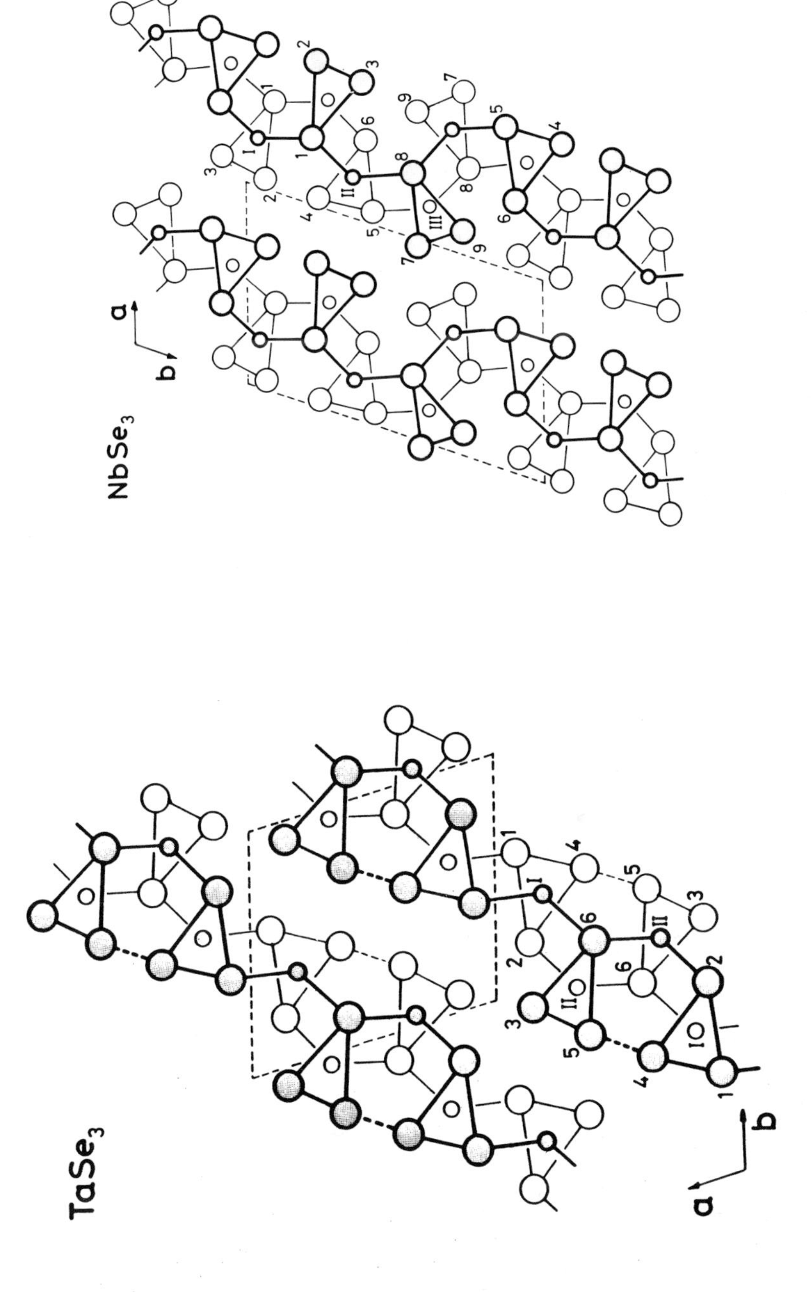

Fig. 3. (a) $TaSe_3$: Projection of the monoclinic structure on the equatorial (**ab**) plane. There are two different prismatic chains I and II. The short anion 4–anion 5 distance contributes to the coupling between the chains (small circles: Ta, large circles: Se). (b) $NbSe_3$: Projection of the monoclinic structure on the (**ab**) plane perpendicular to the c-chain axis. The structure is built by three different prismatic chains I, II and III. The short distance between the anions 2 and 4 comparable with the interchain distances demonstrates the strong interaction between the chains (small circles: Nb, large circles: Se).

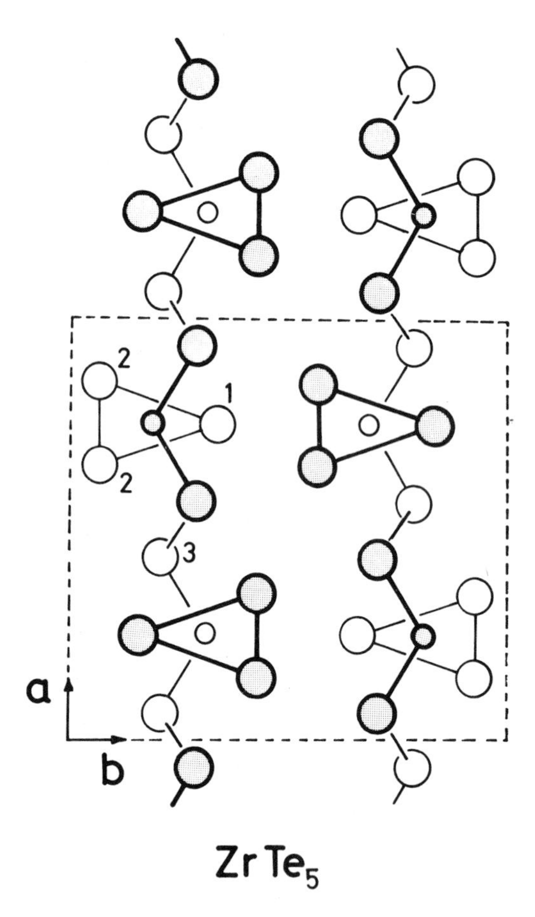

Fig. 4. $ZrTe_5$: Projection of the orthorhombic structure on the (**ab**) plane perpendicular to the chain axis c. The wedged shaped trigonal prismatic columns are linked together by Se(3) zig-zag chains in between (small circles: metal, large circles: anions).

with the widest possible angular aperture at nearly perpendicular incidence. This configuration allows one to avoid the reflected beam and to separate the selection rules of symmetry A_g (x^2, y^2, z^2, xy) from B_g (xz, yz); that is, to distinguish between vibrations perpendicular to the chain and those parallel with the chains [39].

3. Group Theoretical Investigation of the Dynamical Properties

The group theoretical analysis of the lattice vibration symmetries gives the qualitative and geometrical information on the dynamical properties of the crystalline solid. The information that may be obtained includes: symmetry, number and degeneracy of the normal modes of vibration, their optical selection rules, and the direction of their atomic displacements.

The normal modes of the lattice can be described as the irreducible representations of the symmetry factor group. This factor group is isomorphic to the point group, which consists of all operations of the space group without the primitive translations [49].

It has been shown that the correlation method is a simple and convenient means for determining the irreducible representations and the selection rules for layered [50], and, furthermore, for chain-like crystal structures. In applying the correlation method to quasi-one-dimensional materials such as the trichalcogenides, we may take advantage of the individual chain symmetry.

If the primitive unit cell in the structure contains more than one chain as it is the case for all the trichalcogenides described here, we shall define a pseudolattice (non-physical pseudo-crystal).

The pseudolattice in our case is constructed by a primitive unit cell containing the atoms of only one chain. It allows a symmetry analysis of the adjacent chain. As the site symmetry of certain atoms are higher in the pseudolattice than in the real crystal; the correlation between the chain factor group and the crystal factor group is useful for the identification of the crystal normal modes.

The correlation diagram for the $ZrSe_3$ structure is represented in Figure 5, considering both chain and crystal structure. The symmetry of $ZrSe_3$ is characterized by the space group $C_{2h}^2(P2_1/m)$. The chain factor group is C_{2v}. In applying the correlation method to the pseudolattice (structure constituted by isolated chains), the site groups of the atoms in the chain symmetry are determined first. The Zr atom and one Se atom are located in sites which have the C_{2v} symmetry, while for both other Se atoms the site group is C_s. The representations that describe the vibrational modes at a particular site must have translation transformation properties, they are listed in the first left-hand column of Figure 5.

The correlations between the representations of the chain factor group C_{2v} and of the site groups are given by the correlation table of the point group C_{2v} and its subgroups [51].

The irreducible representations labelling the lattice vibrations of the isolated chains are therefore:

$$A_2 + 3B_1 + 4A_1 + 4B_2 \tag{3.1}$$

With help of a similar procedure, but starting from the atoms in the whole crystal structure we obtain, by correlation, the irreducible representation associated with the lattice vibrations in the real crystal:

$$4A_u + 4B_g + 8B_u + 8A_g \tag{3.2}$$

The linear and bilinear transformation properties are given in Figure 5 underneath each irreducible representation. The number of times an irreducible representation occurs in the decomposition of the total representation is directly related to the number of vibrational degrees of freedom associated with the specific vibrational mode. The correlations between the representations of the chain factor group and the crystal factor group are finally obtained through their common subgroup C_s.

This symmetry analysis yields a first description of the vibrational normal modes occuring in the $ZrSe_3$ structure.

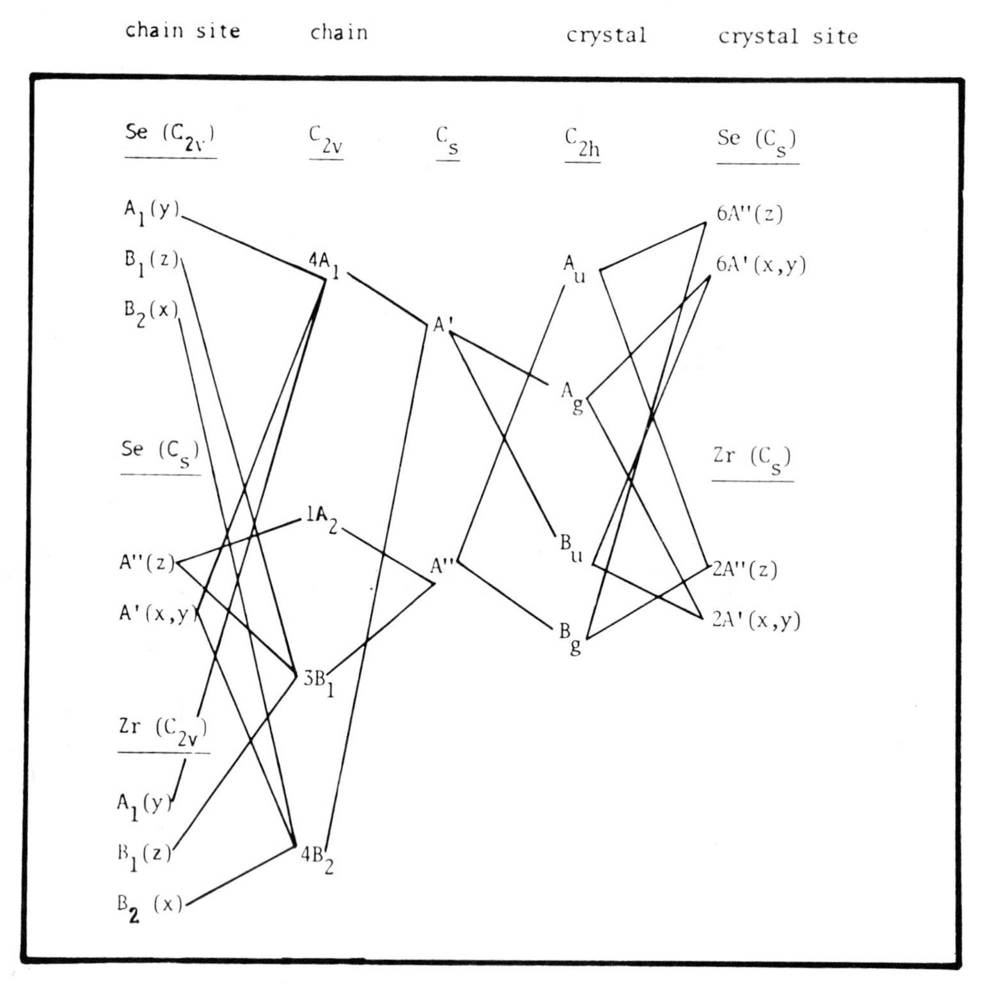

$$\begin{array}{llllllll} \Gamma_{\text{chain}} = & A_2 & + & 3B_1 & + & 4A_1 & + & 4B_2 \\ & xz & & z;yz & & y;x^2,y^2,z^2 & & x;xy,R_z \end{array}$$

$$\begin{array}{lllll} \Gamma = & 4A_u & 4B_g & 8B_u & 8A_g \\ & z & xz,yz & x,y & x^2,y^2,z^2,xy \end{array}$$

Fig. 5. Correlation diagram for the monoclinic $ZrSe_3$ structure.

Since 8 atoms are contained in the primitive unit cell, 24 vibrational normal modes may occur, among which 3 are acoustical modes.

Because of the inversion center situated between the two chains in the middle of the primitive unit cell, the atoms of one chain will vibrate in phase or in counter-phase with respect to the atoms of the second chain. The vibrational modes are then gathered in odd–even pairs in which one mode is IR active and other Raman active.

Each vibrational mode of the isolated chain (C_{2v}) then gives rise to two modes of the (C_{2h}) real crystal structure. The atomic displacements of the crystal vibrational modes as well as their Raman and IR activities will then be analysed through the knowledge of the chain modes and the correspondence mentioned before.

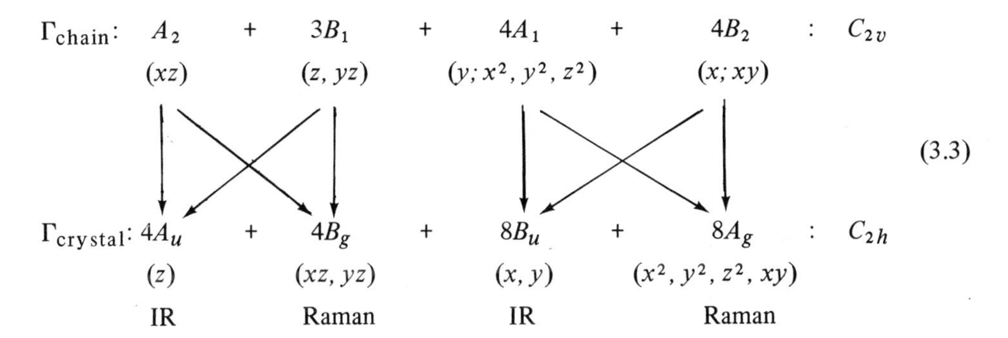

(3.3)

The $4A_u + 4B_g$ modes, on the one hand, have their displacements along the chain axis and according to the correlation rules will then be described by the A_2 and B_1 modes. The $8B_u + 8A_g$ modes, on the other hand, have their atomic displacements in the plane perpendicular to the chains.

Because of the inversion center in the C_{2h} symmetry, the antisymmetric vibrational modes are IR active and the symmetric ones are Raman active exclusively.

From the character tables [52] the optical selection rules for each of the representation appearing in (3.3) can be determined. We find that A_u and B_u representations transform as z and (x, y) respectively.

Except for the three long-wavelength acoustical modes corresponding to rigid translations of the lattice, these modes are associated with an electric dipole transition responsible for the IR activity.

With the z-axis parallel to C_2, according to the first crystallographic setting the IR selection rules are expressed by:

$$A_u \quad : \quad z \qquad\qquad B_u \quad : \quad x, y \tag{3.4}$$

The A_g and B_g Raman active modes, on the other hand, have their selection rules coming from the components of the polarizability tensor [53]

$$A_g \quad : \quad \begin{bmatrix} a & d & 0 \\ d & b & 0 \\ 0 & 0 & c \end{bmatrix} \qquad\qquad B_g \quad : \quad \begin{bmatrix} 0 & 0 & e \\ 0 & 0 & f \\ e & f & 0 \end{bmatrix}$$

which is also expressed by:

$$A_g \quad : \quad x^2, y^2, z^2, xy \qquad\qquad B_g \quad : \quad xz, yz \tag{3.5}$$

In order to describe the displacements of the normal modes of the isolated chain and of the whole crystal in the same reference axis the orthorhombic (C_{2v}) symmetry elements have to be expressed in the crystal reference system.

The C_2 axis in the C_{2v} symmetry is then perpendicular to the chains; its direction, which we call y' makes a small angle (7.48° in $ZrSe_3$) with the y axis in the crystal.

The symmetry elements are:

$$E, \quad C_2(y'), \quad \sigma_h(x, y'), \quad \sigma_v(y', z)$$

and the vibrational modes of the isolated chain are characterized by the following irreducible representations: $A_1(y')$, A_2, $B_1(z)$, $B_2(x)$, of which the corresponding polarizability tensors are:

$$A_1 : \begin{pmatrix} a & . & . \\ . & b & . \\ . & . & c \end{pmatrix} \quad A_2 : \begin{pmatrix} . & . & d \\ . & . & . \\ d & . & . \end{pmatrix} \quad B_1 : \begin{pmatrix} . & . & . \\ . & . & e \\ . & e & . \end{pmatrix} \quad B_2 : \begin{pmatrix} . & f & . \\ f & . & . \\ . & . & . \end{pmatrix}$$

from which we deduce the selection rules:

$$A_1 \quad : \quad x^2, y'^2 z^2, y' \qquad \text{R, IR} \tag{3.5}$$

$$A_2 \quad : \quad xz \qquad \text{R} \tag{3.6}$$

$$B_1 \quad : \quad y'z, z, \qquad \text{R, IR}$$

$$B_2 \quad : \quad xy', x, \qquad \text{R, IR}$$

Within this description, both crystal and chain modes are classified according to the atomic displacements either parallel to the chains (z): B_g, A_u for the crystal; A_2, B_1 for the chain, or perpendicular to them (x, y) A_g, B_u for the crystal and A_1, B_2 for the chain.

For this group theoretical study, the first setting is used here according to our first reference [54]. The second crystallographic setting is often used [11, 55] where the chain axis is along **b** instead of **c** in the above description.

C_2 in C_{2h} is parallel to y in the second setting and C_2 in C_{2v} is then parallel to z' (nearly $\parallel z$).

In the second setting description, the decomposition of the representation associated with the isolated chain can be $\Gamma(C_{2v}) \equiv 4A_1 + A_2 + 4B_1 + 3B_2$. B_1 and B_2 being inverted because of the definition of the symmetry plans (σ_h and σ_v). The essential correspondence between both settings is to replace the z axis by the y axis in the definition [11].

In conclusion, the interpretation of the Raman and IR data will be mainly achieved by the optical selection rules obtained by symmetry analysis: they are those deduced from the representations associated with the crystal symmetry group. The symmetry analysis connected with the isolated chains, however, gives more detailed information on the internal motions of the atoms in the unit cell. It also permits us to distinguish between internal and external vibrational modes with respect to the chains and to give qualitative ideas for preferential IR and Raman activities of the internal modes.

4. Model of Lattice Dynamics: Valence Force Field Calculation

The vibration frequencies of the trichalcogenide lattices have been calculated on the basis of central forces and valence forces models [54, 18, 25, 56]. Such a semi-empirical model brings, on the one hand, a support for the interpretation of the vibrational spectra. On the other hand, it contributes to represent the nature of the chemical bonding within the crystal structure. In particular it can emphasize how both one- and two-dimensional interactions are involved in the lattice dynamics of these trichalcogenide compounds.

A first approach has been tried using a central force model [54]. However, it has been clearly demonstrated that the introduction of angular forces were necessary to fit the experimentally observed vibration frequencies.

Two valence force field models have been discussed [39, 56] on the bases of experimental data. The chronologically first model gives a description for the whole isostructural family of the studied compounds $ZrSe_3$, ZrS_3, HfS_3, $ZrTe_3$, TiS_3 and $HfSe_3$. This model, illustrated in Figure 6, includes the stretching forces described by the c_i ($i = w, b, v$) force constants and the angular bending forces characterized by the K_i ($i = w, b$) force constants.

Six different force constants have been introduced representing bonding forces in the single crystal lattice. They have been calculated in order to fit the experimental vibrational frequencies (Tables IV and V). Within each chain, central forces act between the metal atom and its six nearest chalcogen neighbours; these (M—X) bonds are characterized by one force constant C_w. The chalcogen–chalcogen (X—X) pairing

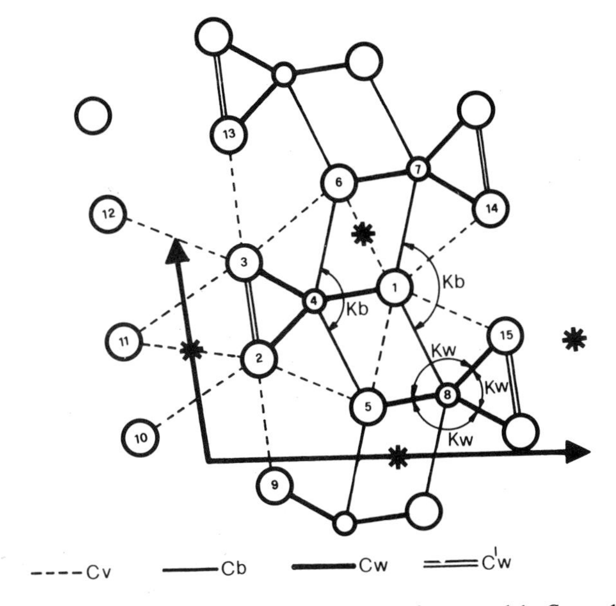

Fig. 6. Force constants used as parameters in the valence force model: C_w and $C_{w'}$ represent the stretching force constants within the chain, C_b is a stretching force constant between the chains, C_v corresponds to an anion–anion interaction between the chains which acts also between the layers. K_w and K_b represent the bending force constants within and between the chains respectively.

within one chain corresponds nearly to the diatomic molecule and is characterized by C'_w. The bond angle restoring forces occurring in the chain are represented by the constant K_w.

The central forces between the chains are characterized by C_b for the M—X bonds and C_v for the X—X bonds. Finally, K_b describes the bond bending forces which occur between the chains.

The potential energy of the valence force model in the internal coordinates can be expressed as

$$\begin{aligned} 2V = {} & 2C_w(\delta R_{41}^2 + \delta R_{42}^2 + \delta R_{43}^2) + C'_w \delta R_{23}^2 + \\ & + C_b(\delta R_{17}^2 + \delta R_{18}^2 + \delta R_{45}^2 + \delta R_{46}^2) + \\ & + 2K_w(R_{41}R_{42}\delta\theta_{412}^2 + R_{41}R_{43}\delta\theta_{413}^2 + R_{42}R_{43}\delta\theta_{423}^2) + \\ & + K_b(R_{17}R_{18}\delta\theta_{178}^2 + R_{45}R_{46}\delta\theta_{456}^2) + \\ & + 2C_v(\delta R_{15}^2 + \delta R_{16}^2 + \delta R_{1,14}^2 + \delta R_{1,15}^2 + \delta R_{25}^2 + \delta R_{2,10}^2 + \\ & + \delta R_{2,11}^2 + \delta R_{3,11}^2 + \delta R_{3,12}^2 + \delta R_{36}^2) \end{aligned} \tag{4.1}$$

where δR_{ij} and $\delta\theta_{ijk}$ are the incremental changes in bond length and bond angle associated with the atomic displacements in the lattice vibrations, and i, j and k label the specific atoms, as shown in Figure 6. In the angular internal coordinates the first index indicates the central atom of the considered angle.

It is obvious that the force constants defined above are not independent with respect to their effect on the frequencies of the normal modes.

Since the considered structure has C_{2h}^2 symmetry (see Section 3), the symmetric and antisymmetric lattice vibrations can be treated separately. The secular determinant for the symmetric modes is therefore of order twelve (8 atoms per unit cell) and contains six unknown force constants.

The values of the force constants have been determined by minimizing the fractional deviation of the calculated frequencies from the seven Raman frequencies observed for each compound in [18, 25, 39] (see Section 5).

The force constants obtained by this procedure are listed in Table III. C'_w is essentially determined by the diatomic mode resulting from the X—X pairing within the chains, while C_w influences every internal mode. According to the experimental results (Section 5), a first adjustment of these two constants is necessary to fit the most visible Raman active modes.

C_b and C_v have been introduced to represent the stretching forces between the chains and between the layers, respectively. They are adjusted mainly by means of the 'rigid chain' modes. It has been demonstrated [57] that the seven Raman frequencies and the three observed IR active modes of the set of six compounds have found better calculated values by introducing the two bending force constants K_w and K_b. One may remark that the stretching and bending forces are not really independent with respect to their influence on the vibrational frequencies, they had to be adjusted mainly considering specific modes like the 'rigid sublattice' modes for the K_w constant (see Figure 7).

As shown in Tables IV and V the agreement between observed and calculated wavenumbers appears to be generally good. The comparison between the described isostructural

TABLE III

List of the force constant values for the different trichalcogenides compounds used in the valence force model and adjusted to fit the vibrational frequencies according to [18, 39]. C_v/C_w gives a quantified approximation of the anisotropic bonding within and between layers while C_b/C_w compares force constants within chains and between the chains in one layer.

		Force constants (dyn/cm)						Ratios	
Type	Crystal	C_w	K_w	C'_w	C_b	K_b	C_v	C_v/C_w	C_b/C_w
A	ZrS_3	4.3×10^4	1.6×10^4	2.3×10^5	7.1×10^4	1.4×10^2	5.7×10^2	0.013	1.65
A	HfS_3	4.0×10^4	1.1×10^4	2.4×10^5	5.2×10^4	1.7×10^4	3.3×10^2	0.008	1.30
A	$ZrSe_3$	3.2×10^4	3.8×10^3	2.0×10^5	3.9×10^4	1.7×10^4	7.2×10^2	0.022	1.22
B	TiS_3	2.9×10^4	1.3×10^4	2.7×10^5	7.1×10^4	1.8×10^4	5.2×10^2	0.018	2.44
B	$HfSe_3$	3.2×10^4	1.3×10^4	1.8×10^5	6.2×10^4	1.5×10^4	4.5×10^2	0.014	1.93
B	$ZrTe_3$	3.4×10^4	3.0×10^3	1.6×10^5	3.1×10^4	4.7×10^3	5.3×10^2	0.015	0.91

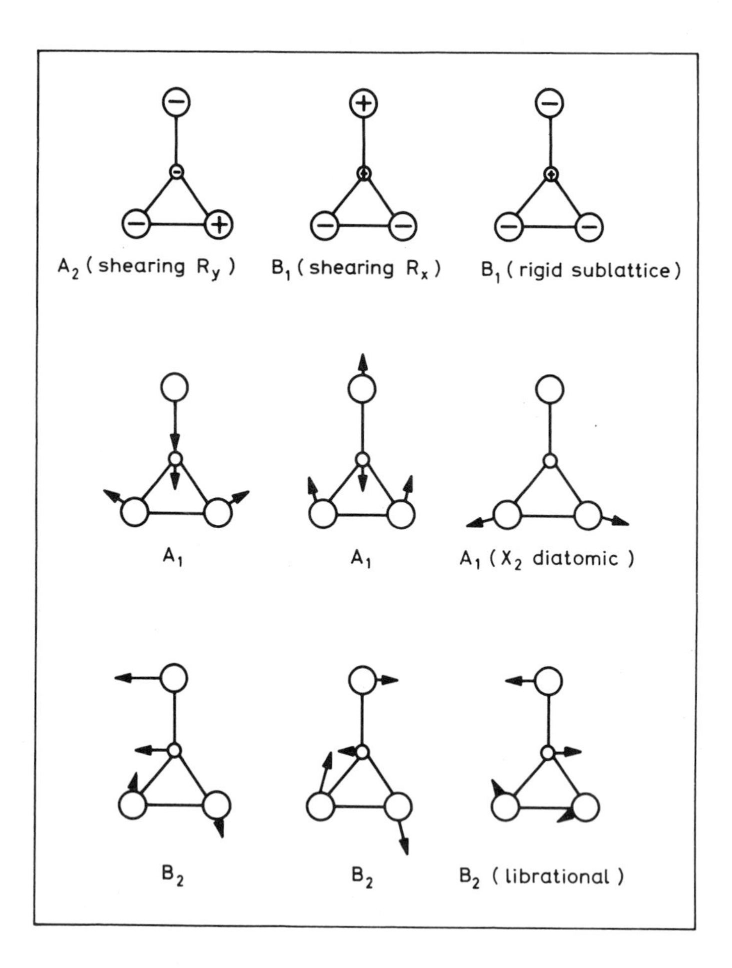

Fig. 7. Schematic representation of the atomic displacements in one chain for the internal vibrational normal modes of the $ZrSe_3$ single crystal.

compounds shows that one single model seems to describe, in the frame of the approximation of the dynamical properties, the whole class of these trichalcogenides. However, it has to be noticed that for specific cases ($ZrSe_3$ and $ZrTe_3$ for example) quite different values of the force constants had to be introduced. This fact suggests that some properties of the chemical bonding are specific to each compound which are already visible in the lattice dynamical properties. The nature of the chemical bonding can still be different depending on the different elements involved in the compounds. Also the long range Coulomb force constants have not been introduced so far, and the electronic structure has not been taken into account.

With help of the described valence force model, the atomic displacements, as well as the frequencies are easily calculated for each normal vibrational modes. Actually,

TABLE IV

Raman and infrared frequencies of ZrS_3, HfS_3, and $ZrSe_3$ compared with the calculated frequencies of the valence-force model. The letters Tr enclosed in parentheses indicate transmission data. The 16 cm^{-1} value given for the lowest frequency rigid-chain vibration of $ZrSe_3$ was calculated by mass scaling (ZrS_3 was the reference).

Vibrational symmetry			ZrS_3 (cm^{-1})		HfS_3 (cm^{-1})		$ZrSe_3$ (cm^{-1})	
Chain	Crystal or Layer	Transf. properties	Raman or IR	Calc.	Raman or IR	Calc.	Raman or IR	Calc.
A_1	A_g	x^2, y^2, z^2	R 530	529	R 524	530	R 302	303
(X_2-diatomic)	B_u	y	IR 515(TO) 522(LO)	529	IR 525(Tr)	530	–	303
B_1 (rigid-	B_g	yz	–	302	–	235	–	177
sublattice)	A_u	z	IR 247(TO) 306(LO)	301	IR 215(TO) 226(LO)	234	IR 200(TO) 223(LO)	176
A_1	A_g	x^2, y^2, z^2	R 322	333	R 322	329	R 236	240
	B_u	y	–	356	IR 286(TO) 300(LO)	343	–	265
B_2	A_g	xy	–	310	–	250	–	176
	B_u	x	IR 305(TO) 334(LO)	315	IR 272(TO) 280(LO)	251	IR 224(TO) 248(LO)	178
A_1	A_g	x^2, y^2, z^2	R 282	206	R 261	171	R 177	140
	B_u	y	–	207	–	174	–	106
B_2	A_g	xy	–	176	–	164	–	96
	B_u	x	IR 247(TO) 254(LO)	186	IR 249(TO) 252(LO)	171	IR 144(TO) 146(LO)	99
A_2	B_g	xz	R 152	173	R 140	160	R 77	86
(shearing)	A_u	(z)	IR 180(Tr)	173	IR 160(Tr)	160	–	86
B_1	B_g	yz	–	159	–	150	–	85
(shearing)	A_u	z	IR 128(Tr)	159	IR 130(Tr)	150	–	84
A_1 (acous./	A_g	xy	R 152	135	R 127	107	R 78	71
rigid-chain)	B_u	x	–	0	–	0	–	0
B_2 (librational)	A_g	xy	R 110	107	R 73	74	R 50	50
	B_u	x, R_t	IR 103(Tr)	57	IR 104(Tr)	46	–	29
B_2 (acous./	A_g	x^2, y^2, z^2	R 21	21	R 15	15	(16)	16
rigid-chain)	B_u	y	–	0	–	0	–	0
B_1 (acous./	B_g	yz	–	23	–	15	–	20
rigid-chain)	A_u	z	–	0	–	0	–	0

TABLE V

Raman and infrared vibrational frequencies of TiS_3, $HfSe_3$, and $ZrTe_3$ compared with the calculated frequencies of the valence force model. The values in parentheses were determined by mass-scaling the rigid-chain frequency of TiS_3.

Vibrational symmetry			TiS_3 (cm^{-1})		$HfSe_3$ (cm^{-1})		$ZrTe_3$ (cm^{-1})	
Chain	Crystal or Layer	Transf. properties	Raman or IR	Calc.	Raman or IR	Calc.	Raman or IR	Calc.
A_1 (diatomic)	A_g	x^2, y^2, z^2	R 559	561	R 296	299	R 216	216
	B_u	y	–	561	–	299	–	217
B_1 (rigid-	B_g	yz	–	314	–	180	–	162
sublattice)	A_u	z	IR 294.5(TO) 328(LO)	313	IR 160(TO) 181.5(LO)	179	–	162
A_1	A_g	x^2, y^2, z^2	R 371	385	R 209	215	R 144	161
	B_u	y	–	411	–	230	–	180
B_2	A_g	xy	–	339	–	189	–	146
	B_u	x	IR 333(TO) 348.5(LO)	347	IR 182(TO) 193(LO)	193	–	147
A_1	A_g	x^2, y^2, z^2	R 300	210	R 171	125	R 108	108
	B_u	y	–	197	–	122	–	83
B_2	A_g	xy	–	149	–	98	R 86	76
	B_u	x	IR 221(TO) 227.5(LO)	159	IR 140(TO) 144.5(LO)	104	–	78
A_2 (shearing)	B_g	xz	R 102	145	R 69	96	R 64	68
	A_u		–	145	–	95	–	68
B_1 (shearing)	B_g	yz	–	129	–	85	–	63
	A_u	z	–	129	–	85	–	63
A_1 (rigid-chain/	A_g	x^2, y^2, z^2	R 176	134	R 101	82	R 62	52
acous.)	B_u	y	–	0	–	0	–	0
B_2 (librational)	A_g	xy	–	112	R 69	67	R 38	39
	B_u	x, R_t	–	53	–	34	–	21
B_2 (rigid-chain/	A_g	xy	R 20	20	(12)	12	(11)	11
acous.)	B_u	x	–	0	–	0	–	0
B_1 (rigid-chain/	B_g	yz	–	25	–	14	–	14
acous.)	A_u	z	–	0	–	0	–	0

they cannot be determined by symmetry arguments only. The direction and length of the eigenvectors are then a result from both symmetry analysis and model calculation.

The displacements vectors calculated in [39] are schematically drawn in Figure 7, where the nine optically normal vibrations of the isolated chains are represented for the $ZrSe_3$ compound.

Referring to Section 3, the crystal vibrational modes are the combination of the two chains vibrating in phase and in counter-phase. The rigid chain modes (they would be acoustic if there were a single chain in the unit cell) are optical modes, as drawn in Figure 8.

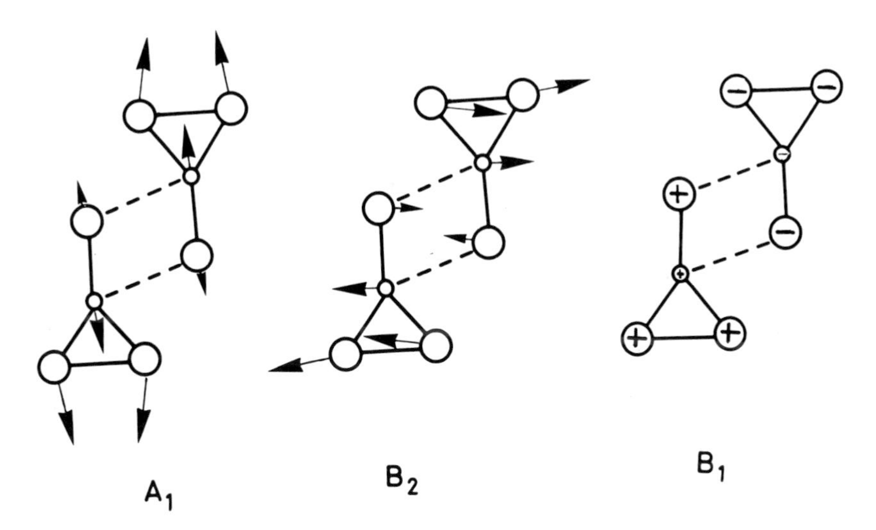

Fig. 8. Schematic representation of the atomic displacements in the 'so-called' rigid-chain vibrations also described as external modes.

From this analysis it is possible to draw the following particular conclusions:

(1) For the A_2 and B_1 modes, the atomic motions are actually directed along the chain axis z.

(2) The existence of the B_1 mode where the anions vibrate as a rigid sublattice in counter-phase with the cations shows the importance of the angular restoring force constant K_w with respect to the pure stretching force constant C_w.

(3) In the vibrational modes of A_1 symmetry the displacements are, as expected, mainly directed along the y axis. However, the x component is appreciable and is actually observed in the Raman and in the IR spectra (see Section 5).

(4) The B_2 modes show the importance of the interchain bonding forces since the associated complex atomic motions cannot be deduced by a simple picture of the isolated chain.

Among the three optical external modes or 'rigid-chain modes' drawn in Figure 8 a real 'rigid-chain' characteristic is well respected only in the z direction.

The interactions between the chains are clearly demonstrated in the A_1 modes by the stretching of the chains and by the rocking behaviour in the B_2 modes.

In [56] a similar valence force field model has been introduced whose results have been compared to the model before described. The main conclusions on the intra-chain forces are very similar; there is however a disagreement about the interpretation of the interchain coupling. Sourisseau *et al.* [56] actually did not interpret the Raman low frequency lines as the 'rigid chain' modes. However, according to Wieting *et al.* [18, 25]

these modes have been observed in most of the compounds, where they do follow a simple mass scaling which tend to confirm their interpretation as rigid chain modes.

5. Experimental Investigation of the MX_3 Lattice Vibration Spectra

The Raman scattering spectroscopy and the far-infrared absorption or reflection are the first main experimental techniques to investigate the lattice vibrations spectra of a crystalline solid.

Several authors have reported Raman and infrared spectra. There are 21 expected optical phonons which are exclusively Raman or infrared active. A first study of ZrS_3 appeared in 1972 [70]. More recently many experimental investigations have been carried out, starting with the type A crystals ZrS_3, HfS_3 and $ZrSe_3$ [10, 11, 12]. Grisel *et al.* have reported comparative IR and Raman data in both type A and type B structures: ZrS_3, HfS_3, $ZrSe_3$, TiS_3, $ZrTe_3$, and $HfSe_3$ [18, 25, 39]. At the same time Zwick *et al.* described quite similar Raman spectra on $ZrSe_3$ [11] and ZrS_3, $ZrTe_3$, $HfSe_3$ [63], while Jandl *et al.* published the Raman spectra of ZrS_3 and HfS_3 [10, 55]. The latter authors also published infrared spectra of the same compounds including TiS_3 [41, 59, 67, 71]. The mixed crystals $HfS_{3-x}Se_x$ [27] and $ZrS_{3-x}Se_x$ [57] have been investigated by Raman scattering experiments.

In spite of the significant experimental effort, all the optical phonons have not been observed and the interpretations are still controversial. In order to understand the spectra, a model is moreover necessary, (Section 4) where the force constants appear as adjustable parameters. A comprehensive confirmation of the force constants and the vibrational frequencies needs complementary experiments like neutron scattering measurements.

The available data do, however, enable some assumptions to be formulated concerning the type and the strength of the bonding forces in the crystals. The discussion will be restricted to a comparison of the experimental data sets with each other and with the calculated frequencies obtained by the valence force model which has been mentioned in Section 4 and which is described in detail in ref [39].

5.1. Raman Spectra

The Raman effect is the inelastic scattering of the light incident on a sample of matter. For a crystalline solid, the observed shift of frequency corresponds to the creation of a phonon (Stokes), or its annihilation (anti-Stokes), which are the simplest energy quantities related to the normal vibrational frequencies of the crystal lattice.

For opaque samples the experimental configurations commonly used in Raman spectroscopy are the so-called 'back-scattering' or 'right-angle' scattering geometries (Ref. [50], p. 356).

In the right-angle scattering geometry, represented in Figure 9, the incident light makes an angle θ with the perpendicular to the surface, in order to avoid the reflected beam and the elastic Rayleigh scattering in the field of the Raman scattering light collected within a cone of angle ϕ.

In the present description of the MX_3 compounds the z axis has been chosen along the chain axis (**c** in the first setting). Therefore, the right-angle scattering allows the

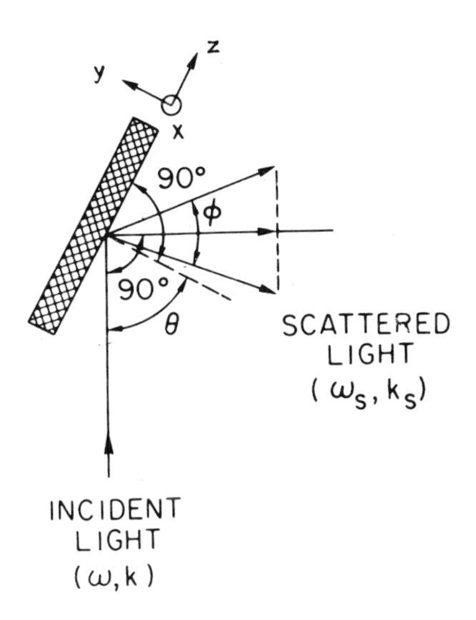

Fig. 9. Right-angle experimental configuration for Raman investigations of opaque samples. θ is a convenient but arbitrary angle between the incident light and the normal to the surface of the sample; ϕ defines the aperture angle of the observed scattered light.

incident polarization vector to be oriented along x or in the yz plane; consequently the A_g and B_g mode symmetries can already be distinguished, since they correspond to the Raman selection rules (x^2, y^2, z^2, xy) and (xz, yz), respectively.

The Raman spectra represented in Figure 10 have been obtained at room temperature with an incident Argon laser beam at the wavelength 5145 Å [39]. They show, for the $ZrSe_3$, ZrS_3 and HfS_3 compounds, the typical selection rules between the A_g and the B_g symmetries.

In the A_g symmetry six of the eight allowed modes have been observed. One mode in the B_g symmetry has clearly been determined.

Similar spectra have been obtained for the type A and for the type B compounds.

The comparison with the experimental Raman investigation by the other authors mentioned above is summarized in Table VI.

5.2. IR SPECTRA

The infrared reflectivity and transmission spectra of the Group IVB trichalcogenides have been investigated over the frequency range 30–600 cm^{-1}. The absorption and Reststrahlen band due to the long wavelength lattice vibrations have then been identified. In [18] and [39] (Figure 11a, b), typical reflectivity spectra have been obtained on single crystal samples with cleaved (010) surfaces (first crystallographic setting).

These spectra for both polarization **E** || **c** and **E** || **a** were obtained at room temperature.

By fitting Lorentzian oscillators to the experimental data the different authors [11,

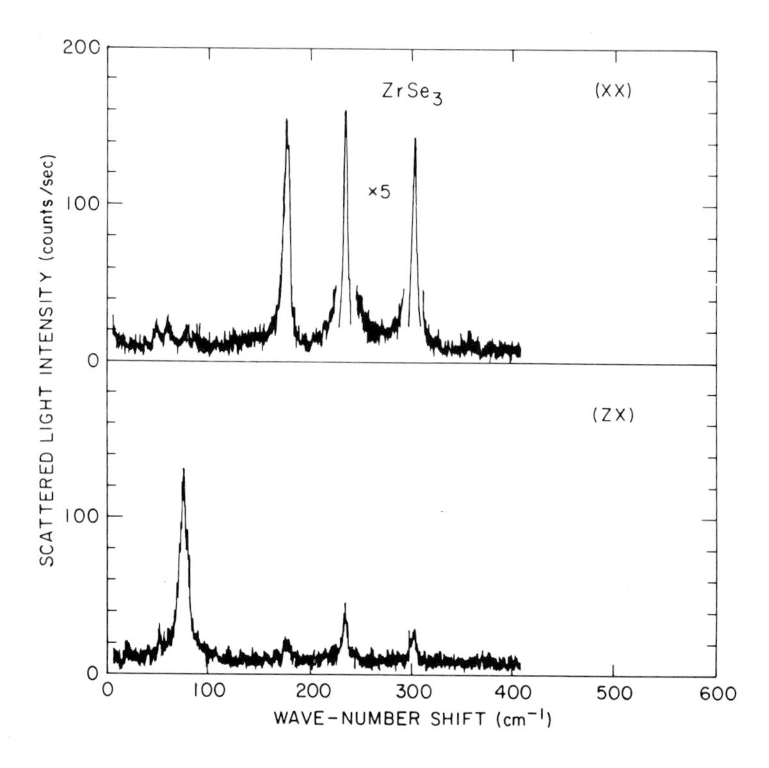
ZrSe3
(XX)
×5
(ZX)
SCATTERED LIGHT INTENSITY (counts/sec)
WAVE-NUMBER SHIFT (cm⁻¹)
0
100
200
300
400
500
600

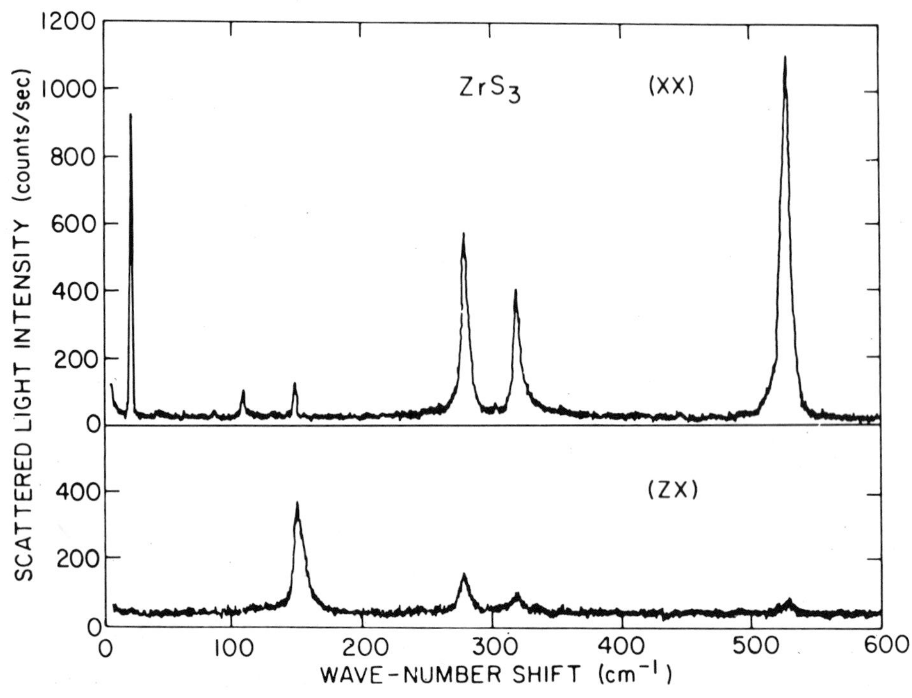
ZrS3
(XX)
(ZX)
SCATTERED LIGHT INTENSITY (counts/sec)
WAVE-NUMBER SHIFT (cm⁻¹)
1200
1000
800
600
400
200
0
100
300
500

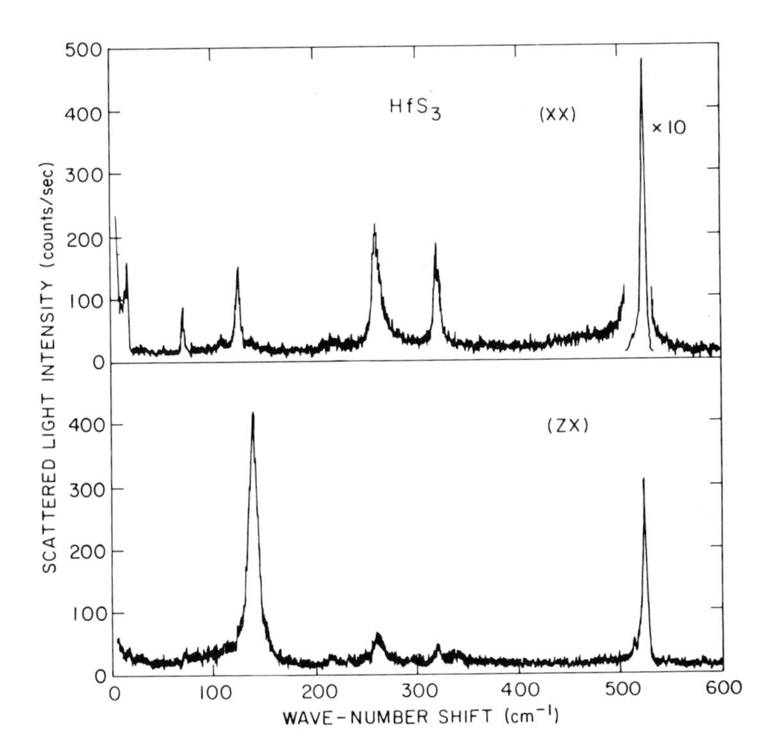

Fig. 10. Raman spectra of (a) $ZrSe_3$, (b) ZrS_3 and (c) HfS_3 compounds in the $(XX)A_g$ and $(ZX)B_g$ configurations.

18, 41] obtained the dielectric constant as well as dispersion parameters of the TO and LO phonons. The dashed lines in Figure 11a, b are damped oscillator fits to the experimental spectra obtained by the factorized form of the dielectric function [18].

$$\epsilon(\omega) = \epsilon_\infty \prod_{j=1}^{M} \frac{\omega_j^2 \mathrm{LO} - \omega^2 - i\gamma_j \mathrm{LO}\omega_j \mathrm{LO}\omega}{\omega_j^2 \mathrm{TO} - \omega^2 - i\gamma_j \mathrm{TO}\omega_j \mathrm{TO}\omega}$$

where ω_jTO and ω_jLO are the transverse-optical (TO) and longitudinal-optical (LO) frequencies respectively, of the jth oscillator. γ_i is a dimensionless damping constant and M is the number of oscillators involved.

From the interpretation of the IR reflectivity spectra, the predominant symmetry of the compounds is the c_{2v} (orthorhombic) symmetry of the chain. It implicates that the $\mathbf{E} \parallel \mathbf{a}$ spectrum is mainly influenced by the three B_2 resonances. The c_{2h} (monoclinic) symmetry of the crystal, however, allows the $3A_1$ modes of the chain to appear in the $\mathbf{E} \parallel \mathbf{a}$ spectrum, although the A_1 resonnances are expected to be weak. Figure 11 shows the reflectance spectra of ZrS_3 and $ZrSe_3$ for $\mathbf{E} \parallel \mathbf{c}$ and $\mathbf{E} \parallel \mathbf{a}$. The two $\mathbf{E} \parallel \mathbf{c}$ spectra clearly exhibit broad Reststrahlen bands which have been indentified with the $A_u(B_1)$

TABLE VI

Experimental Raman frequencies [cm^{-1}] observed by different groups in the Group IVb transition-metal trichalcogenides ZrS_3, HfS_3, $ZrSe_3$, TiS_3, $HfSe_3$ and $ZrTe_3$. (At room temperature). The values between brackets are originally not attributed.

	ZrS_3					HfS_3		$ZrSe_3$		TiS_3		$HfSe_3$		$ZrTe_3$	
	ref [18]	ref [10, 68]	ref [56]	ref [63]	ref [58]	ref [18]	ref [55]	ref [18]	ref [63]	ref [25, 39]	ref [69]	ref [25, 39]	ref [63]	ref [25, 39]	ref [63]
A_g	530	530	527	529	529	524	527	302	302	559	562.0 (555)	296	295	216	215
B_g	–	–	330	–	(334)	–	–	–	–	–	–	137	–	–	–
A_g	322	320	320	320	322	322	322	236	234.5	371	375.4 (366)	209	209	144	143.5
A_g	–	–	360	–	–	–	275	–	–	–	–	–	–	–	–
A_g	282	282	280	281	282	261	262	177	178	300	303.6 (297)	171	172	108	111
A_g	–	277	275	–	277	–	–	–	106.5	–	(277)	153	–	86	84.5
B_gyz	–	–	243	243	244	–	246 (221)	–	–	–	–	69	69	64	66.5
B_gxz	152	152	234	237	237 (152)	140	140 (130)	77	76.5	137	135.6 (A_g)	66	–	–	–
A_g	152	–	150	150	152	127	–	78	80	176	178.0	101	101	62	61.5
A_g	110	125 (77K)	122 (84)	123 (83)	124	73	–	50	50	102	103.8 (B_g)	69	69.5 51	38	37.5
A_g	21	–	19	–	84	15	–	–	–	20	–	19	–	–	–
B_g	–	–	108	–	110	–	–	–	–	–	–	–	–	–	–

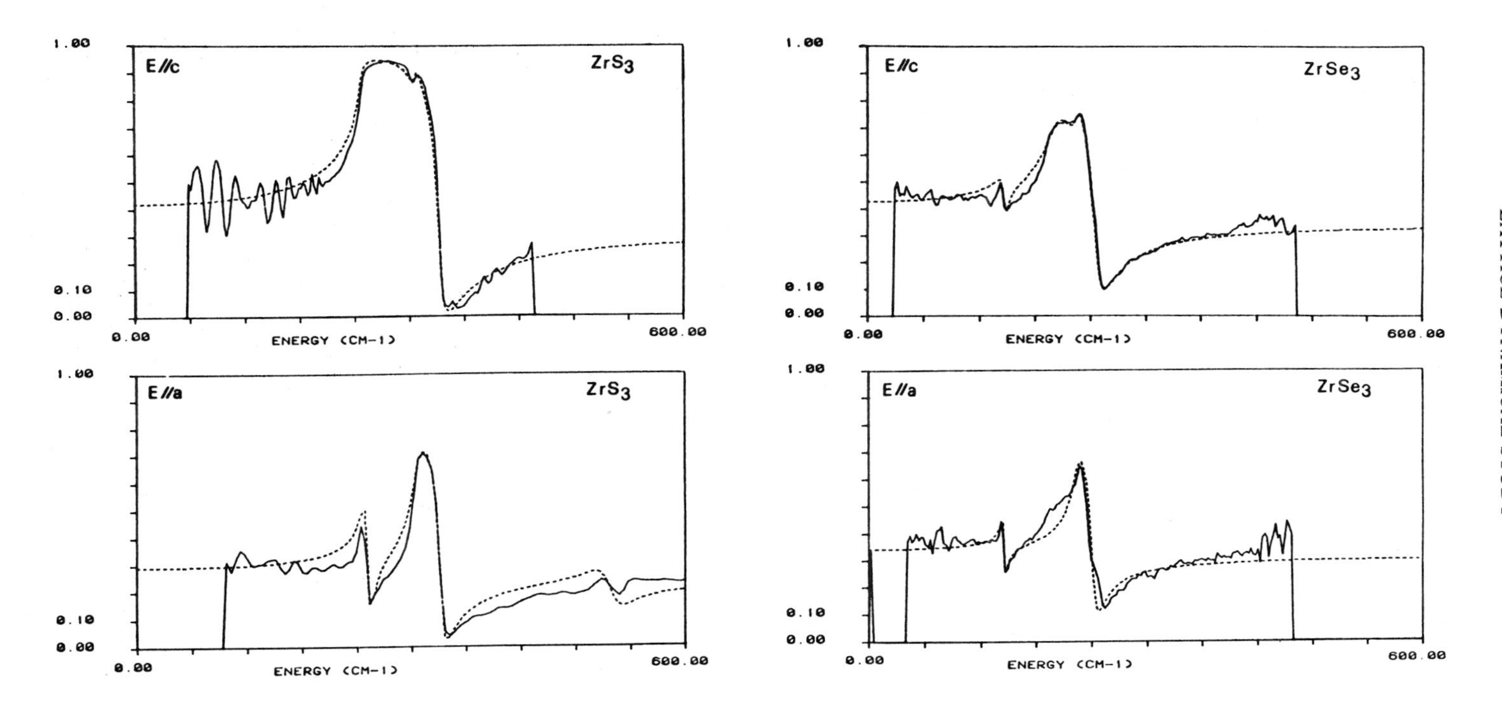

Fig. 11. Room-temperature reflectance spectra of (A) ZrS_3 and (B) $ZrSe_3$ for **E** ∥ c and **E** ∥ a. The dashed curves are damped oscillator fits to the data using the dielectric function in the damped oscillator factorized form.

TABLE VII

Experimental IR frequencies [cm^{-1}] obtained by reflectivity or transmission (Tr) measurements in the ZrS_3, HfS_3, $ZrSe_3$, TiS_3 and $HfSe_3$ compounds. Three of the twelve phonon modes listed are acoustical (*ac*): $B_u(x)$, $B_u(y)$, $A_u(z)$. The original attributions are indicated between brackets when they do not correspond.

		ZrS_3			HfS_3		$ZrSe_3$			TiS_3		$HfSe_3$
		ref [18]	ref [67]	ref [56]	ref [18]	ref [41]	ref [18]	ref [11]	ref [71]	ref [25]	ref [59]	ref [25]
$B_u(y)$	LO	522	–	–	–	–	–	–	300	–	–	–
	TO	515	LO 349(A_u)	526	525(Tr)	527(Tr)	–	–	290	–	–	–
			TO 356(A_u)	310(A_u)		322(Tr)						
$A_u(z)$	LO	306	303	–	226	225	223	255	245(B_u)	328	359	181
	TO	247	253	249	215	220	200	215	230(B_u)	294	295	160
$B_u(y)$	LO	–	272	–	300	300	–	–	219	–	311	–
	TO	–	270	270	286	278	–	–	205	–	309	–
$B_u(x)$	LO	334	332	–	280	276	248	242	232(A_u)	348	346	193
	TO	305	304	320	272	271	224	222	212(A_u)	333	333	182
				(361)								
$B_u(y)$	LO	–	285	–	–	–	–	–	350(A_u)	–	278	–
	TO	–	284	275	–	–	–	–	340(A_u)	–	276	–
$B_u(x)$	LO	254	249	–	252	247(A_u)	146	–	150	227	224	144
	TO	247	245	–	249	242(A_u)	144	–	145	221	217	140
$A_u(z)$	LO	–	–	–	–	–	–	–	166	–	187	–
	TO	180(Tr)	–	227	160(Tr)	136(Tr)	–	–	157	–	177	–
$A_u(z)$	LO	–	–	–	–	–	–	–	–	–	–	–
	TO	128(Tr)	–	–	130(Tr)	128(Tr)	–	–	–	–	–	–
(*Ac*) $B_u(x)$	LO	–	–	–	–	–	–	–	–	–	–	–
	TO	–	–	–	–	–	–	–	–	–	–	–
$B_u(x)$	LO	–	–	–	–	–	–	–	–	–	–	–
	TO	103(Tr)	–	103	104(Tr)	100(Tr)	–	–	–	–	–	–
(*Ac*) $B_u(y)$	LO	–	–	–	–	–	–	–	–	–	–	–
	TO	–	–	–	–	–	–	–	–	–	–	–
(*Ac*) $A_u(z)$	LO	–	–	–	–	–	–	–	–	–	–	–
	TO	–	–	–	–	–	–	–	–	–	–	–

rigid-sublattice mode polarized along the chain axis. The **E** || **a** spectra exhibit two narrower bands of B_u symmetry which have been attibuted to the principal B_2 internal modes of the chain. The weak features in the ZrS_3 **E** || **a** spectrum between 515 and 522 cm^{-1} corresponds to the $B_u(A_1)$ diatomic mode.

It should be noted that the $ZrSe_3$ and HfS_3 spectra are similar to those of ZrS_3 and that the $A_u(B_1)$ modes polarized along the chain axis follow a simple mass-scaling rule. The available experimental data are listed in Table VII.

6. Raman Scattering in the Group VB Transition Metal Trichalcogenides $TaSe_3$ and $NbSe_3$

Referring to Section 2, $TaSe_3$ and $NbSe_3$ have a prismatic chain structure similar to the $ZrSe_3$ structure. However the number of chains in the unit cell is two and three times as large as in $ZrSe_3$, respectively.

There are 16 atoms for $TaSe_3$ and 24 for $NbSe_3$ in their respective primitive unit cells. This fact yields a quite complex Raman spectrum for each of these compounds. Numerous closely spaced lines appear in the Raman spectra, as is shown in the Figures 12 and 13 [60]. It is, however, expected that the 'intra chain' modes will give rise to distinguishable multiplets (doublets for $TaSe_3$ and triplets for $NbSe_3$).

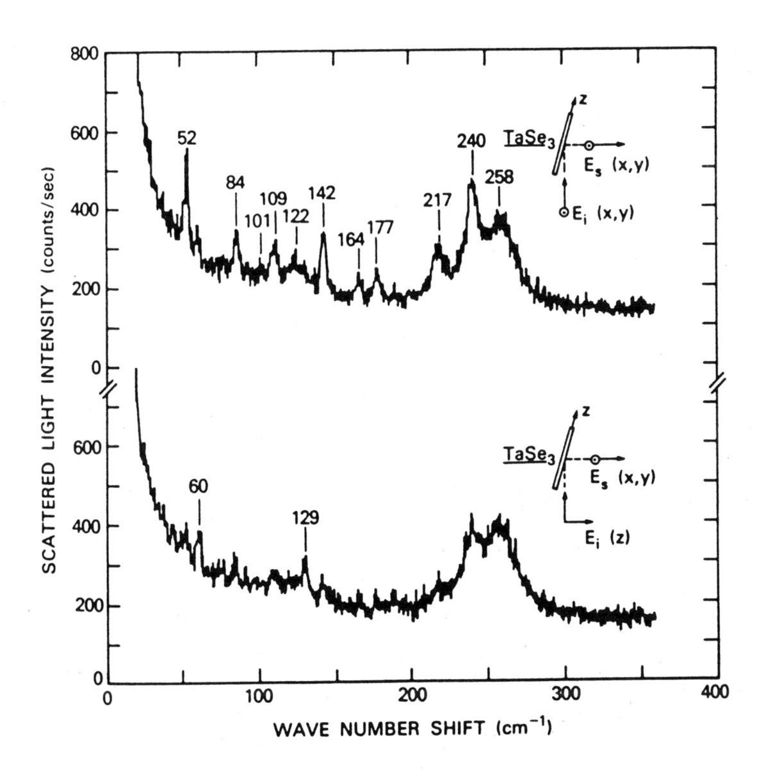

Fig. 12. Raman spectrum of $TaSe_3$ at room temperature.

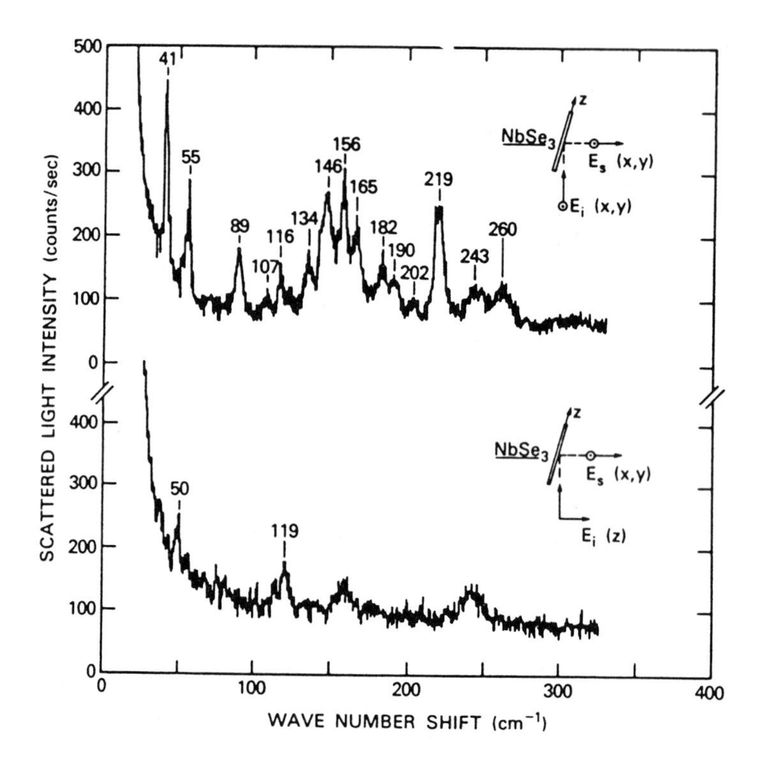

Fig. 13. Raman spectrum of $NbSe_3$ at room temperature.

The complex side linkages between the chains, represented by the different metal–selenium and selenium–selenium bonds complicate this interpretation.

It has been found, however, that the key to interpreting the Raman data on $TaSe_3$ and $NbSe_3$ is the lattice dynamical model developed for the simple $ZrSe_3$ structure. Since the chain, layer and crystal symmetries (C_{2v}, C_{2h} and C_{2h} respectively) are identical to those of $ZrSe_3$, the correlation diagram previously described (Figure 5) can be used.

The vibrational decompositions of the total representations of the chain and the crystals are therefore given by:

$$\begin{array}{lllllllll} \Gamma_{\text{chain}} & = A_2 & + & 3B_1 & + & 4A_1 & + & 4B_2 \\ & xz & & z; yz & & y; x^2, y^2, z^2 & & x, R_z; xy \\ \Gamma_{\text{TaSe}_3} & = 8A_u & + & 8B_g & + & 16B_u & + & 16A_g \\ & z & & xz; yz & & x, y & & R_z; x^2, y^2, z^2, xy \\ \Gamma_{\text{NbSe}_3} & = 12A_u & + & 12B_g & + & 24B_u & + & 24A_g \end{array}$$

where the transformation properties are written below each irreducible representation.

In [60] the experimental results have been reported. Room temperature Raman spectra have been obtained on $TaSe_3$ and $NbSe_3$ for both polarizations, discriminating the A_g from the B_g vibrations.

The Figures 12 and 13 illustrate one of the possible geometries in which, however, the B_g spectra also contain A_g features, since the electric vector of the incident laser beam has a component lying in the xy plane of the crystal.

A tentative attribution of the Raman lines has been discussed on the basis of the valence force model previously worked out for $ZrSe_3$. For this purpose, the comparison of several bond lengths in the three structures ($ZrSe_3$, $TaSe_3$, $NbSe_3$) proved to be useful. In particular the Se—Se bond length within the chain is longer and more variable in $TaSe_3$ and $NbSe_3$. Taking into account three distances (Table II), it was possible to predict the set of high frequency lines. The change in bond length of the metal–selenium bond within and between the chains as well as the much shorter Se—Se bonds between the chains have been found to correlate well with changes in the force constants when the $ZrSe_3$ model is applied to $TaSe_3$ or $NbSe_3$.

For simplicity in the first approach of the [60] the vibrational frequencies have been calculated, neglecting the bond-angle restoring forces originally introduced in the $ZrSe_3$ model since it is assumed that they affect the low frequency vibrations only.

For comparison with the $ZrSe_3$ structure the same stretching force constants have been introduced.

As reported in Section 4, C_w represents the metal–selenium force constant within the chains and C'_w characterizes the selenium–selenium pairing. C_b is the metal–selenium force constant between the chains and C_v links every pair of Se atoms between the chains. Beyond this basic model a second force constant within the chains C''_w has been introduced for $TaSe_3$, to distinguish the strength of Se—Se pairing in chain II from that in chain I. A Se—Se interchain force constant C'_b has also been added to take into account the different arrangement of the chains in $TaSe_3$. It has been found to be zero.

For $TaSe_3$, Table VIII summarizes the best fit obtained with the following force constants:

$$\begin{aligned} C_w &= C_b = 9.8 \times 10^4 \quad \text{[dyn/cm]} \\ C'_w &= 1.1 \times 10^5 \\ C_v &= 7.2 \times 10^2 \\ C'_b &= 0 \\ C''_w &= 4.0 \times 10^4 \end{aligned}$$

One conclusion of this work was that the strength of the intrachain Se—Se bond is different in the two chains of $TaSe_3$, that the metal–selenium bond is stronger in $TaSe_3$ than in $ZrSe_3$ and that the B_g modes are almost independent of the coupling introduced between the chains. A comparison between the displacement vectors of the B_g modes in $TaSe_3$ and those of an isolated chain shows that they are nearly identical.

The layered character of the $TaSe_3$ structure is reinforced by the Ta—Se bond ($C_b = C_w \simeq C'_w$) which links the chains into layers along (110).

In $NbSe_3$ since there are three different chains in the unit cell the doublets in $TaSe_3$

TABLE VIII

Raman data for $ZrSe_3$ and $TaSe_3$ compared with the calculated frequencies respectively from the valence force model and from a central force model.

Vibrational symmetry		$ZrSe_3$ (cm^{-1})		$TaSe_3$ (cm^{-1})	
Chain	Crystal	Raman	Calc.[a]	Raman	Calc.
A_1 (diatomic)	A_g	302	303	258 240	257 240
B_1 (rigid-sublattice)	B_g	–	177	– –	208 207
A_1	A_g	236	240	– 217	230 218
B_2	A_g	–	176	– 177	191 187
A_1	A_g	177	140	164 142	169 153
B_2	A_g	106.5[b]	96	122 109	135 128
A_2	B_g	77	86	129 –	138 137
B_1	B_g	–	85	– –	137 136
A_1 (acous./ rigid-chain)	A_g	78	71	101 84	77 73
B_2 (rot./libration)	A_g	50	50	60 52	24 23
B_2 (acous./ rigid-chain)	A_g	–	16	– –	18 11
B_1 (acous./ rigid-chain	B_g	–	20	– –	17 11

[a] Reference [18]
[b] Reference [11]

will then become triplets. Preliminary results tend to show that the layered character is enhanced in $NbSe_3$ with respect to the trichalcogenides.

7. Vibration Spectra of the Pentachalcogenides $ZrTe_5$ and $HfTe_5$

The strong similarity of the $ZrTe_5$ structure to the $ZrSe_3$ type is well demonstrated in the respective lattice vibration spectra. The likeness is emphasized by the chosen approach to interpret the experimental spectra, starting from the single chain modes towards the comprehensive representation of the modes in the crystal symmetry. The valence force model established for $ZrSe_3$, however, cannot be applied to the pentatellurides since, in these phases, the basic structural trigonal prismatic slabs are linked together by chains of tellurium atoms running between them. Nevertheless, features common to the two structures have been used as a starting point for the interpretation of the available vibration frequencies measured in Raman scattering experiments. In some cases, these steps even allowed confirmation of the original assignment of the modes in the trichalcogenides ($ZrTe_3$). Up to now, it has not actually been possible to grow single crystals large enough to allow reflectivity measurements in the infrared.

The trigonal prismatic slabs are typical building units in $ZrSe_3$ and in $HfTe_5$ types of structures. They are linked together to form layers. Internal modes are expected to be of the same order of magnitude. In particular, the characteristic short and strong bond between two Te atoms in the trigonal prism is nearly the same in the pentatellurides as in $ZrTe_3$, as may be deduced from the respective bondlengths: Te(2)—Te(2) = 2.762 ± 1 Å.

The Raman spectra have been measured at room and at low temperature in $ZrTe_5$ by Zwick *et al.* [61] and in $HfTe_5$ by Taguchi *et al.* [62]. The factor group analysis have been carried out by the two groups of authors with the help of correlation diagrams. A comptability diagram has been established to relate the single chain vibration to the crystal modes. In Zwick's investigation, the primitive unit cell containing 12 atoms is set according to the original crystallographic data, with the **a** axis along the chain axis. In Taguchi's work, the **c** axis is chosen parallel with the chain axis and corresponds with the displacements of the vibration modes along the z direction. In keeping with the first sections on the trichalcogenides, this setting will be retained in the following.

The 36 normal modes of the crystal with D_{2h} symmetry correspond to the 18 normal modes classified according to the single chain symmetry C_{2v}. The irreducible representation of the orthorhombic D_{2h}^{17} point symmetry group at the center of the Brillouin zone is

$$\Gamma_{\text{crystal}} = \underset{x^2,\,y^2,\,z^2}{6A_g} + \underset{y}{6B_u} + \underset{xz}{2B_{2g}} + 2A_u + \underset{yz}{4B_{3g}} + \underset{z}{4B_{1u}} + \underset{xy,\,R_z}{6B_{1g}} + \underset{x}{3B_{3u}}$$

The correlation diagram between the reduction of the chain factor group for the 6 atoms in one chain unit and the representation related to the crystal symmetry is shown in Figure 14. Because of the inversion center in the crystal, there are 18 Raman active modes and 18 odd parity modes with 15 infrared active and 3 acoustic phonons.

The modes can also be classified according to the respective displacements: the $2B_{2g} + 4B_{3g}$ and $4B_{1u}$ modes have displacements parallel with the chain axis and the $6B_{2u} + 6B_{3u}$ infrared modes have their displacements perpendicular to the **c**-axis. Since there is no center of symmetry for the chain, the modes are not divided into odd and even symmetry types in this reduction scheme. The inversion center in the crystal correlates the two chains in the unit cell, so that each chain mode splits into an odd–even pair in the crystal. But the crystal retains the symmetry elements of the chain. Consequently, there is no mixing of the chain modes symmetries in the crystal, in contrast with the case of the monoclinic structure in the $ZrSe_3$ type crystals.

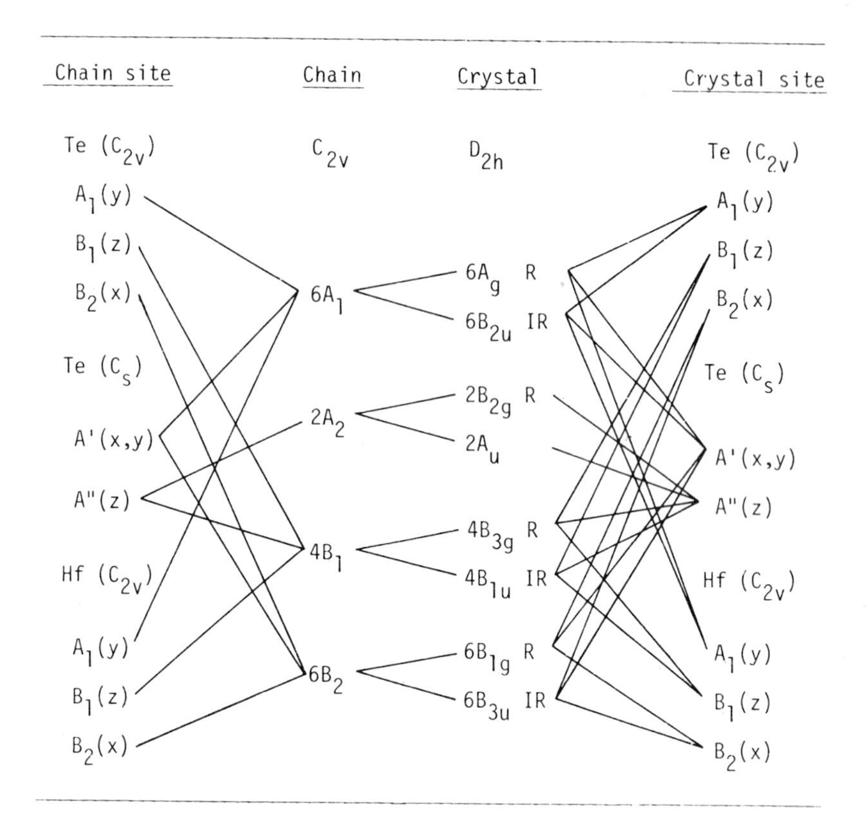

Fig. 14. Correlation diagram for the transition metal pentatellurides $HfTe_5$ and $ZrTe_5$ [62].

In Table IX the measured Raman frequencies are listed for the two compounds $ZrTe_5$ and $HfTe_5$. The assignments of the modes are those proposed by the two different authors according to their original interpretation. The single chain representation is also given. The agreement between the two sets of results is quite acceptable. For the case of $HfTe_5$, a series of experimental Raman spectra measured at various temperatures is shown in Figure 15. The polarized Raman spectra of $ZrTe_5$ for various geometries are given in Figure 16.

In $HfTe_5$, 7 Raman modes have been observed in the $YZ(XX)YZ$ geometry, in contradiction with the theoretical prediction of $6A_g$ modes correlated with the $6A_1$ chain modes. The six modes are observed in $ZrTe_5$ in a backscattering configuration.

The line with the lowest energy at 37 cm^{-1} ($HfTe_5$) and 39 cm^{-1} ($ZrTe_5$) is assigned to the external rigid chain vibration with displacement perpendicular to the plane of the layers, along the **b** axis. This attribution is corroborated by the comparison with the established rigid chain mode in $ZrTe_3$ at 37 cm^{-1} [63].

As in the tritelluride, the stretching mode of the Te_2—Te_2 short-bonded pairs in the trigonal prism should approach the frequency typical of the bimolecular mode. In $ZrTe_5$, its higher value of 239 cm^{-1} indicates a stronger bond than in $ZrTe_3$ [61].

TABLE IX

Raman frequencies in the transition metal pentachalcogenides. (a) Vibration frequencies for $HfTe_5$ and $ZrTe_5$ measured by Raman spectroscopy [61, 62] and calculated with help of a simple central force model [66]. (b) Optimalized forces constants used in the central force model [66].

(a)

$HfTe_5$ [62]			$ZrTe_5$ [61]					Central force model
Chain	Crystal	ν[cm^{-1}] (295 K)	Chain	Crystal	ν[cm^{-1}] (295 K) [61]	[62]	(77 K) [61]	$\bar{\nu}$[cm^{-1}] [66]
A_1	A_g	204	A_1	A_g			239	
A_1	A_g	182	A_1	A_g	181.5	179	183.5	179
A_1	A_g	166						
A_1	A_g	140	A_1	A_g	147	143	152	149
A_1	A_g	119	A_1	A_g	121	119	123	97
A_1	A_g	115	A_1	A_g	116	115	117	93
A_2	B_{2g}	84	A_2	B_{2g}	86.5	84	89.5	84
A_2	B_{2g}	66	A_2	B_{2g}	72	69	74	69
A_1	A_g	37	A_1	A_g	39	39	40	–

(b)

Bond	Force constants used in the calculation [dyn/cm] [66]	
Te_2-Te_2	$C_{w\perp}$	9.8×10^4
$Zr-Te_2$	C_w	3.9×10^4
$Zr-Te_3$	C_b	3.9×10^4
Te_3-Te_3	C_v	2.9×10^4

In the *YZ*(*YX*)*YZ* configuration, two B_{2g} modes are measured. They correlate with the A_2 modes for the single chain. In $HfTe_5$, the line at 66 cm^{-1} has been assigned to the A_2 shearing mode of the Te_2 pairs along the **c**-axis. The high frequency of the correspond mode in $ZrTe_5$ (72 cm^{-1}) agrees with the tendency of the bonds to strengthen when going from Hf to the more polarizable Zr ion. The second B_{2g} mode has been tentatively attributed to the motion of the Te_3 atoms in the zig-zag chains between the trigonal prismatic slabs.

Zwick's interpretation [61] of the $ZrTe_5$ spectrum also attributes the two lines measured at 74 and 89.5 cm^{-1} and considered as a doublet, to the mixed shearing modes of the $Zr-Te_2$ and $Zr-Te_3$ vibrations.

The temperature dependence of the Raman spectra does not show any strong anomaly (cf. Figure 15). The only noticeable pecularity is the strong increase in intensity of the A_g mode at 166 cm^{-1} with decreasing temperature in contrast with the diminution of the lowest lying peak at 37 cm^{-1}. However, no new peak could be detected which would be connected with any structural transition as observed in the transition metal dichalcogenides.

The experimental investigation of the Raman spectra, however, provides information for the phonon vibration frequencies at the center of the Brillouin zone $\mathbf{k} = 0$ only.

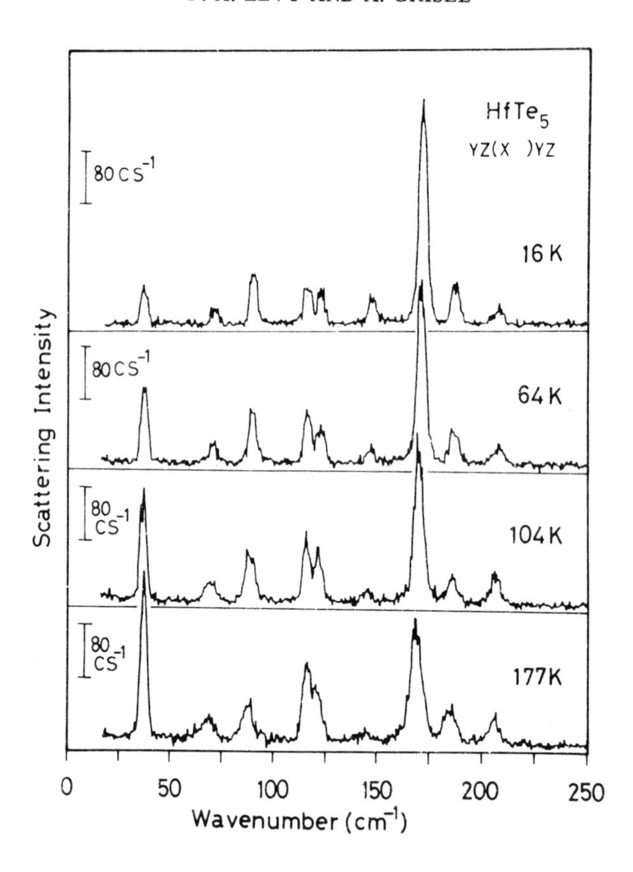

Fig. 15. Raman spectra of $HfTe_5$ measured at various temperatures.

Anomalies in the dispersion relations are often possible for finite **k** values, so that it is not allowed, at this stage, to exclude any phonon anomaly at low temperature. The occurrence of such structural and vibrational anomalies has actually been questioned in relation to the results of the investigation of the electrical transport properties. In the dependence on the temperature of the resistivity, unexplained maxima have been found around $T = 141$ K for $ZrTe_5$ [37] and around $T = 83$ K for $HfTe_5$ [64]. No evidence for a structural phase transition could be found up to now [65], so that the situation appears to be different from the case of $NbSe_3$.

In order to verify the interpretation of the experimental lines, Taguchi [66] has recently tried to calculate the vibration frequencies on the basis of a central force model with four central forces. The provisory results are summarized in Table IX. Even with help of such a simple model it is surprising how well the frequencies (except two) can be reproduced.

8. Final Remarks

In crystals with strongly anisotropic structures, the low dimensionality influences the

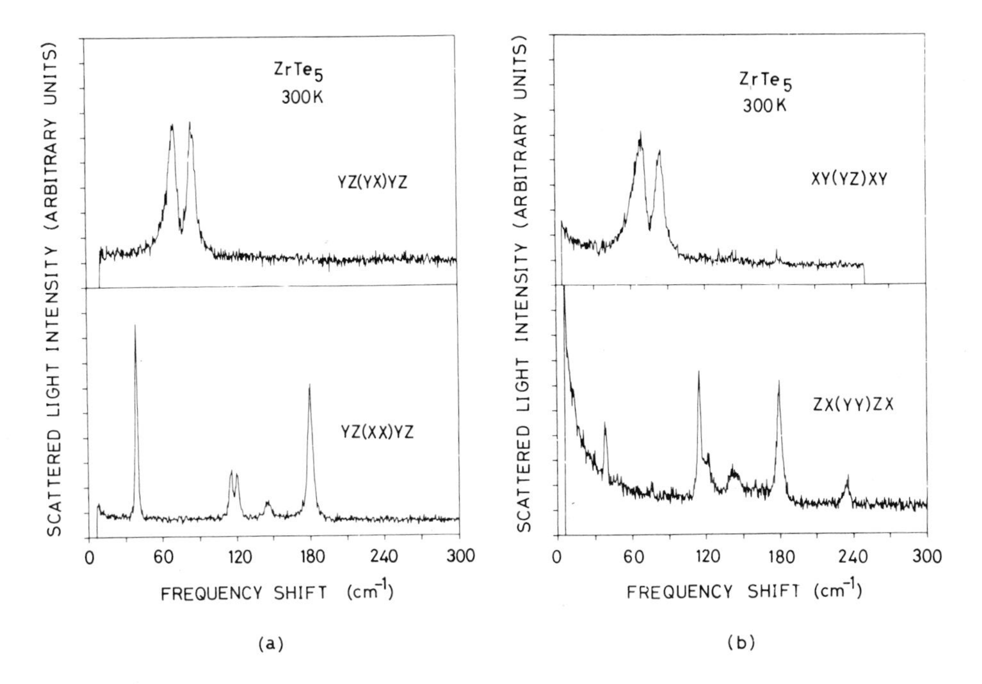

Fig. 16. Raman spectra of $ZrTe_5$ at room temperature, measured for various geometries.

lattice vibration spectra. The interpretation of the vibration mode frequencies in a family of compounds is based on assumptions about the relative strength of the atomic bonds. The description of the dynamical properties then gives a critical and reliable picture of the lattice. In particular, it gives a fair representation of the one- and two-dimensional building features of the structure.

The most typical results of the lattice dynamical properties of crystals of the $ZrSe_3$ type are summarized as follows.

In the transition metal trichalcogenides, the internal modes A_2 and B_1 of the single chains have displacements along the chain axis. They reflect the quasi-one-dimensional character of the trigonal prismatic slabs. This is due to the alignment of the transition metal cations and to the significant number of bonds with the chalcogen atoms along the direction of the chain.

In $HfTe_5$ and $ZrTe_5$, the B_{2g} crystal modes correlate with the A_2 chain modes and similarly reveal the one dimensional nature of the structure. A particularity of this structure is the unique antiphased motion, parallel to the chain axis, of the Te atoms in the zig-zag chains between the prismatic units.

In the plane perpendicular to the chain axis, the external modes A_1, with displacements perpendicular to the layers have significantly higher frequencies than the external B_2 modes parallel with the layers. The frequency of these A_1 modes is mainly related to the interaction between the chains in the layer planes. Consequently, this anisotropy demonstrates the layered character of the structure and the strong correlation between the chains which build layers.

The last feature common to all the vibration spectra in the tri- and pentachalcogenides is the high frequency of the nearly diatomic mode corresponding to the vibration of the chalcogen pairs in the prismatic unit. The strong bonds between these anion pairs are so confirmed, as the typical wedge shape of the trigonal prismatic slabs.

The particular distorted trigonal prism around the cation and the appreciable correlation between the chains in the layer planes lead to a different representation of the structure in relation to the transition metal dichalcogenides in conformity with the more usual coordination of the Group IV transition metals. As is outlined in Figure 17(a) for the $ZrSe_3$ type of structure, each nearly diatomic chalcogen molecule can be considered at a single apex of a distorted octahedron around the cation. Two further equivalent apexes are occupied by the two opposite single anions of the prismatic unit and the last two apexes coincide with anions belonging to the neighbouring chain. In this sketch, the layer appears to be built of edge-sharing distorted octahedra. The structure appears

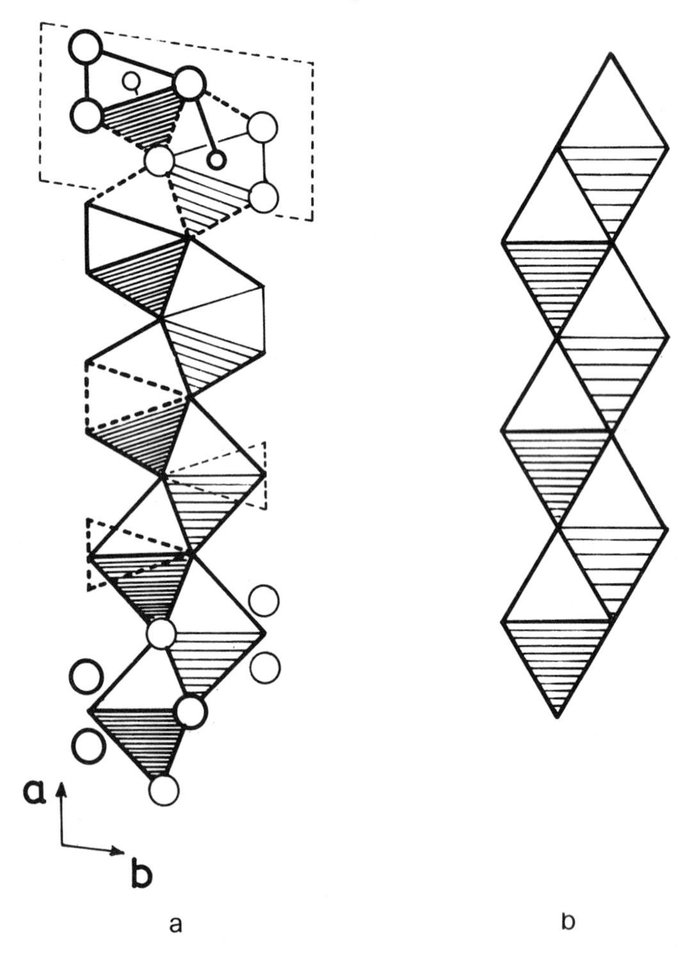

Fig. 17. $ZrSe_3$ structure represented as a stepped double-layer of edge sharing distorted octahedra: (a) $ZrSe_3$ structure; (b) idealized layer similar to the layers in the FeOCl structure [23].

to be a loose arrangement of pseudo-octahedra like the stepped double-layer structure for MX_2 compounds [23] FeOCl or γ-AlO(OH). Such a representation accentuates the layered character of the structure, in agreement with many of the physical properties.

References

1. H. Temkin, D. B. Fitchen, *Solid State Commun.* **19**, 1181 (1976).
2. H. J. Stolz, H. Wendel, A. Otto, L. Pintschovius, H. Kahlert, *Phys. Stat. Sol.* **b78**, 277 (1976).
3. H. Wendel, *J. Phys. C* **10**, L1 (1977).
4. E. F. Steigmeier, R. Loudon, G. Harbeke, H. Auderst, G. Scheiber, *Solid State Commun.* **17**, 1447 (1975).
5. K. Carneiro, G. Shirane, S. A. Werner, S. Kalser, *Phys. Rev.* **B13**, 4258 (1976).
6. H. Kuzmany, H. J. Stolz, *J. Phys. C: Sol. State Phys.* **10**, 2241 (1977).
7. H. Kuzmany, B. Kundu, H. J. Stolz, *Intern. Conf. Lattice Dynamics*, M. Balkanski Ed. Paris 1978, p. 584.
8. H. Temkin, D. B. Fitchen, F. Wudl, *Solid State Commun.* **24**, 87 (1977).
9. J. C. Tsang, C. Hermann, M. W. Shafer, *Phys. Rev. Lett.* **40**, 1528 (1978).
10. S. Jandl, C. Deville Cavellin, J. Y. Harbec, *Solid State Commun.* **31**, 351 (1979).
11. A. Zwick, M. A. Renucci, *Phys. Stat. Sol.* **b96**, 757 (1979).
12. T. J. Wieting, A. Grisel, F. Lévy, Ph. Schmid in *Quasi-One-Dimensional Conductors I*, Lecture Notes in Physics **95**, S. Barišic *et al.* ed. Springer 1979 pp. 354.
13. T. Sambongi, M. Yamamoto, T. Tsutsumi, Y. Shiozaki, K. Yamaya, Y. Abe, *J. Phys. Soc. Japan Letters* **42**, 1421 (1977).
14. S. Kurita, J. L. Staehli, M. Guzzi, F. Lévy, *Physica* **105** B+C, 169 (1981).
15. A. Meerschaut, J. Rouxel, P. Haen, P. Monceau, M. Nuñez Rugueiro, *J. Physique Lettres* **40**, L157 (1979).
16. A. Meerschaut, L. Guémas, J. Rouxel, *J. Solid State Chem.* **36**, 118 (1981).
17. P. Monceau, N. P. Ong, A. M. Portis, A. Meerschaut, J. Rouxel, *Phys. Rev. Lett.* **37**, 602 (1976).
18. A. Grisel, F. Lévy, T. J. Wieting, *Physica* **99B**, 365 (1980).
19. H. Haraldsen, A. Kjekshus, E. Røst, A. Steffensen, *Acta Chem. Scand.* **17**, 1283 (1963).
20. L. Brattås, A. Kjekshus, *Acta Chem. Scand.* **26**, 3441 (1972).
21. S. Furuseth, L. Brattås, A. Kjekshus, *Acta Chem. Scand.* **29**, 623 (1975).
22. W. Krönert, V. Plieth, *Z. Anorg. Allg. Chem.* **336**, 207 (1965).
23. F. Hulliger: 'Structural Chemistry of Layer Type Phases', in *Physics and Chemistry of Materials with Layered Structures*, Vol. 5, F. Lévy ed. D. Reidel Dordrecht, 1976.
24. A. Meerschaut, J. Rouxel in *Physics and Chemistry of Materials with Quasi-One-Dimensional Structures*, to be published.
25. T. J. Wieting, A. Grisel, F. Lévy, *Physica* **105B**, 366 (1981).
26. S. Jandl, R. Provencher, *J. Phys. C: Sol. State Phys.* **14**, L461 (1981).
27. A. Zwick, G. Landa, M. A. Renucci, R. Carles, A. Kjekshus, *Phys. Rev.* **B26**, 5694 (1982).
28. P. Monceau, N. P. Ong, A. M. Portis, A. Meerschaut, J. Rouxel, *Phys. Rev. Lett.* **37**, 602 (1976).
29. F. Kadijk, F. Jellinek, *J. Less-Common Metals* **19**, 421 (1969).
30. J. Rijnsdorp, F. Jellinek, *J. Sol. State Chem.* **25**, 325 (1978).
31. T. Cornelissens, G. Van Tendeloo, J. Van Landuyt, S. Amelinckx, *Phys. Stat. Sol.* **a48**, K5 (1978).
32. E. Bjerkelund, A. Kjekshus, *J. Less-Common Metals* **7**, 231 (1964).
33. T. Sambongi, K. Tsutsumi, Y. Shiozaki, M. Yamamoto, K. Yamaya, Y. Abe, *Solid State Commun.* **22**, 729 (1977).
34. A. Meerschaut, L. Guémas, J. Rouxel, *C. R. Acad. Sci.* (France) **290**, C215 (1980).
35. K. Yamaya, Y. Abe, *Mol. Cryst. Liq. Cryst.* **81**, 133 (1982).
36. S. Furuseth, L. Brattås, A. Kjekshus, *Acta Chem. Scand.* **25**, 2783 (1971).
37. S. Okada, T. Sambongi, M. Ido, Y. Tazuke, R. Aoki, O. Fujita, *J. Phys. Soc. Japan* **51**, 460 (1982).

38. F. Lévy, H. Berger, *J. Cryst. Growth* **61**, 61 (1983).
39. A. Grisel: Thèse 1981, Ecole Polytechnique Fédérale, Lausanne.
40. S. Kikkawa, M. Koizumi, S. Yamanaka, Y. Onuki, S. Tanuma, *Phys. Stat. Sol.* **a61**, K55 (1980).
41. S. Jandl, J. Deslandes, *Phys. Rev.* **B24**, 1040 (1981).
42. S. Kikkawa, N. Ogawa, M. Koizumi, Y. Onuki, *J. Sol. State Chem.* **41**, 315 (1982).
43. J. L. Hodeau, M. Marezio, C. Roucau, R. Ayroles, A. Meerschaut, J. Rouxel, P. Monceau, *J. Phys. C: Solid State Phys.* **11**, 4117 (1978).
44. C. Roucau, R. Ayroles, P. Monceau, L. Guémas, A. Meerschaut, J. Rouxel, *Phys. Stat. Sol.* **a62**, 483 (1980).
45. E. Bjerkelund, J. H. Fermor, A. Kjekshus, *Acta Chem. Scand.* **20**, 1836 (1966).
46. E. Bjerkelund, A. Kjekshus, V. Meisalo, *Acta Chem. Scand.* **22**, 336 (1968).
47. S. Furuseth, L. Brattås, A. Kjekshus, *Acta Chem. Scand.* **27**, 2367 (1973).
48. A. Meerschaut, J. Rouxel, *J. Less Common Metals* **39**, 197 (1975).
49. H. Poulet, J. P. Mathieu *Spectres de vibrations et symétrie des cristaux*, Gordon and Breach, New York, 1970.
50. T. J. Wieting and J. L. Verble, *Electron and phonons in layered crystal structures*, T. J. Wieting and M. Schlüter Eds. Reidel, Dordrecht, 1979.
51. W. G. Fateley, F. R. Dollish, N. T. McDevitt and F. F. Bentley, *Infrared and Raman selection rules for molecules and lattice vibrations: The correlation methods*, Wiley-Interscience New York, 1972, p. 201.
52. S. Califano *Vibrational States*, J. Wiley & Sons (1976) Appendix I.
53. Tenseurs de polarisabilité (R. Loudon *Adv. Phys.* **13**, 423 (1964)).
54. A. Grisel and T. J. Wieting, *Helv. Phys. Acta* **52**, 365 (1979).
55. C. Deville Cavellin and S. Jandl, *Solid State Commun.* **33**, 813 (1980).
56. C. Sourisseau and Y. Mathey, *Chemical Physics* **63**, 143 (1981).
57. A. Zwick, G. Landa, R. Carles, M. A. Renucci and A. Kjekshus, *Solid State Commun.* **45**, 889 (1983).
58. C. Deville Cavellin, G. Martinez, O. Gorochov and A. Zwick, *J. Phys. C: Solid State Phys.* **15**, 5371 (1982).
59. S. Jandl, J. Deslandes and M. Bauville, *Infrared Phys.* **22**, 327 (1982).
60. T. J. Wieting, A. Grisel and F. Lévy, *Mol. Cryst. Liq. Cryst.* **81**, 117 (1982).
61. A. Zwick, G. Landa, R. Carles, M. A. Renucci, *Solid State Commun.* **44**, 89 (1982).
62. I. Taguchi, A. Grisel, F. Lévy, *Solid State Commun.* **45**, 541 (1983), and **46**, 299 (1983).
63. A. Zwick, M. A. Renucci, A. Kjekshus, *J. Phys. C: Solid State Phys.* **13**, 5603 (1980).
64. M. Izumi, K. Uchinokura, E. Matsuura, *Solid State Commun.* **37**, 641 (1981).
65. D. W. Bullet, *Solid State Commun.* **42**, 691 (1982).
66. I. Taguchi, Personal Communication, Meeting of the Japan Phys. Soc. 1983.
67. S. Jandl, M. Bonville, J. Y. Harbec, *Phys. Rev. B* **22**, 5697 (1980).
68. J. Y. Harbec, C. Deville Cavellin and S. Jandl, *Phys. Stat. Sol. B* **96**, K 117 (1980).
69. D. W. Galliardt, W. R. Nieveen and R. D. Kirby, *Solid State Commun.* **34**, 37 (1980).
70. C. Perrin, A. Perrin et J. Prigent, *Bull. Soc. chem. France B*, 3086 (1972).
71. S. Jandl and J. Deslandes, *Canadian Journal of Physics* **59**, 7,936 (1981).

INDEX OF NAMES

INDEX OF SUBJECTS